MARITIME DELIMITATION AND INTERIM ARRANGEMENTS IN NORTH EAST ASIA

Publications on Ocean Development

Volume 40

A Series of Studies on
the International, Legal, Institutional and Policy Aspects of Ocean Development

General Editor: Vaughan Lowe
Chichele Professor of Public International Law and Fellow of All Souls College, Oxford University

The titles published in this series are listed at the end of this volume.

Maritime Delimitation and Interim Arrangements in North East Asia

SUN PYO KIM

MARTINUS NIJHOFF PUBLISHERS
THE HAGUE/LONDON/NEW YORK

A C.I.P. Catalogue record for this book is available from the Library of Congress.

ISBN 90-04-13669-X

Printed on acid-free paper.

Brill Academic Publishers incorporates the imprint Martinus Nijhoff Publishers.
http://www.brill.nl

Printed and bound in The Netherlands

PREFACE

This work will be a useful guide for those who look for rules and practice on the relations between neighbouring States in the absence of maritime boundaries. The main question the author is trying to tackle is how to handle the relations between neighboring coastal States when there is no maritime boundary in place. This book attempts to clarify the legal issues of exploitation of oil, gas and fisheries resources, and jurisdictional conflicts with regard to marine scientific research and protection of the marine environment in disputed areas. This book shows numerous instances of provisional arrangements in disputed areas around the globe together with as many as 45 valuable maps. The author, a scholar and diplomat of Korea, gives an up-to-date and in-depth analysis of the complicated legal issues of maritime delimitation and provisional arrangements in North East Asia. The English texts of the provisional arrangements in the region annexed to the book are also valuable materials.

CONTENTS

ACKNOWLEDGEMENTS

The idea of writing a book on provisional arrangements first occurred to me when I was involved in the negotiations with Japan and China for maritime delimitation and fisheries agreements in the mid-1990s. The question of which law governs the relations between neighbouring States in the absence of maritime boundaries occupied my mind and kept bothering me. Then I had an opportunity to research this topic as a Ph.D. student at the Faculty of Law, at the University of Edinburgh under the supervision of Professor Alan Boyle and Professor Bill Gilmore. It has always been a rewarding challenge to research under their supervision. I also had a good occasion when Professor Lowe of the University of Oxford came to Edinburgh as an external examiner for my Ph.D. degree and made valuable comments. I thank Professor Lowe for his insightful comments and his encouragement to publish the thesis.

Most of this book was written when I stayed in the Holland House Annex and the Churchill House in Edinburgh where I found good companions from all over the world. I cannot forget the wonderful outings I had with them for drinks, talks and movies. I hope that every one of my Edinburgh friends will achieve his or her dream.

I wish to avail this opportunity to express my indebtedness to many who helped me in the course of my life. I thank God who helped me overcome what appeared to be insurmountable obstacles in my life. I owe a considerable debt of gratitude to Judge Choon-ho Park at the International Tribunal for the Law of the Sea for his invaluable teaching and encouragement. I must express my sincere appreciation to those who guided me in learning international law at the Korea University and the University of Edinburgh.

I wish to extend my deep gratefulness to those who guided and helped me in the course of my career at the Korean Foreign Ministry and the Korea

Maritime Institute. I wish all the best for Mr. Hyung Ki Lee and Miss Soo-jeong Choi at the Korea Maritime Institute who have been good acdemic companions to me. I also thank those who helped me in proofreading of this book.

I express my apologies to my family for the time I had to spend alone in Edinburgh writing this book. My little daughter, Min-chung, did not understand why her father had to stay abroad for such a long time. I hope Min-chung and Min-sun will be interested in this book when they are grown-ups.

Lastly, I would like to dedicate this book to those who look for some guidance on the rules and practice of relations between neighbouring States in the absence of maritime boundaries. I hope this book will shed some lights on their quest.

Seoul — Sun Pyo Kim
15 April 2003

ABBREVIATIONS

A.J.I.L.	*American Journal of International Law*
B.Y.I.L.	*British Yearbook of International Law*
CCSBT	Convention for the Conservation of Southern Bluefin Tuna
Cm. and Cmnd.	Command Paper of the United Kingdom
ECAFE	(United Nations) Economic Commission for Asia and Far East
FAO	(United Nations) Food and Agricultural Organisation
FICMZ	Falkland Islands Conservation and Management Zone
G.Y.I.L.	*German Yearbook of International Law*
H.I.L.J.	*Harvard International Law Journal*
ICJ	International Court of Justice
I.C.L.Q.	*International and Comparatively Law Quarterly*
ICNT	Informal Composite Negotiation Text
I.J.E.C.L.	*International Journal of Estuarine and Coastal Law*
I.J.M.C.L.	*International Journal of Marine and Coastal Law*
ILC	*Yearbook of the International Law Commission*
I.L.M.	International Legal Materials
IPOA	International Plan of Action
ISNT	Informal Single Negotiation Text
IUU	Fishing Illegal, Unreported and Unregulated Fishing
ITLOS	International Tribunal for the Law of the Sea
J.A.I.L.	*Japanese Annual of International law*
JRA	Joint Regime Area
M.N.	Nautical Mile
M.P.	Marine Policy
MOU	Memorandum of Understanding

NBL	Northern Boundary Line
NLL	Northern Limits Line
N.Y.I.L.	*Netherlands Yearbook of International Law*
N.I.L.R	*Netherlands International Law Review*
O.D.I.L.	Ocean Development and International Law
OFCZ	Outer Fishery Conservation Zone
PCIJ	Permanent Court of International Justice
PFSEL	Provisional Fisheries Surveillance and Enforcement Line
Recueil des Cours	*Recueil des Cours de l'Académie de droit international*
R.I.A.A.	Reports of International Arbitration Awards (United Nations)
RSNT	Revised Single Negotiation Text
SAFC	South Atlantic Fisheries Commission
S.D.L.R.	*San Diego Law Review*
UKMIL	United Kingdom in International Law (a section in the *B.Y.I.L.*)
UNCLOS	United Nations Conference on the Law of the Sea
U.N.T.S.	*United Nations Treaty Series*
V.J.I.L.	Virginia Journal of International Law

TABLE OF MAPS

TABLE OF CASES

Aegean Sea Continental Shelf Case *(Interim Measures)*, 1976 *ICJ Reports.*

Anglo-French Continental Shelf Arbitration (1977), 18 *I.L.M.* (1979), pp. 397-494.

Anglo-Norwegian Fisheries Case, 1951 *ICJ Reports.*

Eritrea-Yemen Arbitration (First Phase, 1998), Permanent Court of Arbitrator's <http:pca-cpa.org.>

Eritrea-Yemen Arbitration (Second Phase, 1999), Permanent Court of Arbitrator's <http:pca-cpa.org.>

Fisheries Jurisdiction Case *(Federal Republic of Germany v. Iceland)*, 1973 *ICJ Reports.*

Fisheries Jurisdiction Case *(UK v. Iceland)*, 1974 *ICJ Reports.*

Jan Mayen Maritime Delimitation Case *(Denmark v. Norway)*, 1993 *ICJ Reports.*

Jan Mayen conciliation (Iceland v. Norway), 1981, 20 I.L.M. (1981), pp. 797-842.

Guinea/Guinea Bissau Maritime Boundary Arbitration (1985), 25 *I.L.M* 251 (1986), pp. 251-307.

Gulf of Fonseca Case *(El Salvador v. Honduras)*, 1992 *ICJ Reports.*

Gulf of Maine Case *(Canada v. U.S.A.)*, 1984 *ICJ Reports.*

Libya/Malta Continental Shelf Case, 1985 *ICJ Reports.*

Maritime Delimitation between Guinea-Bissau and Senegal, 1995 *ICJ Reports.*

Maritime Delimitation and Territorial Questions between Qatar and Bahrain, 2001 *ICJ Reports.*

The Minquiers and Ecrehos Case *(France v. UK)*, 1953 *ICJ Reports.*

The Monte Confurco Case (*Seychelles v. France*, prompt release), 2000 ITLOS.

The M/V Saiga Case No. 1 (*St. Vincent and the Grenadines v. Guinea*, prompt release), 1997.

The M/V Saiga Case No. 2 (*St. Vincent and the Grenadines v. Guinea*, Provisional Measures), 1998 ITLOS.

The M/V Saiga Case No. 2 (*St. Vincent and the Grenadines v. Guinea*, Judgement on the Merits), 1999 ITLOS.

Southern Bluefin Tuna Cases (*Australia v. Japan* and *New Zealand v. Japan*, Provisional Measures), 1999 ITLOS.

Southern Bluefun Tuna arbitration (*Australia and New Zealand v. Japan*), 2000, Arbitral Tribunal constituted under Annex VII of the LOS Convention, at International Centre for Settlement of Investment Dispute's, <www.worldbank.org/icsid/>

St.Pierre and Miquelon Arbitration (Canada v. France), 31 *I.L.M.* (1992) pp. 1145-1220.

North Sea Continental Shelf Cases *(Federal Republic of Germany v. Denmark, Federal Republic of Germany v. The Netherlands)*, 1969 *ICJ Reports*.

Tunisia/Libya Continental Shelf Case, 1982 *ICJ Reports*.

Introduction

The law of delimitation has been one of the major research topics in the modern law of the sea. It mainly covers ways of delimiting overlapping areas of the sea between States with opposite or adjacent coasts.

The task of making maritime boundaries is, however, time-consuming and sometimes even politically impossible. The following examples show how difficult it usually is. The process of delimitation lasted from 1964 to 1971 in the cases of the continental shelf in the North Sea between the Federal Republic of Germany and Denmark and the Netherlands.[1] Canada and the United States submitted the dispute concerning the maritime boundary in the Gulf of Maine to a Chamber of the International Court of Justice (ICJ) in 1981 after five years of intensive high-level negotiations and the Chamber rendered its judgment in 1984.[2] The delimitation agreement between Argentina and Chile concerning the Beagle Channel was concluded in 1984, 17 years after Chile had submitted the issue to arbitration in 1967.[3] The Aegean Sea continental shelf delimitation issue between Greece and Turkey over which the ICJ declared that it had no jurisdiction in 1978 still awaits solution.[4]

According to Professor Gerald Blake at the University of Durham, the number of potential maritime boundaries in the world is 434 and only 150 of them had been agreed in whole or in part as of mid-1999.[5] This means that only 34.5% of the potential maritime boundaries have been solved at least in part and that the remaining 65.5% still await delimitation.

Thus follows the important question of how to handle the relations between States with opposite or adjacent coasts when there is no maritime boundary in place. This book focuses on the legal problems arising from the absence of maritime boundaries between States with adjacent or opposite coasts and on how to deal with the respective rights and duties of the States concerned before the ultimate delimitation of the exclusive economic zone (EEZ)/continental shelf can take place.

Unfortunately, the legal problems which might arise before the ultimate boundaries of EEZ/continental shelf are formalised were given only limited attention in the third UN Conference on the Law of the Sea (UNCLOS III). Articles 74(3) and 83(3) of the resulting UN Convention on the Law of the Sea (LOS Convention) deal with the legal problems pending delimitation of the EEZ/continental shelf. They provide, in identical terms, that "pending agreement as provided for in paragraph 1, the States concerned, in a spirit of understanding and co-operation, shall make every effort to enter into provisional arrangements of a practical nature". As these are the only provisions in the LOS Convention which explicitly purport to deal with the legal relations between States with adjacent or opposite coasts pending maritime delimitation of EEZ/continental shelf, their analyses become considerably important.

However, it must be underlined that even though no illustrations of provisional arrangements are given in the provisions, there have been many actual cases of them in practice even before the conclusion of the LOS Convention in 1982. For example, joint development zones of the continental shelf, joint fishing zones, and provisional maritime boundaries were adopted from time to time by coastal States as provisional measures.

There are a dozen or so cases of joint oil and gas development zones over the world which attracted much scholarly attention.[6] For instance, the British Institute of International and Comparative Law organized a team to study joint developments of offshore oil and gas and completed a model agreement with a commentary in 1989.[7]

It is worthwhile to note that fisheries are also an important area which needs to be managed before the EEZ is delimited. Hence, it is normal that some joint fishing zones can be found throughout the world which are intended as provisional arrangements before the final delimitation. How to classify a joint fishing zone into a Grey or White Zone was mentioned in 1978 in UNCLOS III.[8] This classification is now popular among international lawyers though it had not been reflected in the text of the LOS Convention. The fisheries agreement between Sweden and the USSR in 1977 is a typical example of the White Zone Approach while the fisheries agreement between Norway and the USSR in 1978 is frequently cited as a good example of the Grey Zone Approach.[9] It is regrettable that few academic articles have thus far addressed the legal nature of provisional fisheries arrangements.

Besides issues of oil and gas and fisheries, there are other complex issues that call for clarification before the final delimitation of the EEZ/continental shelf. For example, it is difficult to answer which country can give its consent to a marine scientific research plan of a third country when such a research is to be conducted in an area of overlapping claims. Similarly, the question of which country is entitled to exercise its jurisdiction over the protection of the marine environment in overlapping areas raises complicated legal issues.

In this book, I first intend to show why maritime delimitation is so difficult. The legal nature of the rights and duties between States with opposite and adjacent coasts pending the ultimate conclusion of maritime boundaries will then be defined. In this context, the implications of Articles 74(3) and 83(3) of the LOS Convention will be explored through an objective interpretation of the provisions and by examining the drafting history of the articles. As many relevant cases of provisional arrangements as possible, which exist in practice throughout the world, will be analysed in order to shed some light on this issue. Finally, the provisional maritime regimes of North East Asia, where very complicated legal issues are currently arising, will be explored in the hope that some valuable lessons can be extracted for the readers in others parts of the globe.

NOTES

1. The ICJ rendered its judgement on 20 February 1969 and the final delimitation agreements were concluded on 28 January 1971.

2. David Colson, "Report Number 1-3: Canada-United States", *International Maritime Boundaries* (J.I. Charney and L.M. Alexander, Martinus Nijhoff Publishers, 1993), p. 401.

3. B.H. Oxman, "Political, Strategic, and Historical Considerations", *International Maritime Boundaries*, p. 12.

4. Douglas M. Johnston, *The Theory and History of Ocean Boundary-Making*, McGill-Queen's University Press (1988), pp. 154-159.

5. G. Blake *et al., IBRU Workshop on International Maritime Boundaries*, University of Durham, 4-9 April 1999.

6. This number includes the instances of joint development zones and comprehensive exploitation zones: see Chapter Three.

7. See H. Fox *et al., Joint Development of Offshore Oil and Gas: A Model Agreement*

for States for Joint Development with Explanatory Commentary, London: British Institute of International and Comparative Law (1989).

8. Statement made by the Chairman of Negotiating Group 7 on 12 September 1978, Conf.Doc.NG7/23.

9. For example, see also R. Churchill, "Fisheries Issues in Maritime Boundary Delimitation", 17 *M.P.* (January 1993).

CHAPTER ONE

What Makes Maritime Delimitation Difficult?

1. *MARE CLAUSUM* IN DISPUTE

When Grotius wrote the treatise *Mare Liberum* in March 1609, he opened what has proceeded to be the major controversy on the law of the sea. For Grotius, the sea as a whole is too immense to be appropriated by a nation or a race and thus should be seen as *res communis* and thus should be open to all.[1] Even though Grotius's main objective in this renowned treatise was especially directed against the Portuguese monopoly over the passage round the Cape of Good Hope and trade with the East Indies, the freedom of fishing was also an important element in his concept of the freedom of the sea.[2]

However, when King James I of England issued, within less than two months of the publication of *Mare Liberum*, a proclamation forbidding unlicensed fishing by foreigners off the British coast,[3] the distinction between the high seas and maritime zones under the jurisdiction of a coastal State began to emerge. A Scottish lawyer, Welwood, in his book *An Abridgement of all Sea-Lawes* published in 1613,[4] endeavoured to refute the arguments advanced in *Mare Liberum*, and argued that strangers should be stayed from scattering and breaking the shoals of fish on the coast of Scotland. It appeared that Welwood intended to protect Scottish interests from the unregulated herring fishing by the Dutch fishermen off the Scottish coast and to support King James I.[5] Welwood agreed with the principle of complete freedom of the sea. However, for him the freedom of the sea was to be applied as far as the "main Sea or great Ocean" is concerned, which was "*farre removed from the just and due bounds above mentioned properlie perteyning to the neerest Land of everie Nation*".[6] Later, in 1636, John Selden, an English lawyer joined the argument and asserted the rights of the Crown of England to the Dominion of the British seas. In his famous treatise *Mare*

Clausum in 1636, Selden challenged the arguments by Grotius, aiming to restrict foreign fishermen's fishing in the waters off the English coast. However, when the extensive claims by King James I and the pretensions of other States to a similar dominion altogether fell to disuse in the 18th century, *Mare Liberum* seemed to have won the battle against *Mare Clausum*.[7]

At the beginning of the 18th century, however, the question of the appropriation of the sea was placed on another footing with the emerging concept of the territorial sea. A Dutch lawyer Bynkershoek, in his book *Dominion of the Sea* published in 1704, asserted that the open ocean could not be wholly brought under dominion but admitted that various nations had at different times enjoyed such dominion. He laid down the proposition that: "*terrae dominium finitur, ubi finitur armorum vis*".[8] The cannon shot rule, which was understood to mean to limit the territorial waters to 3 nautical miles (N.M.)[9] or one league emerged strongly in the practice of the international community until the first half of the 20th century.[10]

However, from the second half of the 20th century, a movement to extend jurisdiction over the sea emerged. Then, the battle between *Mare Liberum* and *Mare Clausum* entered a new round. Sometimes the movement to extended maritime jurisdiction was coupled with the desire to extend the territorial sea or to establish exclusive fisheries zones or the continental shelf beyond the territorial sea.

Ironically, the United States, which had been a strong patron of 3 N.M. for the maximum limit of the territorial waters,[11] heralded a new epoch of extended maritime zones after World War II by issuing the two Truman Proclamations.[12] The Truman Proclamations on fisheries and on the continental shelf in September 1945 set the movement for extending maritime zones in motion. The Proclamation on the continental shelf made it clear that "the Government of the United States regards the natural resources of the subsoil and sea bed of the continental shelf beneath the high seas but contiguous to the coasts of the United States as appertaining to the United States, subject to its jurisdiction and control".[13] And the Proclamation with regard to fisheries claimed a fisheries-conservation zone in waters contiguous to the territorial sea.[14] Along with the United States' move to extend the maritime zones, the United Kingdom laid claim to submarine areas off its dependent territories by extending their boundaries to include the continental shelf contiguous to their coast.[15]

Following the Truman Proclamations, coastal States began to claim extensive areas of the continental shelf, and in some instances, waters above the continental shelf. These claims came firstly from Latin America. The movement to extend maritime jurisdiction up to 200 N.M. swept Latin America like an epidemic from 1945 to 1951: Mexico, Argentina, Panama,

Chile, Peru, Costa Rica, El Salvador and Honduras claimed their respective 200 N.M. zones in this period.[16] In 1952, Chile, Ecuador and Peru signed the Santiago Declaration. They proclaimed that "as a principle of their international policy that each of them possesses sole sovereignty and jurisdiction over the area of sea adjacent to the coasts of their respective countries, to a minimum distance of 200 N.M. from the said coasts."[17]

When the First United Nations Conference on the Law of Sea (UNCLOS I) was held in Geneva in 1958, there were thus various claims as to the breadth of the maritime zone of a coastal State ranging from 3 N.M. to 200 N.M. The proposal in UNCLOS I by the United States for a 6 N.M. territorial sea plus a further 6 N.M. fisheries zone failed to get a two-third majority.[18] UNCLOS I failed to reach a consensus on the maximum breadth of territorial waters and the contiguous fisheries zone.[19] However, as a result of UNCLOS I four Conventions were adopted. The tendency for extended maritime zones was further stimulated by the Geneva Conventions on the Continental Shelf and on the Fishing and Conservation of the Living Resources of the High Sea in two aspects. Firstly, the Geneva Convention on the Continental Shelf recognised the inherent right of a coastal State to the continental shelf. And secondly, the Convention on Fishing and Conservation of the Living Resources of the High Seas recognised that "a coastal State has a special interest in the maintenance of the productivity of the living resources in any area of the seas adjacent to its territorial seas."[20] The Convention went on to allow a coastal State to "adopt unilateral measures of conservation appropriate to any stock of fish or other marine resources in any area of the high sea adjacent to its territorial sea."[21]

The Second United Nations Conference on the Law of the Sea (UNCLOS II), held in Geneva in 1960, also failed to produce a consensus on the maximum limits of a territorial sea and exclusive fishing zones. It deserves mention that in the intervening period between UNCLOS I and UNCLOS II some countries such as Iraq, Panama, Iran, Libya and Iceland had proclaimed 12 N.M. fisheries zones.[22] Even if UNCLOS II failed to reach a consensus, the joint proposal of the United States and Canada for the formula of 6 N.M. territorial sea plus 6 N.M. exclusive fishing zone lacked only one vote to gain formal approval.[23] The failure in UNCLOS II to fix the maximum extent of territorial sea and contiguous fishing zone tells us that the battle between *Mare Liberum* and *Mare Clausum* was going to enter yet another phase. Following UNCLOS II, there was a movement by coastal States to extend their fisheries jurisdictions. In particular, many countries proclaimed 12 N.M. fishing zones, either by unilateral act or by bilateral agreement.[24] In the *Fisheries Jurisdiction* cases of 1974, where the Icelandic 50 N.M. fisheries zone was challenged by Britain and Germany, the ICJ tried not to render judgement on the possible outcome of the United Nations Conference on the Law of the Sea (UNCLOS III) on extended maritime zones as it would be judgement *sub specie legis ferende*.[25] However, the Court noted that a 12 N.M. fisheries zone had become generally accepted as lawful through the practice of States following UNCLOS II.[26] But, it is to be recalled that the claim of a fisheries zone of 12 N.M. is a modest one considering the fact that there were a number of States that claimed even 200 N.M. zones at that time.[27] In the late 1970s, the move to extend maritime claims tended to be expanded further. For instance, between 1969 and 1982, 62 States proclaimed 200 N.M. exclusive economic zones, 17 States claimed 200 N.M. exclusive fishing zone, and 7 States claimed 200 N.M. territorial seas.[28] It appeared that this movement was also influenced by UNCLOS which lasted from 1973 to 1982.

The LOS Convention, which was adopted in 1982 after 11 years of negotiations in the Third United Nations Conference on the Law of the Sea (UNCLOS III), adopted institutions of the 12 N.M. territorial sea and the 200 N.M. exclusive economic zone. These institutions are a compromise between the desire by some to extend national maritime zones and the desire of naval powers to keep a territorial sea within 12 N.M.[29] From 1982 to 1994, about 40 States claimed their respective 200 N.M. exclusive economic zones.[30] This trend continued. For instance, lastly before the new millennium, Canada, South Korea, Japan, and Denmark claimed their respective 200 N.M. exclusive economic zones in 1996 and China claimed its 200 N.M. exclusive zone in 1998.[31]

Thus, it can be said that the provisions of the LOS Convention which give a coastal State the right to establish 200 N.M. exclusive economic zone is not new to the practices of international community. As we will see later, the provision on the institution of a 200 N.M. zone is now a rule of customary international law.[32]

At present, more than 100 States have claimed a 200 N.M. EEZ or fisheries zone. The extension of maritime jurisdiction by the coastal States up into 200 N.M. or even further in case of the continental shelf has been a source of dispute between States with opposite or adjacent coast. The International Court of Justice has dealt so far with 10 cases of maritime delimitation, and there are two more such cases before the Court.[33] As mentioned earlier, as of mid-1999, only 34.5% of the potential maritime boundaries have been solved in whole or part.[34] However, delimiting the maritime boundaries is hard to achieve, whereas claiming the extended maritime zone is easily done. This means that disputes are bound to arise in the overlapping claims areas of the world.

2. INHERENT DIFFICULTIES IN DELIMITING BOUNDARIES OF THE EEZ/CONTINENTAL SHELF

2.1. *Various Factors*

It seems that many factors come into play to make the delimitation of maritime boundaries difficult in practice. For example, a coastal State can simply choose a strategy to delay the delimitation in order to keep the traditional fishing pattern in the disputed area. Or it can choose a "do-nothing policy", fearing possible criticism from the public after the delimitation is done however good the result might be. Even if two governments are determined to delimit the overlapping area, confrontation on applicable principles for delimitation can happen and they would find that Articles 74(1)/83(1) of the LOS Convention, which provide principles of delimitation of exclusive economic zone and continental shelf respectively, are not clear-cut. Even if they agreed that the principle of equity should be applied in their delimitation the following problem would come up: What are the relevant factors? Even if they agreed on the relevant factors, the next question would be how the relevant factors can be balanced off.[35]

If the two countries agreed on the principle of equidistance to be applied in their delimitation, there is also a possibility of a dispute arising regarding the validity of some base points in the delimitation or on the degree of the effect of base points.[36]

If there is a sovereignty dispute over an island in the disputed area, then the delimitation might be politically insoluble. Among the various factors which make delimitation difficult, the focus of exploration will be given to (a) the lack of clear rules, (b) the political nature of the maritime delimitation, (c) sovereignty disputes over islands, and (d) inefficiency of the dispute settlement.

2.2. *Lack of Clear Rules for Maritime Delimitation*

As we have seen earlier, there is a constant tension between the equidistance principle and the equitable principles, and each principle failed to win the titling battle against the other in the course of UNCLOS III. The provisions on delimitation in the LOS Convention are simply the result of a compromise, avoiding any clear reference either to the principle of equidistance or principle of equity. Articles 74(1)/83(1) of the LOS Convention provide that:

> The delimitation of the exclusive economic zone/continental shelf between States with opposite or adjacent coasts shall be effected by agreement on the basis of international law, as referred to in Article 38 of the Statute of the International Court of Justice, in order to achieve an equitable solution.

From this new formula, only one thing is clear: the maritime boundaries should be effected by agreement, which was enunciated also in Article 6 of the Geneva Convention and upheld in the *North Sea Continental Shelf* Cases.[37] But what else is clear? There can arise an argument, in the course of negotiations, on the meaning of the above provisions between a State which upholds the equidistance principle and another State which argues on the basis of the principle of equity. However, it is plain that the provisions of the LOS Convention leave room for such argument. Judge Oda has commented on the inefficiency of the provisions. He said that:

> Given, however, the difficulty of deriving any positive meaning from these provisions, it would seem that the satisfaction must be essentially of a negative kind, i.e., pleasure that the opposing school has not been expressly vindicated.[38]

The reason for the obscurity of the provisions is that they are the result of a compromise in UNCLOS III, intended to satisfy the two opposing groups. The issue of delimitation of maritime boundaries was one of the hard-core issues of UNCLOS III for which seven small negotiating groups were established at the seventh session (1978), and Negotiation Group 7 (NG7) was charged with the issues of delimitation and dispute settlement procedure.[39] The negotiations in NG7 were dominated by the conflict between a pro-equidistance group and a pro-equitable principle group.[40] The suggestion of the pro-equidistance group read:

> The delimitation of the Exclusive Economic Zone/Continental Shelf between adjacent or opposite States shall be effected by agreement employing, ***as a general principle, the median or equidistance line, taking into account any special circumstances*** where this is justified (emphasis added).[41]

However, a counter-proposal was made by the pro-equitable principles group. It read:

> The delimitation of the Exclusive Economic Zone/Continental Shelf between adjacent or/and opposite States shall be effected by agreement, in accordance with ***equitable principles taking into account all relevant***

> ***circumstances*** and employing any methods, where appropriate, to lead to an equitable solution (emphasis added).[42]

The deadlock in the negotiations on the rule of delimitation continued even in the resumed tenth session in 1981, and NG7 failed to solve the dilemma. Then the President of UNCLOS III, Ambassador Tomy Koh from Singapore, took the initiative. He managed to work out a compromise formula by avoiding any particular reference either to the equidistance principle or equitable principle.[43] The provisions were intended to please the opposing groups in a negative way: "the pleasure that the opposing school has not been expressly vindicated," as Judge Oda later commented.[44] Ambassador Koh himself appealed to the delegations of UNCLOS III "to avoid making interpretative statements, since such statements might undermine what had been achieved after difficult negotiations".[45]

Even if there is no reference to any particular principle in the provisions of Articles 74(1)/83(1), there is a reference to be found to "international law, as referred to in Article 38 of the Statute of the International Court of Justice". On its face the provisions beg the question of why such a reference is needed because it goes without saying that Article 38 of the Statute is generally regarded as an indication of sources of international law and an international court or a tribunal would apply international law as defined in Article 38 of the Statute.[46] With regard to this, there was a relevant remark by Manner, who was the Chairman of the NG7. According to him, the reference was intended:

> to indicate that international law, as a basis of delimitation agreements, does not differ from the law applied by the Hague Court (the ICJ). That kind of consistency is, of course, quite appropriate, but it may be questioned whether the reference to the Statute has any significance in the practical application of the relevant Articles in the Convention.[47]

The fact that Articles 74(1)/83(1) simply refer to international law which can be applied by the ICJ means that the provisions do not entail any change to the current content of customary international law on delimitation, and States should turn to customary international law developed by cases of the ICJ and other international tribunal for the rules of delimitation.[48] Evans has observed that: "Although this provision is almost entirely vacuous, it has been endorsed as an accurate statement of customary international law, although that in itself does not take the matter very much further".[49] The ICJ has remarked in the *Greenland/Jan Mayen* Case on the provision that:

> That statement of an "equitable solution" as the aim of any delimitation process reflects the requirements of customary law as regards the delimitation both of continental shelf and of exclusive economic zones.[50]

In a similar context, Professor Charney of the University of Vanderbilt emphasised that in international maritime boundary law, the judgments and awards take on even greater salience because of, *inter alia*, "the absence of clearer guidance from codification efforts".[51]

As we have discussed earlier, the two identical rules do not necessarily mean that the two boundaries should be the same. Rather the application of the identical formula can lead to two boundaries different from each other because an equitable solution for a continental shelf boundary might not be an equitable one for the delimitation of the water column.[52] It was rightly observed that:

> Although the provisions of the Convention governing continental shelf and EEZ delimitation are the same, their wording is sufficiently imprecise for the provisions to be capable of application in more than one way, and thus for the continental shelf and EEZ boundaries to differ.[53]

2.3. *Political Nature of Maritime Delimitation*

"There is no boundary which is not political".[54] The maritime boundary, like the land boundary, is the fruit of hard negotiations between neighbouring States, or the respect of a decision of an international court. The ICJ observed that: ". . . a delimitation, whether of a maritime or of a land boundary, is a legal-political operation . . ."[55]

Some important political decisions have to be made stage by stage in the process of maritime delimitation. Oxman identified four political decisions which should be made in connection with maritime boundaries: (i) the decision to negotiate, (ii) the decision to propose a particular boundary, (iii) the decision to make concessions with a view to reaching agreement, and (iv) the decision to agree on a particular boundary.[56]

Johnston observed that there are 11 diplomatic options regarding maritime delimitation: Do-Nothing Policy, Agreement to Disagree, Agreement to Designate, Agreement to Consult, Agreement on Access, Preliminary Joint Enterprise Policy, Operational Joint Development Policy, Agreement on Sharing of Services, Agreement on Limited Joint Management Arrangement, Agreement on Permanent Joint Management, Final Boundary Treaty.[57]

When faced with each decision and each option, governments tend to be cautious at every step. When a government chooses a "Do-Nothing

Policy" the delimitation of maritime boundaries will remain unsolved as long as the policy is maintained. There is on occasion a possibility that the parliament would not give its consent to the ratification of the maritime boundaries agreement. For example, the United States Senate has yet to approve the maritime boundary agreement between the United States and Cuba signed in 1977.[58] Also, it took 4 years for the government of Japan to obtain the consent from its national assembly for the ratification of a joint development of continental shelf treaty signed in 1974 with South Korea. Sometimes delimitation can become a heated national and diplomatic issue. For example, in the Beagle Channel delimitation between Argentina and Chile there were even fears of an armed conflict when Argentina rejected the award from the arbitration court in 1977 and then the Vatican intervened to mediate.[59] The dispute between Greece and Turkey, regarding the Aegean Sea continental shelf has developed into a very serious one with significant international implications. It is notable that the Security Council in the United Nations, which has the primary responsibility for the maintenance of international peace and security, dealt with the dispute while the ICJ was considering the dispute.[60] It is remarkable that the delimitation dispute in the Aegean Sea, which erupted in 1973 is still awaiting settlement.

Maritime delimitation issues can become even more intractable because of security concerns. For example, strategic considerations influenced both Sweden and the Soviet Union to take conflicting positions on the effect of the Swedish Gotland island in the negotiations on delimitation in the Baltic Sea. Sweden argued and the Soviet Union denied that the island should have a full effect. For Sweden, Gotland has a crucial role in its national defence system, and similarly the Baltic Sea is traditionally important for the Soviet Union. The negotiations became more and more politicised by the Karlskrona incident, which concerned the stranding of a Soviet submarine in the vicinity of the Swedish naval base Karlskrona on 27 October 1981.[61]

Sometimes political and security considerations are openly presented in the international court dealing with the maritime delimitation issue. For example, in the *Gulf of Maine* Case the United States stressed that the line claimed by Canada was further to the south of the line put forward by United States. It also pointed out that "if Canada's line were adopted, the entire Atlantic coast north of Philadelphia – where one-quarter of the entire population of the United Sates lives – would face not Europe, not the high seas of the open Atlantic, but Canadian waters," and then argued that "This issue raises the most fundamental question of sovereignty".[62] In this case, the United States noted the possibility of the exclusive economic zone turning into a zone of a political character in the future when it mentioned that "no State knows with any certainty what the future of the economic zone regime will hold".[63]

2.4. *Sovereignty Disputes over Islands*

There is no doubt that sovereignty disputes over islands between States make the delimitation of maritime boundaries much more complicated, because there tends to be nationalism and passionate public opinion associated with the sovereignty dispute. The reason for the extraordinary length of time taken for the delimitation of the Beagle channel between Argentina and Chile was partly, if not wholly, due to a sovereignty dispute over islands between them.[64] However, at the same time, it cannot be denied that the LOS Convention has indirectly contributed to the intensification of the sovereignty disputes over the islands, because the Convention manifestly entitles islands except rocks to generate full 200 N.M. exclusive economic zones and even farther-reaching continental shelves. The fact that a small island in the ocean can generate a 200 N.M. exclusive economic zone and continental shelf may well make the disputant States more concerned about the destiny of the disputed island than ever before. Paragraphs 2 and 3 of Article 121 of the LOS Convention provide that:

> 2. Except as provided for on paragraph 3, the territorial sea, the contiguous zone, the exclusive economic zone and the continental shelf of an island are determined ***in accordance with the provisions of this Convention applicable to other land territory***.
> 3. Rocks which cannot sustain human habitation or economic life of their own shall have no exclusive economic zone or continental shelf. [emphasis added]

Smith and Thomas from the State Department of the United States rightly observed that:

> It is fair to say that maritime resources jurisdiction is at least an implicit issue in almost all island disputes, whether or not openly stated in the publicity surrounding those disputes, and regardless of when and over what issue the dispute originated. This aspect of island disputes is clearly more pronounced than it was three or four decades ago.[65]

Some techniques have been developed to deal with the issues of disputed islands in maritime delimitation. One way is to terminate a boundary line at a point before it reaches the location where a disputed island would influence the boundary. France and Mauritius used this approach in the Indian Ocean to avoid the issue regarding the island of Tromelin.[66] Similarly, Canada and the United States agreed on the landwards terminus of the boundary at sea, in a manner which avoided the issue of sovereignty over Machias Seal Island and North Rock, when submitting their delimitation issue in the Gulf of Maine to a Chamber of the ICJ.[67] When Denmark (Greenland) and Canada agreed on a continental shelf boundary line in 1973, they left out a small area around the disputed Hans Island, which lies in mid-channel and measures less than a mile north-south.[68]

It appears that there are 27 cases of sovereignty disputes over islands in the world,[69] and it goes without saying that the sovereignty issue over islands is an important factor making maritime delimitation more difficult, however smart the delimitation technique might be.

2.5. *Inefficiency of the Dispute Settlement Procedure*

As was seen earlier, there are some inherent difficulties in maritime boundary delimitation. If bilateral negotiations on delimitation get into a stalemate then the States concerned might look into the possibility of their dispute being resolved by a third party dispute settlement procedure.[70] The LOS Convention provides for dispute settlement procedures in paragraph 2 of Articles 73 and 83. The identical paragraphs provide that if no agreement can be reached within a reasonable period of time, the States concerned shall resort to the procedures provided for in Part XV [Settlement of Dispute].

As there is no definition in the LOS Convention on the meaning of "a reasonable period of time", the definition remains an open question.[71] Once the condition of "after a reasonable period of time" is met, then the complicated provisions of Part XV (Settlement of Dispute) will be applied.

Maritime boundary disputes are basically subject to compulsory binding settlement because these are not part of the general exceptions from the compulsory binding procedure, which are provided for in Article 297. However, the Parties are free to choose, under Article 298, to opt out of the compulsory binding procedure.[72] Even if the LOS Convention is probably the most important development in the settlement of international disputes since the adoption of the UN Charter and the Statute of the International Court of Justice,[73] this opt-out provision will be disappointing to those who expected that many of the maritime boundaries disputes would be settled by the LOS Convention. But one of the main reasons for this leakage in the mechanism of compulsory binding procedures of the LOS Convention came from the fact that the provisions on the delimitation in the LOS Convention are not clear-cut. Professor Brown has observed that:

> For those who believe that the law should offer a reasonable degree of precision and predictability in exchange for the acceptance of compulsory settlement procedures, Article 83(1) leaves a great room to be desired and it is hardly surprising, therefore, that no formula embodying such an obligation [to accept compulsory binding procedures] was able to provide a basis for consensus at the Conference.[74]

However, a Party is obliged, under Article 298(1)(a)(i), to submit the boundary dispute to "compulsory conciliation" if the other Party requests it. It means that the LOS Convention adopts compulsory dispute settlement procedures on maritime delimitation disputes. The compulsory conciliation procedure for delimitation disputes is the very essential nature of the boundary dispute settlement mechanism in the LOS Convention, and it is a compromise formula to solve the confrontation between the two positions: Position I excluding all delimitation disputes from compulsory settlement procedures, and Position II subjecting all delimitation disputes to compulsory settlement procedures.[75] It was noted by the Chairman of the NG 7 that:

> . . . only a proposal based upon the procedure of compulsory conciliation is consistent with a realistic view of the possibilities, if any, to reach a compromise on this controversial issue.[76]

Why was "compulsory conciliation" viewed as the only possible compromise? It is because conciliation is not a form of "judicial settlement" such as ICJ proceedings.[77] The function of a conciliation commission is limited to investigating the dispute and suggesting the terms of a possible settlement.[78] It is "a kind of institutional negotiation".[79] The final statement of the conciliation commission is not a "judgement" or "award" which is legally binding, but a mere "report" which contains "recommendations" for the settlement of a dispute. Note here that the term "compulsory" must not be misunderstood as to mean that it is compulsory for Parties to obey the report of the conciliation commission. The term "compulsory" simply means that a dispute is to be referred to a conciliation procedure if one of the Parties wants it so.[80]

Even the application of compulsory conciliation is restricted by certain conditions. Firstly, any dispute that necessarily involves the concurrent consideration of any unsettled dispute concerning sovereignty or other rights over continental or insular land territory shall be excluded from such submission.[81] And secondly, the compulsory conciliation procedure does not apply to disputes which have arisen before the entry into force of the LOS Convention.[82] This is also the result of a long debate in UNCLOS III.[83]

Once the report of the conciliation commission is made, then the Parties negotiate an agreement on the basis of that report.[84] What if the negotiation does not result in an agreement? Article 298(1)(a)(ii) of the LOS Convention provides that:

> . . . if these negotiations do not result in an agreement, the parties shall, ***by mutual consent***, submit the question to one of the procedures provided for in section 2 (Compulsory Procedures Entailing Binding Decisions), unless the parties otherwise agree . . . [emphasis added].

Therefore, it should be noted that the obligation of Parties to submit their boundary dispute is loosened by the condition of "by consent," which means that it is no more than a *pactum de contrahendo*.[85] Therefore, a party to a boundary dispute can refuse to submit the dispute to the compulsory binding procedure.

From the examination of the dispute settlement procedure of the LOS Convention, it can be foreseen that the provisions of the LOS Convention will not be instrumental in reducing the number of disputes on maritime delimitation.

3. OBSERVATIONS

There have been constant battles between the concepts of *mare liberum* and *mare clausum* through the history of the law of the sea. Sometimes it appeared that *mare liberum* had won the battle. However, since the second half of the 20th century *mare clausum* began to dominate as coastal States proclaimed extended maritime zones such as the EEZ, exclusive fisheries zones and the continental shelf. As we have seen, the space of the high seas has significantly dwindled after World War II.

The institutions of the continental shelf and the EEZ are rules under customary international law. Although many provisions on the continental shelf and the EEZ in the LOS Convention codify customary law, some of the provisions seem to develop customary law. In particular, it remains to be seen what the future of the unattributed rights in the EEZ might be.

It was also noted that there are some inherent difficulties in delimiting boundaries of the continental shelf and the EEZ. First of all, the task of delimitation has political characteristics, involving a number of political decisions. Second, the unsolved dilemma between the principle of equity and the principle of equidistance is also an important factor. And the provisions of the LOS Convention, which are the result of a compromise between the two opposing schools and thus simply contain reference to international law and an equitable solution would not be helpful in making the delimitation law clearer. Thirdly, it was pointed out that policy makers tend to be cautious in negotiating maritime boundaries, largely due to the stability of maritime boundaries and the possibility of more solidity in the future. Fourthly, where there are sovereignty disputes over islands, maritime delimitation has another hurdle to surmount before it can ever take place. The dispute settlement procedures in the LOS Convention on maritime boundaries would not play a significant role in reducing the number of maritime boundaries disputes, as they do not provide for compulsory judicial procedures.

In a circumstance where overlapping claims are made but delimitation of the area of overlapping claims is not made, an obvious need arises to search for rules applicable between neighbouring coastal States pending ultimate maritime boundaries which may never take place.

NOTES

1. Hugo Grotius, *Mare Liberum*; Grotius wrote, "Can the vast, the boundless sea be the appendage of one kingdom alone, and it not the greatest?"; See *The Freedom of the Sea or The Right which Belongs to the Dutch to Take Part in the East Indian Trade*, translated by Ralph Van Deman Magoffin, Oxford University Press, New York (1916), p. 4.

2. Thomas Wemyss Fulton, *The Sovereignty of the Sea*, William Blackwood and Sons, Edinburgh and London (1911), pp. 340-377; The opinions and reasoning of Grotius in *Mare Liberum* as to the free use of the sea were repeated more concisely and with some modification in his greatest work, *The Rights and War and Peace*, which was published in 1625.

3. *Proclamation of James I for the Restraint of Foreigners Fishing on the British Coasts*, the text is reproduced in Thomas Wemyss Fulton, *ibid.*, pp. 755-756.

4. William Welwood, *An Abridgement of all Sea-Lawes, gathered forth of all*

Writing and Monument, which are to be found among any people or Nation upon the coasts of the great Ocean and Mediterranean Sea: And specially ordered and disposed for the use and benefit of all benevolent Sea-farers, within his Maisties Dominions of Great Britanne, Ireland, and the adjacent Isles thereof, London (1613); T.W. Fulton, *ibid.*, p. 362.

5. T.W. Fulton, *ibid.*, pp. 345-346.

6. T.W. Fulton, *ibid.*, p. 353.

7. T.W. Fulton, *ibid.*, p. 536: Joannis Seldenus, *Mare Clausum seu De Dominio Maris*, 1636, pp. 150-161.

8. It means that "the dominion of the land ends where the range of fire-arms ends"; Bynkershoek wrote that: "Wherefore on the whole it seems a better rule that the control of the land over the sea extends as far as cannon will carry; for that is as far as we seem to have both command and possession". See Cornelius van Bynkershoek, *De Dominio Maris Dissertatio* (1744), translated into English by Ralph Van Deman Magoffin, Oxford University Press, 1923, p. 44; P.C. Jessup, *The Law of Territorial Waters and Maritime Jurisdiction*, G.A. Jennings Co. (1927), p. 7.

9. Note that one nautical mile is 1,852 metres, whereas a mile is 1,600 metres. However, some writers on the law of the sea inadvertently use the term "mile" to denote "nautical mile".

10. See Wyndham L. Walker, "Territorial Waters: The Cannon Shot Rule," 22 *B.Y.I.L.* (1945), pp. 20-231. It is to be noted, however, Spain claimed 6-mile limits in 1820, and Scandinavian countries maintained a 4-mile rules established in 1812.

11. For the background see Ann L. Hollick, *US Foreign Policy and the Law of the Sea*, Princeton University Press (1981), pp. 234-237.

12. Presidential Proclamation No. 2667 concerning the Policy of the United States with Respect to the Natural Resources of the Subsoil and Seabed of the Continental Coastal Shelf High Seas, 28 September 1945; Presidential Proclamation No. 2668 concerning the Policy of the United States with Respect to Coastal Fisheries in Certain Areas of the High Seas, 28 September 1945. The Proclamations are reproduced in S. Houston Lay *et al.* eds., *New Directions in the Law of the Sea*, Oceana Publication (1973), Vol. 1, pp. 95-98 and pp. 106-109.

13. See the text of the Proclamation. It will be seen later that the Proclamation influenced the consideration of the ICJ in the *North Sea Continental Shelf* Cases, because the Proclamation mentioned "equitable principles" for delimitation of boundary of the continental shelf.

14. However, the second Proclamation was never applied: see David Joseph Attard, *The Exclusive Economic Zone*, Oxford Clarendon Press (1987), pp. 1-3.

15. See Vallat, "The Continental Shelf," 23 *B.Y.I.L.* (1946), pp. 333-338.

16. Young, "Recent Development with Regard to the Continental Shelf," 42 *A.J.I.L.* (1948), pp. 851-857; Ann L. Hollic, "The Origin of 200-mile Offshore Zones, 71 *A.J.I.L.* (1977), pp. 492-498.

17. For the Santiago Declaration, see Churchill, Simmond, Welch eds., *Op. cit.* (1973), p. 247.

18. D.W. Bowett, *The Law of the Sea*, Manchester University Press (1967), pp. 19-11. It is to be noted that according to the proposal, the rights of such States as had an established fishing practice in the outer-six zone-established for a period of five years prior to 1 January 1958 – can preserve these fishing rights in perpetuity. Canada proposed a six-mile territorial sea with an additional six-mile exclusive fisheries zone, where traditional fishing was not be recognised. See Ann L. Hollick, "US Foreign Policy and the Law of the Sea", Princeton University Press (1981), p. 141.

19. See *Official Records of the First United Nations Conference on the Law of the Sea* (1958) Vol. 3.

20. Article 6(1) of the Convention on Fishing and Conservation of the Living Resources of the High Seas.

21. Article 7(1) of the Convention on Fishing and Conservation of the Living Resources of the High Seas.

22. Whiteman, *Digest of International Law* (1965), pp. 117-120.

23. D.W. Bowett, *op. cit.* (1967), p. 11. It is to be noted that the joint proposal differed from the US proposal of 1958 in that existing fishing rights in the outer-six zone were to be preserved not in perpetuity but for a limited period of 10 years. See also Ann L. Hollick, *op. cit.* (1981), pp. 156-157.

24. For example, in 1965, South Korea and Japan concluded a fisheries agreement which entitled the parties to establish 12 N.M. exclusive fishing zone. The London Convention of 1964 provided a 3 plus 9 formula, with phasing out provisions, the Parties being Austria, Belgium, Denmark, France, Ireland, Italy, Luxembourg, the Netherlands, Portugal, the UK, and Sweden. (*U.N.T.S.*, vol. 581, p. 57.) The countries which enacted 12 N.M. fisheries zones were Iceland (1958), Faroes (1958), Morocco (1962), Tunisia (1962), Canada (1964), Ireland (1964), Turkey (1964), USA (1966), Australia (1967), Spain (1967).

25. 1974 *ICJ Reports* (*UK v. Iceland*), para. 3; 1974 *ICJ Reports* (*Germany v. Iceland*), para. 45.

26. 1974 *ICJ Reports* (*UK v. Iceland*), para. 52; 1974 *ICJ Reports* (*Germany v. Iceland*), para. 44. However, the Court found that the Icelandic 50 N.M. fisheries zone was not opposable to the Applicants, i.e., UK and Germany.

27. R. Jennings and A. Watts, ed., *Oppenheim's International Law* (9th ed.), Longman (1992), vol. 1, p. 788.

28. Source: "National Claims over Maritime Zones", *Law of the Sea Bulletin* No. 23 (June 1993), p. 67.

29. Attard wrote that "the EEZ represents an endeavour to solve the classic *mare clausum/mare liberum* dilemma"; Attard, *op. cit.* (1987), p. 1. Churchill and Lowe noted that "the EEZ could be seen as something of a compromise between those States that claimed a 200-mile territorial sea and those developed States which were hostile to extended coastal State jurisdiction; R.R. Churchill and A.V. Lowe, *The Law of the Sea*, Manchester University Press (1999), p. 161. However, it should be also noted that the agreement of the maritime States on the breadth of 12 nautical miles for the territorial waters is obtained, in part, for the assurance of enhanced navigational rights – the so-called "transit passage" – in the straits used for international navigation because the increase to 12 miles in the breadth of the territorial sea meant that many straits through which there was previously a belt of high sea would become territorial sea straits; see E.D. Brown, *The UN Convention on the Law of the Sea, 1982: A Guide for National Policy Making Legislation and Administration*, Book 1, UK Government Commonwealth Secretariat, 1987, p. 32.

30. Churchill and Lowe, *ibid.*, Appendix I.

31. The People's Republic of China announced its intention to proclaim a 200 N.M. EEZ in 1996, but legislated the Act on EEZ and continental shelf in 1998.

32. *Tunisia/Libya Continental Shelf* Case, 1985 *ICJ Reports*, para. 34.

33. Pending cases are Maritime Delimitation between Nicaragua and Honduras in the Caribbean (1999-) and Territoral and Maritime Dispute between Nicaragua and Colombia (2001-). Closed cases are Maritime Delimitation in the Area between Greenland and Jan Mayen (1988-1993); Land, Island and Maritime Frontier Dispute (El Salvador/Honduras: Nicaragua intervening 1986-1992); Continental Shelf (Libyan Arab Jamahiriya/Malta, 1982-1985); Delimitation of the Maritime Boundary in the Gulf of Maine Area (Canada/United States of America, 1981-1984); Continental Shelf (Tunisia/ Libyan Arab Jamahiriya, 1978-1982); Aegean Sea Continental Shelf (Greece/Turkey, 19760-1978); and North Sea Continental Shelf (Federal Republic of Germany/Denmark; Federal Republic of Germany/Netherlands, 1967-1969); Maritime Delimitation and Territorial Questions between Qatar and Bahrain (1991-2001); and Land and Maritime Boundary between Cameroon and Nigeria (1994-2002). It is to be noted that the ICJ held that it had no jurisdiction on the merits of the Aegean Sea Continental Shelf dispute.

34. G. Blake *et al., IBRU Workshop on International Maritime Boundaries* (4-9 April 1999), University of Durham.

35. For example, in the *Tunisia/Libya Continental Shelf* Case, both parties agreed

that the delimitation was to be effected in accordance with equitable principles taking into account of all the relevant circumstances. See 1982 *ICJ Reports*, para. 15-16.

36. For example, in negotiation on the delimitation in the Baltic Sea, Sweden and the Soviet Union both agreed that the boundary should be based on principle of equidistance. However, they disagreed on the baselines from which the equidistant line is drawn; see Alex G. Oude Elfernik, *The Law of Maritime Boundary Delimitation: A Case Study of the Russian Federation*, Martinus Nijhoff Publishers (1994), p. 207.

37. See 1969 *ICJ Reports*, para. 85.

38. Dissenting Opinion of Judge Oda, 1982 *ICJ Reports*, p. 246.

39. *UNCLOS III Official Records*, Vol. X, p. 6.

40. M.H. Nordquist *et al.* eds., *Op. cit.* (1993), pp. 801-810.

41. UN Doc NG 7/2 of 20 April 1978, reproduced by Platzöder, *Third United Nations Conference on the Law of the Sea: Document*, Vol. IX, p. 392.

42. UN Doc NG 7/10 of 1 May 1978, reproduced by Platzöder, *ibid.*, p. 402.

43. M.H. Nordquist *et al.* eds., *op. cit.* (1993), pp. 801-810.

44. Dissenting Opinion of Judge Oda, 1982 *ICJ Reports*, p. 246.

45. *UNCLOS III Official Records* Vol. XV, p. 40.

46. See Ian Brownlie, *Principles of International Law*, Oxford University Press (1998), pp. 3-10.

47. E.J. Manner, "Settlement of Sea-Boundary Delimitation Dispute According to the Provisions of the 1982 Law of the Sea Convention," J. Makarczysk, ed., *Essays in International Law in Honour of Judge Lachs*, Martinus Nijhoff Publishers (1984), p. 639.

48. E.D. Brown, *The Sea-Bed Energy and Mineral: The International Legal Regime, Vol. 1: The Continental Shelf*, Martinus Nijhoff Publishers (1992), p. 360.

49. M.D. Evans, *op. cit.* (1993/*B.Y.I.L.*), p. 294.

50. 1993 *ICJ Reports*, para. 48.

51. J.I. Charney, "Progress in International Maritime Boundary Delimitation Law," 88 *A.J.I.L.* (1994), p. 227.

52. R.R. Chirchill and A.V. Lowe, *op. cit.* (1999), p. 196.

53. R.R. Chirchill and A.V. Lowe, *ibid.*

54. P. Lapradelle, "La Frontiere: Etude de droit international," *Les Editions Internationales* (1928), p. 11.

55. 1984 *ICJ Reports*, para. 56.

56. Bernard H. Oxman, "Political, Strategic, and Historical Considerations," *International Maritime Boundaries*, pp. 10-11.

57. Douglas M. Johnston, *op. cit.* (1988), pp. 262-263; "Do-Nothing Policy" is wholly passive strategy of inaction; "Agreement to Disagree" is to be followed by a good-faith effort by the neighbours to adhere to a policy of non-provocation; "Agreement to Designate" is entering into an agreement designating the boundary area acknowledged by them to be at issue; "Agreement to Consult" is an agreement on consultation or a limited degree of co-operation within the defined area at issue; "Agreement on Access" is an agreement on access to the designated boundary area for specific purpose; "Preliminary Joint Enterprise Policy" is to negotiate a preliminary joint undertaking that contemplates future production based on some unspecified form of co-operation; "Operational Joint Development Policy" permit the negotiation of a joint development agreement for designated production purposes within the disputed area; "Agreement on Sharing of Services" is agreement on sharing of specified state services relayed to ocean uses in the boundary area; "Agreement on Limited Joint Management Arrangements" are limited to the purposes of their existing or projected joint ocean development projects; "Agreement on Permanent Joint Management" is for the establishment of a full-scale joint management system; "Final Boundary Treaty" is a formal treaty.

58. Robert W. Smith, "Report Number 1-4: Cuba-United States," *International Maritime Boundary*, pp. 416-425; the Agreement entered into force provisionally on 1 January 1978 through an exchange of notes.

59. The boundary delimitation issue in the Beagle Channel was also associated with a territorial dispute over islands and islets. Efforts by the two governments to deal with the dispute can be traced back as far as 1904. In December 1967, Chile submitted the dispute to arbitration under the General Treaty of *Arbitration* signed in 1902 between Chile and Argentina. But Argentina denied the applicability of the Treaty of 1902. Following the disagreement on dispute settlement procedure and negotiation, the two countries signed in 1972 a treaty whereby the dispute was submitted to *ad hoc* arbitration. The award was promulgated in May 1977. However, public and governmental reaction was extremely negative in Argentina and Argentina rejected the award on 25 January 1978. A tense situation was created between the two countries. Pope John Paul II intervened to mediate on 26 December 1978. Under the auspices of the Pope, the two countries solved their dispute by signing the Treaty of Peace and Friendship in 1984. It is notable that this treaty was submitted to a plebiscite in Argentina; see D.M. Johnston, *op. cit.* (1988), pp. 192-196; E. Jimenez de Aréchaga, "Report Number 3-1: Argentina – Chile," *International Maritime Boundaries*, pp. 719-721; Oxman, *op. cit.* (in *International Maritime Boundaries*), p. 10.

60. One of the main issues between Greece and Turkey in the Aegean sea was the question of entitlement of Greek islands to continental shelf in the Aegean sea. The dispute entered a critical stage in November 1973, when the Turkish government granted petroleum exploration permits for submarine areas close to the Greek islands. After a series of unsuccessful negotiations, the Greek government referred the matter in August 1976 simultaneously to the ICJ and the Security Council. Several days later the Security Council passed a resolution, calling upon the two governments to resume negotiations. In September 1976, the ICJ denied Greece's application for interim measures. In 1978 the ICJ decided that it had no jurisdiction over the dispute; see 1976 *ICJ Reports*, and Report 1978; see also Jon M. Van Dyke, "The Aegean Sea dispute: option and avenues," 20 *M.P.* (1996), pp. 397-403.

61. Alex G. Oude Elferink, *op. cit.* (1994), pp. 207-213; Oxman, *op. cit.* (in *International Maritime Boundaries*), p. 24; Sweden and the Soviet Union signed a maritime delimitation agreement in 1988 by giving 75% effect to the Swedish island of Gotland.

62. Mr. Robinson, the Agent for the Government of the United States of America stated that: "The coast of the Gulf of Maine is the principal coast of New England, facing the Atlantic Ocean and the outside world. Under the United States proposals, the maritime boundary will proceed in a southeasterly direction perpendicular to the general direction of the coast. Yet Canada persists in advocating a boundary that moves still further south across the front of the United States east coast, claiming nearly half of the last great fishing bank in the north west Atlantic Ocean to which United States fishermen have access. If Canada's line is adopted, the entire Atlantic coast north of Philadelphia – where one-quarter of the entire population of the United Sates lives – would face not Europe, not the high seas of the open Atlantic, but Canadian waters. This issue of cut-off of which the United States has made so much in this case, will be replicated in maritime boundary disputes around the world. This issue raises the most fundamental question of sovereignty. The question of cut-off posed in this case, not only in a geographic sense, but in a political sense as well, affects the interests of all States. For all our hopes that the customary law of coastal-State rights will now stabilise, no State knows with any certainty what the future of the economic zone regime will hold;" The *Gulf of Maine* Case, *ICJ Pleading*, Vol. VII, pp. 265-266.

63. *ICJ Pleading, idem.*

64. B.H. Oxman, *op. cit.* (in *International Maritime Boundaries*), p. 12. Similarly, Qatar and Bahrain settled their territorial disputes over islands, Zubarah, Hawar, and Janan, along with their delimitation issues of the territorial water, continental shelf and EEZ by the judgment of ICJ in 2001 after 10 years since Qatar asked the ICJ to adjudicate the issues in 8 July 1991: see *Qatar v. Bahrain* Case, 2001 *ICJ Reports*.

65. Robert W. Smith and Bradford L. Thomas, *Island Disputes and the Law of the Sea: An Examination of Sovereignty and Delimitation Disputes*, Maritime Briefing Vol. 2 No. 4, University of Durham (1998), p. 15.

66. Victor Prescott, "Report Number 6-5: France (Reunion)-Mauritius", *International Maritime Boundaries*, pp. 1354-1358.

67. 1984 *ICJ Reports*, para. 211; David Colson, "Report Number 1-3: Canada-United States", *International Maritime Boundaries*, p. 408.

68. Lewis M. Alexander, "Report Number 1-1: Canada-Denmark (Greenland)", *International Maritime Boundaries*, p. 372.

69. See Robert W. Smith and Bradford L. Thomas, *op. cit.* (1998), p. 3.

70. The answer to a question of whether a dispute is a maritime boundary dispute can be different depending on the formulation of the dispute. Depending on the formulation of the dispute, the dispute can be a dispute on fisheries, entitlement or delimitation; see A.E. Boyle, "Problems of Compulsory Jurisdiction and the Settlement of Disputes Relating to Straddling Fish Stocks", 14 *I.J.M.C.L.*, vol. 14 (1999), pp. 13-15.

71. A.O. Adede, *The System for Settlement of Disputes under the United Nations Convention on the Law of the Sea*, Martinus Nijhoff Publishers (1987), p. 280.

72. Argentina, Belarus, Iceland, Germany, Italy, Norway, Russia, Tunisia, and Ukraine made declaration for opt-out for the disputes on delimitation or for the all categories of disputes listed in Article 298 of the LOS Convention; see E.D. Brown, "Dispute Settlement and the Law of the Sea: the UN Convention regime", 21 *M.P.* (1997), p. 32.

73. See Alan E. Boyle, "Dispute Settlement and the Law of the Sea Convention; Problems of Fragmentation and Jurisdiction", 46 *I.C.L.Q.* (1997), p. 36.

74. E.D. Brown, *op. cit.* (1997/*M.P.*), p. 23.

75. A.O. Adede, *op. cit.* (1987), p. 182.

76. *Report of the Chairman on the Work of Negotiation Group 7*, UN Doc. NG 7/4 (1979); *UNCLOS III Official Records*, Vol. 12, p. 107.

77. Article 1 of the Regulation of the Procedure of International Conciliation adopted by the Institute of International Law in 1961 defined "conciliation" as "a method for the settlement of international disputes of any nature according to which a Commission set up by the Parties, either on a permanent basis or an *ad hoc* basis to deal with a dispute, proceeds to the impartial examination of the dispute and attempts to define the terms of a settlement susceptible of being accepted by them or of a affording the Parties, with a view to its settlement, such aid as they may have requested."

78. J.G. Merrills, *International Dispute Settlement* (3rd ed.), Cambridge University Press (1998), p. 70.

79. J.G. Merrills, *ibid.*, p. 70.

80. Norway and Iceland agreed in 1980 to refer their maritime boundary dispute around Jan Mayen to conciliation and settled their boundary dispute in 1981 following the recommendation of the conciliation commission.

81. Article 298(1)(a)(i) of the LOS Convention.

82. Article 298(1)(a)(i) of the LOS Convention.

83. A.O. Adede, *op. cit.* (1987), pp. 177-178.

84. Article 298(1)(a)(ii) of the LOS Convention.

85. E.D. Brown, *op. cit.* (1997/*M.P.*), p. 25. For the exposition on the *Pactum de Contrahendo* see, Anthony Aust, *Modern Treaty Law and Practice*, Cambridge University Press (2000), p. 25. Aust wrote that: "It has been used to refer to an agreement to conclude a treaty; agreement to include certain clauses in future agreements between the same parties; and an agreement to become party to a treaty which has already been concluded, such as the celebrated undertaking by Poland in the Treaty of Versailles to accede to the Berne Convention on Railway Transport".

CHAPTER TWO

Which Law Governs in the Absence of Maritime Boundaries?

1. CAN A UNILATERAL EQUIDISTANCE LINE BE A SOLUTION?

1.1. *Delimitation of Boundaries and Provisional Lines*

At present, many countries over the world are facing the question of how best to handle the sensitive area of overlapping claims. There have been attempts to settle the problems arising from the absence of maritime boundaries by unilaterally providing a median or equidistant line as a provisional boundary line in the absence of maritime boundary.[1] In this regard, it is clear that a unilateral boundary cannot be a valid solution under international law because of its unilateral character. According to Articles 74(1)/83(1), boundaries should be delimited by agreement. In the *Anglo-Norwegian Fisheries Jurisdiction* Case the ICJ held that:

> The delimitation of sea areas has always an international aspect; it cannot be dependent merely upon the will of the coastal State as expressed in its municipal law. Although it is true that the act of delimitation is necessarily a unilateral act, because only the coastal State is competent to undertake it, the validity of the delimitation with regard to other States depends upon international law.[2]

Similarly, the ICJ in the *North Sea Continental Shelf* Cases pointed out that:

> On a very general precept of justice and good faith, actual rules of law are here involved which govern the delimitation of adjacent continental shelves- that is to say, rules binding upon States for all delimitation – . . . the parties are under an obligation to enter negotiations with a view to arriving at an agreement . . .[3]

Having said that the maritime boundaries should be drawn by agreement through negotiations, there still remain questions: is the same rule applicable also to a "provisional boundary"? Should even a "provisional line" be drawn by agreement? These questions seem very important because if a unilateral provisional line is regarded as valid then it can be a useful instrument in preventing problems which might arise from the absence of maritime boundaries. It appears that 27 or so States in the world mention the equidistance/median line in their legislation on the continental shelf and exclusive economic zone and among those about a dozen mention the equidistance/median line as a provisional line pending the settlement of boundaries by agreement or judicial settlement.[4] It would be difficult to argue that the unilateral equidistance line as a provisional measure is a general customary law by relying on the fact that there are 12 or so instances where an unilateral equidistance line is mentioned as a provisional line. Furthermore, it is to be noted that a unilateral equidistance line cannot serve as a valid solution in the disputed areas pending maritime delimitation, unless all the neighbouring littoral States are willing to accept the equidistance line as a provisional line.[5] From the overview of relevant domestic laws in the world, a case where all the neighbouring States provide for the equidistance line as a provisional line in their domestic law could not be identified.[6] This possibly means that the unilateral equidistance line as a provisional measure is not a customary law in any region in the world, nor is it a solution in the disputed areas. Rather the unilateral equidistance line by one coastal State in a disputed area can cause a dispute if the other coastal State does not recognise it. For the above reasons, it seems appropriate to examine the positions of the Geneva Convention on the Continental Shelf and the LOS Convention to see whether they allow a unilateral equi-distance line as a provisional solution pending maritime delimitation by agreement.

1.2. *Rules under the Geneva Convention on the Continental Shelf*

Article 6 of the 1958 Geneva Convention on the continental shelf provides that:

> . . . ***In the absence of agreement***, and unless another boundary line is justified by special circumstances, ***the boundary is the median line*** . . .

Although the ILC's Commentary on these provisions suggests that the provisions were at first intended to be applied by a third Party institution,[7] the provisions themselves do not say that only a third party institution has a power to justify the median line in the absence of agreement. It is because, as Professor Bowett pointed out, "[t]he question of what are 'special circumstances' and who decides whether they justify a departure from the median line cannot be avoided".[8] Therefore, there is room in the Geneva Convention for a State to justify a provisional unilateral line when the line is the median line. Interestingly, the ICJ gave some thought to this issue, raising a question:

> [W]hat is the relationship between the requirements of Article 6 for delimitation by agreement, and the requirements relating to equidistance and special circumstances that are to be applied in the "absence of" such agreement, i.e., in the absence of agreement on the matter, is there a presumption that the continental shelf boundary between any two adjacent States consists automatically of an equidistance line, – or must negotiations for an agreed boundary prove finally abortive before the acceptance of a boundary drawn on an equidistance basis become obligatory in terms of Article 6, if no special circumstances exist?[9]

We can see in this question raised by the ICJ an important clue to an answer to the question whether or not a provisional unilateral line is permitted in international law. According to the Court's view, the median line can be applied automatically in the absence of agreement and the only question is simply whether it is before or after negotiations are exhausted for maritime boundaries. Although the Court said the determination of the question "is not necessary for the purpose of the present case",[10] it actually gave an answer to the question when it said that:

> the Parties are under an obligation to enter negotiations . . . as a sort of prior condition for ***the automatic application of a certain method of delimitation in the absence of agreement***.[11]

From the above observation, there is great room for a State to justify a unilateral provisional line in the absence of agreement if the line is a median line in terms of the 1958 Geneva Convention and case law.

1.3. *Rules under the LOS Convention*

We can see that the relevant provisions of the LOS Convention do not follow the provisions of the Geneva Convention on the Continental Shelf. Articles 74(3)/83(3) provide that:

> Pending agreement as provided for in paragraph 1, the States concerned, in a spirit of understanding and co-operation, shall make every effort to enter into provisional arrangements of a practical nature . . .

In the above provisions no possible justification for a provisional unilateral line can be found, because there is no mention of the median line which can be applied in the absence of agreement. Furthermore States concerned should make every effort in a spirit of co-operation to conclude a provisional arrangement. It will be seen later in this chapter that a proposal for the median or equidistant line as a provisional line was not accepted in UNCLOS III, because there was a fear that reference to median or equidistance line as a provisional measure might delay the final delimitation when there is no compulsory judicial settlement on maritime delimitation.

There are jurists who hold critical views on Articles 74(3)/83(3). For instance, Professor Brown has mentioned that "it is difficult to see that any great purpose is served by this provision".[12] However, it must be noted that this provision creates new rules which are in not accordance with provisions in the Geneva Convention. Under this new provision justification for a unilateral median line cannot be found, unlike under Article 6 of the Geneva Convention. These provisions can be useful in some respects, at least by prohibiting a unilateral provisional line. We will discuss several important purposes these provisions can serve in the absence of boundaries.

1.4. *The Need for Searching for Rules in the Disputed Areas*

After seeing that a unilateral line cannot be a solution in the absence of maritime boundaries under the provisions of the LOS Convention, a question arises as to what law is applicable in disputed areas. There is no doubt that some fundamental norms which will be applicable in the absence of an agreed boundary between States with opposite or adjacent coasts pending the delimitation of boundaries can be found in general principles of law. As Article 38 of the Statute of the ICJ provides, the general principles of law recognised by civilised nations are an important source of international law applicable by the Court. Principle of *res judicata*, the principle of good faith, certain principles relating to procedure, the principle that what is not forbidden is allowed, the principle proscribing the abuse of rights, the principle according to which, under special circumstances, the stronger takes rightful precedence over the weaker, and the principle *lex specialis generalibus derogat* are examples of the general principles of law that were cited in the *travaux preparatoire* of Article 38 of the ICJ Statute.[13] However, it is to be noted that general principles of law do not consist of specific rules formulated for practical purposes.[14]

Article 2(3) of the Charter of the United Nations Charter, which obliges all Member States to settle their international disputes by peaceful means is also an important rule in relation to disputed areas. The principle of good neighbourliness, which is sometimes explained by reference to the maxim *sic utere tuo, ut alienum non laedas* also seems important for the States with opposite or adjacent coasts pending final delimitation.[15] These are of significance and importance because there is a high possibility of a coastal State exercising its rights or carrying out activities in the disputed area in such a way as to impair the rights of the other coastal States. According to the principle of good neighbourliness, "international law does not allow States to conduct activities within their territories, or in common spaces, without regard for the rights of other States".[16]

If the States concerned are parties to the LOS Convention, then the relevant provisions are applicable. Then what rules in the LOS Convention are applicable between States with opposite or adjacent coasts? Firstly, the general provisions of Part XVI such as good faith and abuse of rights (Article 300), peaceful uses of the seas (Article 301), and disclosure of information (Article 302), are to be applied. There are other relevant rules of the LOS Convention which can be applicable. For example, the States shall co-operate on a regional basis for the protection and preservation of the marine environment of the disputed area.[17]

However, it seems that the provisions which deal, in the most direct form, with the legal relations between States with opposite or adjacent coasts pending ultimate delimitation of EEZ and the continental shelf are Article 74(3) and Article 83(3) of the LOS Convention.

> Pending agreement as provided for in paragraph 1, the States concerned, in a spirit of understanding and co-operation, shall make every effort to enter into provisional arrangements of a practical nature and, during this transitional period, not to jeopardise or hamper the reaching of the final agreement. Such arrangement shall be without prejudice to the final delimitation.

In the analysis of rules applicable in the absence of maritime boundaries of extended maritime zones priority will be given to these provisions, as they form the essential rules in the legal regime between States concerned before final delimitation of the EEZ/continental shelf. In the interpretation of the provisions, we will search for the ordinary meaning of the terms of the articles in the their context and in the light of their object and purpose as provided for in Article 31 of the Vienna Convention on the Law of Treaties.[18]

However, before directly embarking upon an interpretation based upon the ordinary meaning of the provisions, whether it be literal or logical interpretation, it seems desirable to look into the drafting history of the provisions and *travaux preparatoires* in order to ascertain the intentions of the negotiating parties in UNCLOS III.[19] An interpretation of provisions of the LOS Convention, which are a result of 11 years of negotiations, and particularly the provisions of Articles 74(3)/83(3) which were adopted through a complicated process of consideration, without resort to *travaux preparatoires* might lead to an unreasonable interpretation. O'Connell rightly observed that: "To interpret it (a treaty) without reference to the struggle for compromise is gravely to over-simplify the problem of treaty application".[20] In this regard, it is to be recalled that Article 32 of the Vienna Convention on the Law of Treaties allows for resort to *travaux preparatoires* as a supplementary means of interpretation in order either to confirm the result of the ordinary meaning interpretation or to determine the meaning of the provisions when the ordinary meaning interpretation leads to a manifestly absurd or unreasonable result.[21] However, in interpretation of Articles 74(3)/83(3), a balance between the final text and the amorphous mass of documentation which arises under the heading of *travaux preparatoires* will be maintained.

2. DRAFTING HISTORY OF ARTICLES 74(3)/83(3) OF THE LOS CONVENTION

2.1. *General Background*

During the Third United Nations Conference on the Law of the Sea (UNCLOS III), Articles 74 and 83, which deal with problems regarding delimitation of the EEZ and the continental shelf respectively, were negotiated together.[22] Furthermore, the "provisional arrangements pending delimitation" matter was discussed in Committee II along with such issues as delimitation and dispute settlement.[23] Although the problem concerning "provisional arrangements pending delimitation" was not highlighted as a main issue of UNCLOS III or the key factor in regard to boundary delimitation, there were some serious discussions on this issue.

At the 1973 session of the Sea-Bed Committee, proposals regarding the delimitation of overlapping claims were divided into two lines of arguments: one was the principle of equidistance or median line and the other was the principle of equity. At this session, the issue of provisional arrangements that are to be applied before delimitation, was not discussed.[24]

2.2. *The First Discussion at the Second Session of UNCLOS III*

The issue of provisional measures pending maritime delimitation arose for consideration for the first time in the Second Committee at the second session of UNCLOS III (1974). It appears from the drafting history that the Netherlands initially touched upon the issue of the absence of maritime boundaries.[25] It supported the application of equitable principles in determining maritime boundaries between States and went on to argue that, pending an agreement on overlapping claims, neither State could establish its maritime boundaries beyond the median line.[26]

A proposal by Greece during the second session of the Conference stressed the role of the median line in delimitation of the continental shelf/EEZ, and suggested that the median line should be used in the absence of agreement between States.[27] Japan joined the argument and put forward a proposal which was similar to that of Greece.[28]

However, since those proposals made by the Netherlands, Greece and Japan included the median line concept, although as an interim measure, it was difficult to expect the States which supported the equitable principle doctrine to agree to it. Consequently, Ireland, which supported the equitable principles doctrine for the delimitation of the continental shelf/the EEZ, proposed a moratorium of exploration or exploitation activities in any area which is claimed *bona fide* by any other State except with the express consent of that State as a provisional measure.[29]

2.3. *From Median Line (ISNT) to Provisional Arrangements (RSNT)*

At the third session of 1975, an informal consultative group on delimitation was set up within the Second Committee, which in turn, had a serious discussion about the principle of maritime delimitation and provisional measures pending delimitation. After the discussion at the third session, Article 61 of the ISNT (Informal Single Negotiation Text)/Part II supported the median or equidistant line as an interim measure pending delimitation of the EEZ or continental shelf boundaries. Article 61 (3) of the ISNT read as follows:

> Pending agreement, no State is entitled to extend its exclusive economic zone beyond the median line or the equidistance line.[30]

The RSNT(Revised Single Negotiation Text) produced at the fourth session (1976) dropped the reference to the median or equidistant line as an interim arrangement. This was because there were concerns that using the median or equidistant line before boundary delimitation might delay a final solution when there is no compulsory judicial settlement on maritime delimitation.[31] With regard to such concerns, the Chairman of the Second Committee remarked:

> Since the Conference may not adopt a compulsory jurisdictional procedure for the settlement of delimitation disputes, I felt that the reference to the median or equidistant line as an interim solution might not have the intended effect of encouraging agreements. In fact such reference might defeat the main purport of the article as set out in paragraph 1. Nonetheless, the need for an interim solution was evident. The solution was, in my opinion, to propose wording in paragraph 3, which linked it more closely to the principle in paragraph 1.[32]

Even if the reference to the median or equidistant line as a provisional measure was deleted in the RSNT, it still recognised the need for certain interim measures before delimitation. Thus, paragraph 3 of Article 62/71 of the RSNT provided that:

> Pending agreement or settlement, the States concerned shall make provisional arrangements, taking into account the provisions of paragraph 1.[33]

It is to be noted that this requires the States concerned to make provisional arrangements taking into account paragraph 1, pending agreement or settlement. Here we can see the difference between provisions of the ISNT and the RSNT in this regard. In other words, the provision of the RSNT purported to encourage development activities through provisional arrangements pending delimitation, whilst the intention of the article of the ISNT was to prevent States from exercising jurisdiction beyond the median line. It can be said that the provision of the RSNT takes "an incentive approach", whereas the provision of the ISNT is based on "a prohibitive approach".

At the fifth session (1976), the Second Committee discussed various proposals on the delimitation of the continental shelf and EEZ between opposite or adjacent States. However, the direction of the discussion continued to vary between equitable principles and the principle of equidistance.[34]

At the sixth session (1977), there was an attempt by Spain to reintroduce the median or equidistant line as a provisional measure pending delimitation.[35] States which supported delimitation in accordance with equitable principles supported the approach of the RSNT, which provided for "provisional arrangements" without any reference to the median line.

However, those States which supported the median line contended that it should be used as an interim measure. In the ICNT (Informal Composite

Negotiation Text) of 1977, the provisions on delimitation of the EEZ and the continental shelf in the RSNT were renumbered as Articles 74 and 83 respectively as the LOS Convention stands now, but repeated almost verbatim as Articles 62/71 of the RSNT. And the provisions of paragraph 3 of Articles 74 and 83 of the ICNT were the exactly same as those of paragraph 3 of Articles 62/71 of the RSNT which provided for "provisional arrangements" without referring to median lines as provisional lines.[36]

2.4. *Median Line, Moratorium, or Provisional Arrangement?*

At the seventh session in the spring of 1978, the topic of delimitation of the EEZ and the continental shelf and settlement of disputes thereon were classified as hardcore issues and referred to Negotiating Group 7 (NG7) under the chairmanship of Eero J. Manner.[37] The difference between the "Preventive Approach" of States advocating the median line and the "Incentive Approach" of States supporting the use of equitable principle became a principal issue in NG 7. A group of 20 States which endorsed the median line made the following suggestion:

> 3. Pending agreement or settlement in conformity with paragraphs 1 and 2, the parties in the dispute shall refrain from exercising jurisdiction beyond the median or equidistance line unless they agree on alternative interim measures of mutual restraint.[38]

On the other hand, a rival group of 22 States that favoured the equitable principles approach argued for the maintenance of the provisions of the RSNT, which provided a provisional arrangement pending delimitation without any reference to the median or equidistance line.

Under such circumstances a formula to bridge the gap between the two opposing positions had to be pursued. A proposal to this end was made by Morocco:

> Pending the conclusion of an agreement or settlement, the States concerned shall abstain from any measure which could prejudice a final solution or in any way, aggravate their conflict, and shall endeavour to reach mutually acceptable, provisional arrangements, regarding the activities in the *bona fide* disputed area.[39]

By way of contrast, Papua New Guinea reintroduced the idea of a moratorium on economic activities in the area under dispute.[40] An informal paper put forward by Norway contained a number of elements for discussion on provisional arrangements, including (i) whether guidelines for limitation on the exercise of jurisdiction should be subject to functional distinctions; and (ii) whether general guidelines should address the conduct of the States concerned under provisional arrangements.[41]

The Chairman of NG 7 reported that while there was general agreement on including a provision on interim measures to be applied pending agreement in delimitation cases, in view of the link with paragraph 1 there were no formulations that would offer a substantially improved prospect of consensus.[42]

At the resumed seventh session in the autumn of 1978, most of the NG 7's attention was devoted to provisional measures pending definitive delimitation and the settlement of disputes.[43] There was a view that a more modest approach might be better, such as one that focuses on the duties of coastal States throughout their economic zones and continental shelf under the substantive provisions of the Convention and which provides for consultation.[44] Mr. Manner noted that there was no consensus that States should be obliged to make provisional arrangements; instead they should be encouraged to adopt interim measures as appropriate. He also mentioned that the Moroccan proposal could be the basis for compromise.[45]

2.5. *Consensus for Mutual Arrangements*

At the eighth session of UNCLOS III in 1979, it became clear that a moratorium could not form an acceptable part of a provisional regime.[46] Mr. Manner suggested that, in order to avoid the introduction of a "moratorium" on economic exploitation pending agreement, the prohibition on "unilateral actions" might be added.[47] India, Iraq and Morocco put forward a joint proposal. This suggested that provisional arrangements between the States concerned should be entered into "in a spirit of co-operation". According to the proposal, arrangements could be based on mutual restraint or mutual accommodation. Furthermore, the arrangement was not to prejudice the final solution on delimitation. The proposed text read:

> 3. Pending agreement or settlement, the States concerned shall, in a spirit of co-operation, freely enter into provisional arrangements. Accordingly, they shall refrain from activities or measures, which may aggravate the situation or jeopardise the interests of either State, during the transitory period. Such arrangements, whether of mutual restraint or mutual accommodation, shall be without prejudice to the final solution on delimitation.[48]

In the course of discussions at the eighth session, the Chairman of the NG7 offered a suggestion which was in substance the same as the above proposal made by India, Iraq and Morocco.[49] The Chairman's proposal, however, did not gain enough support to justify a revision of the ICNT. With the lack of consensus on the Chairman's proposal, paragraph 3 of Articles 74/83 of ICNT/Rev.1 repeated verbatim the corresponding provisions of the RSNT and ICNT. The text read:

> 3. Pending agreement or settlement, the States concerned shall make provisional arrangements, taking into account the provisions of paragraph 1.[50]

At the resumed eighth session (1979), the discussions on the provisional arrangements were conducted on the basis of the Chairman's proposal which was put forward at the eighth session of 1978. As a result of these discussions, the text of a compromise proposal was prepared by the Chairman of NG 7 and included in his summary report. It read:

> 3. Pending agreement as provided for in paragraph 1, the States concerned, in a spirit of understanding and co-operation, shall make every effort to enter into **provisional arrangements of a practical nature** and, during this transitional period, not to jeopardise or hamper the reaching of the final agreement. Such arrangements shall be without prejudice to the final delimitation [emphasis added].[51]

At the ninth session of 1980, the group of States favouring the use of the median or equidistance line proposed the following alternative text:

> 3. Pending agreement or settlement in conformity with paragraphs 1 and 2, the parties in the dispute shall refrain from exercising jurisdiction beyond the median or equidistance line unless they agree on alternative interim measure of mutual restraint.[52]

The group of States which supported the use of equitable principles as the primary criterion in delimitation argued for provisions on provisional arrangements which are identical to the corresponding provisions in the RSNT, ICNT and ICNT/Rev.1. It read:

> 3. Pending agreement or settlement, the States concerned shall make provisional arrangements, taking into account the provisions of paragraph 1.[53]

Notwithstanding the difference between the proposals of the two groups, the ICNT/Rev.2 which was produced at the ninth session contained the provisions which were identical to the text of the compromise proposal prepared by the Chairman of NG 7 at the resumed eighth session.[54]

The above provisions in the ICNT/Rev.2 remained intact throughout the ninth, the resumed ninth, and the tenth sessions and thus became the framework of the present paragraph 3 of Articles 74/83 of the Los Convention.

3. MEANING OF ARTICLES 74(3)/83(3) OF THE LOS CONVENTION

3.1. *Provisional Measures and Provisional Arrangements*

3.1.1. *Provisional Arrangement as Provisional Measures*

An "interim measure" or a "provisional measure" is generally acknowledged to be a measure which is intended to be implemented before the final decision. In domestic legal proceedings provisional measures or interlocutory injunctions are widely used by courts, at any time during the pendency of litigation for the short-term purpose of preventing irreparable injury to the petitioner prior to the time that the court will be in a position to either grant or deny permanent relief on the merits.[55] In national and international procedural law generally, provisional measures have only a temporary character and leave unaffected the legal positions of the States concerned, particularly those of any parties to the dispute. It can be noted here that the provisional arrangements in paragraph 3 of Articles 74/83 share common characteristics with the "interim measure" which are commonly used in courts, because these arrangements purport to be applied before the final stage, i.e. ultimate delimitation, without affecting the rights of the Parties.

The legal nature and function of provisional measures in international law are well illustrated in the United Nations Charter. Article 40 of the Charter confers on the Security Council of the United Nations, which has the primary responsibility for the maintenance of international peace and security, a power to decide upon provisional measures. It provides that:

> ***In order to prevent an aggravation of the situation***, the Security Council may, before making the recommendations or deciding upon the measures provided for in Article 39, call upon the parties concerned to comply with such provisional measures as it deems necessary or desirable. ***Such provisional measures shall be without prejudice to the rights, claims or position of the parties concerned***. The Security Council shall duly take account of failure to comply with such provisional measures [emphasis added].

As the phrase "in order to prevent the aggravation of the situation" implies, the provisional measures of the Security Council are of a preventive character rather than one of incentive. They are simply a means of preventing an aggravation of the situation.[56] Indicating provisional measures is also one of the important functions of the International Court of Justice. It has competence to indicate a provisional measure before final judgment. Article 41 of its Statute provides that:

> 1. The Court shall have the power to indicate, if it considers that circumstances so require, ***any provisional measures which ought to be taken to preserve the respective rights of either party***.
> 2. ***Pending the final decision***, notice of the measure suggested shall forthwith be given to the parties and to the Security Council [emphasis added].

In the orders made in the *Fisheries Jurisdiction* Cases on 17 August 1972, the ICJ spelled out the essential character of the provisional measures of the Court:

> Whereas the rights of the Court to indicate provisional measures as provided for in Article 41 of the Statute has as its object to preserve the respective rights of the Parties pending the decision of the Court, and presupposes that irreparable prejudice should not be caused to the rights which are the subject of dispute in judicial proceedings and that the Court's judgement should not be anticipated by reason of any initiative regarding the measures which are in issue.[57]

It has also been pointed out that: "The function of interim protection is easily stated. It is to safeguard the rights which are in dispute, pending the Court's decision on the merits".[58]

With regard to obligations between parties to a dispute under interim measures, Rosenne observed that:

> There is sometimes said to exist a generally recognised principle according to which the institution of judicial proceedings itself operates as a provisional measure of protection, since the parties are under an implied obligation, until the Court has reached its decision in the case, to refrain from any steps which might have a prejudicial effect on the execution of the Court's final decision, or which might exacerbate the dispute.[59]

The LOS Convention also provides for provisional measures in regard to its dispute settlement procedures. These provisions are intended to protect the marine environment or to preserve the rights of States concerned before a court or a tribunal reaches the final decision on the merits of the case. Paragraph 1 of Article 290 provides that:

> . . . If a dispute has been duly submitted to a court or tribunal which considers that *prima facie* it has jurisdiction . . . the court or tribunal may prescribe any provisional measures which it considers appropriate under

> the circumstances to preserve the respective rights of the parties to the dispute or to prevent serious harm to the marine environment, pending the final decision.

Paragraph 5 of Article 290 also provides a basis for provisional measures even before the constitution of an arbitral tribunal to which a dispute is being submitted under the LOS Convention.[60]

The International Tribunal for the Law of the Sea ("ITLOS") has ordered provisional measures on two occasions since its establishment in 1996. The first occasion was in the *M/V Saiga* Case No. 2 between St. Vincent and Grenadines and Guinea.[61] The Tribunal ordered provisional measures on the basis of paragraph 1 of Article 290 of the LOS Convention thereby requiring Guinea to comply, without delay, with the judgment of the Tribunal of 4 December 1997.[62] The Tribunal also made it clear that the parties should take no action that might aggravate or extend the dispute.[63] When the Tribunal ordered the provisional measures on 11 March 1998 it examined its rationale for them as follows:

> Considering that the rights of the Applicant would not be fully preserved if, pending the final decision, the vessel, its Master and the other members of the crew, its owners or operators were to be subjected to any judicial or administrative measures in connection with the incidents leading to the arrest and detention of the vessel and to the subsequent prosecution and conviction of the Master; . . . Considering that, in order to prevent aggravation or extension of the dispute, the parties should endeavour to find an arrangement to be applied pending the final decision, without prejudice to their contentions on jurisdiction or merits; Considering that any action or abstention by either party to avoid aggravation or extension of the dispute should not in any way be construed as a waiver of any of its claims or an admission of the claims of the other party to the dispute.[64]

The ITLOS again ordered interim measures on 28 August 1999 on the basis of paragraph 5 of article 290 of the LOS Convention in the *Southern Bluefin Tuna* Cases between Australia, New Zealand and Japan.[65] The Tribunal ordered, *inter alia*, that "Australia, Japan and New Zealand each ensure that no action is taken which might prejudice the carrying out of any decision on the merits which the arbitral tribunal may render, and that each State shall refrain from conducting an experimental fishing program involving the taking of a catch of southern bluefin tuna, except with the agreement of the other parties or unless the experimental catch is counted against its annual national allocation . . ."[66]

When the Tribunal prescribed the provisional measures in the *Southern Bluefin Tuna* Cases, it examined the rationale behind the necessity of the provisional measure and said that:

> . . . Australia and New Zealand contend that further catches of southern bluefin tuna, pending the hearing of the matter by an arbitral tribunal, would cause immediate harm to their rights; . . . that the conservation of the living resources of the sea is an element in the protection and preservation of the marine environment; . . . although the Tribunal cannot conclusively assess the scientific evidence presented by the parties, it finds that measures should be taken as a matter of urgency to preserve the rights of the parties and to avert further deterioration of the southern bluefin tuna stock.[67]

It seems from the above that provisional measures under the LOS Convention are basically the same as those of the ICJ. However, it is to be noted that the provisional measures of the LOS Convention are intended to protect not only the respective rights of the parties but also to prevent serious harm to the marine environment.[68] In other words, the marine environment itself is regarded as an important legal interest in the LOS Convention.

3.1.2. *Provisional Arrangements by Mutual Consent*

As noted above, the provisional arrangements provided for in paragraph 3 of Articles 74/83 share common characteristics with the provisional measures of the Security Council, the ICJ, and the ITLOS. Such arrangements should (i) preserve the respective rights of the parties on the issues of final delimitation and on the resources of the disputed areas; (ii) irreparable prejudice should not be caused to the disputed rights; and (iii) final delimitation should not be anticipated by reason of the provisional arrangements.

Even if the provisional arrangements have similar functions by virtue of being a kind of provisional measure in international law, those which are envisaged in paragraph 3 of Articles 74/83 of the LOS Convention also have their own distinctive characteristics. Firstly, such provisional arrangement is something "mutually" done between two or more States, whereas the provisional measure of the Security Council, the ICJ, or the ITLOS is an order of a third institution. It is to be noted that the Security, the ICJ, or the ITLOS has a discretion, if it has the proper jurisdiction, as to whether a situation requires an interim measure or not, whereas coastal States are obliged to make every effort to enter into provisional arrangement. In this regard, it has been pointed out that the term "arrangement" rather than "measures" was chosen to avoid the implication of assigning the role of making such arrangement to a third party forum under Part XV of the LOS Convention.[69]

Secondly, an important and distinct aspect of the provisional arrangements under consideration in this study is that they are intended not only to preserve the rights of States concerned but also to promote certain activities, i.e., utilising activities of the area to be delimited. As we have seen in the drafting history of common paragraph 3, the provisions of the paragraph are based upon an incentive approach. It is to be recalled here that the proposal of Ireland which supported a moratorium on economic activities in the overlapping areas and the preventive approach of the ISNT which prohibited extending of the EEZ beyond the equidistant or median line were both discarded. In contrast, the nature of the provisional measures of the Security Council, the ICJ, and the ITLOS is basically prohibitive as these are based upon the preventive approach. The provisional measures of those institutions are not intended to promote specified activities but only to prevent the aggravation of situations and to preserve the rights of States concerned.

3.2. *Legal Nature of Arrangements in Terms of the Law of Treaties*

3.2.1. *Practice in Use of the Term "Arrangements"*

As Articles 74(3)/83(3) utilise, for the purpose of referring to the provisional measure, the term "arrangement" rather than "agreement" or "treaty", it will be worthwhile to examine the legal nature of the "arrangement".

The term "arrangement" has been used often to indicate legal instruments less formal than "treaty" and "agreement". For example, Lord McNair picked up the term "arrangements" for describing the inter-departmental agreement, "which concern matters of private law rather than matters of an international legal character".[70] For O'Connell, 'an arrangement' is an agreement of an ephemeral nature".[71]

The 1988 UN Convention against Illicit Traffic in Narcotic Drugs and Psychotropic Substances envisage co-operation between parties through agreements or arrangements. Article 17(4) of the Convention provides that:

> 4. In accordance with paragraph 3 or in accordance with treaties in force between them or in accordance with any agreement or ***arrangement*** otherwise reached between those Parties, the flag State may authorise the requesting State to, *inter alia*:
> (a) Board the vessel;
> (b) Search the vessel;

> (c) If evidence of involvement in illicit traffic is found, take appropriate action with respect to the vessel, persons and cargo on board. [emphasis added].

In this regard, it has been pointed out that Article 17 of the 1988 UN Convention is primarily concerned with making detailed provisions for procedures designed to provide maximum opportunities for Parties, to obtain enforcement jurisdiction with the consent of the flag State.[72] As the term "arrangement" is used along with the terms "treaties" and "agreement" in the above Article 17(4), the term "arrangement" seems to include something less formal than those which treaties and agreements normally denote. In practice, co-operation between the law enforcement agencies of two countries has been taking place through less formal arrangements. It was observed that the United States government has entered into special arrangements with several countries for co-operation in the field of law-enforcement at sea and in most cases the arrangement is oral and operates case by case.[73] The SEARIDER program formulated on 29 January 1986 between the Royal Bahamas Defence Force and US Coast Guard, and Shiprider Memorandum of Understanding concluded on 8 February 1996 between the UK on behalf of the British Virgin Islands and the USA were intended for law enforcement co-operation at sea through bilateral arrangement.[74]

State practices of the United Kingdom and the United States regarding the use of the term "arrangement" shows that an arrangement is usually intended to be informal, i.e. not intended to be legally binding.[75] Regarding this, the International Agreement Regulation made by the US State Department in 1981 in paragraph (a) (2) provides that:

> (2) *Significance of the arrangement.* Minor or trivial undertakings, even if couched in legal language and form, are not considered international agreements. . . . In deciding what level of significance must be reached before a particular arrangement becomes an international agreement, the entire context of the transaction and the expectations and intent of the parties must be taken into account. It is often a matter of degree . . .[76]

However, it is also to be noted that the term "arrangement" has been used sometimes to indicate co-operative mechanism between States or the instruments which provide legal basis for such mechanism regardless of their nomenclatures. When the term "arrangement" is used in this context, then the titles and formalities of instruments providing such arrangement can vary. For example, the Charter of the United Nations adopted the term "arrangement" in Article 52(1). It provides that:

> Nothing in the present Charter precludes the existence of regional **arrangements** or agencies for dealing with such matters relating to the maintenance of international peace and security as are appropriate for regional action, provided that such **arrangements** or agencies and their activities are consistent with the Purposes and principles of the United Nations [emphasis added].

According to a commentary on the Charter of the United Nations, 23 treaties of friendship and mutual assistance between the Soviet Union and the Eastern European Peoples' Democracies, the Treaty of Brussels among Great Britain, France, Belgium, the Netherlands and Luxembourg concluded in 1948, the North Atlantic Pact concluded in 1949 and the Treaty of Rio de Janeiro for the Mutual Defence of America concluded in 1949 are all arrangements under this provision.[77] Thus it is obvious that the term "arrangements" in the U.N. Charter was not used to indicate only MOU or other informal or less formal instruments. Professor Bowett viewed the term "arrangement" in the Charter as a term focusing on action rather than organisation itself. He has argued that the question is not whether a given organisation is a regional arrangement or not, but rather, whether particular action is taken as a regional arrangement or not.[78]

The UN Fish Stocks Agreement also used the term "arrangement" to indicate co-operative mechanism rather than a particular form of instrument. Notice that it obliges coastal States and fishing States "to make every effort to enter into provisional arrangements of a practical nature" "pending agreement on compatible conservation and management measures".[79] Article 1 of the Agreement defines the term "arrangement" as a co-operative mechanism. It defined the term as follows:

> "arrangement" means a co-operative mechanism established in accordance with the Convention and this Agreement by two or more States for the purpose, *inter alia*, of establishing conservation and management measures in sub-region or region for one or more straddling fish stocks or highly migratory fish stocks.[80]

When the terms "arrangement" is used to indicate co-operative mechanism and action, then the titles and formalities of the documents providing such mechanism and action would not matter. We can find that this kind of usage of the term "arrangement" in the *Fisheries Jurisdiction* Case. In this case, the United Kingdom challenged the legality of the Icelandic 50 N.M. exclusive fisheries zone. On 17 August 1972, the ICJ made an Order for provisional measure indicating, *inter alia*, that pending the Court's final decision, Iceland should refrain from taking any measures to enforce its fisheries law against British fishing vessels in the disputed area.[81] However, Iceland did not comply with the Court's Order, denying the jurisdiction of the Court on this issue and began to enforce its new fisheries law against British fishing vessels.[82] Under these circumstances, the British government proposed to the Icelandic government negotiations for an "interim regime".[83] Negotiations for an interim arrangement were carried on intermittently during 1972 and 1973. At the last stage of the negotiations, the discussions between the Prime Ministers of the two countries in 1973 led to the conclusion of an "Interim Agreement in the Fisheries Dispute" constituted by an Exchange of Notes dated 13 November 1973. The terms of the Agreement were set out in the Icelandic Note, which provides that:

> . . . the following ***arrangements*** have been worked out for ***an interim agreement*** relating to fisheries in the disputed area, pending a settlement of the substantive dispute and without prejudice to the legal position or rights of either Government in relation thereto . . . [emphasis added][84]

Then the Court went on to examine the relationship between the interim agreement which had a two-year duration clause and its judgment. It said:

> The judgement of the Court . . . will not completely replace with immediate effect the interim agreement, which will remain ***a treaty in force***. In so far as the judgement may possibly deal with matters which are not covered in the interim agreement, the judgement would . . . have immediate effect . . .[85]

From the dicta of the ICJ in the *Fisheries Jurisdiction* Case, we can see that the term "arrangements" was used interchangeably with the term "agreement" to indicate the Exchange of Notes, which is a treaty.

3.2.2. *Legal Nature of Arrangements in the Light of the Drafting History*

Now, let us turn to the drafting history of Articles 74(3)/84(3) in order to see what was the intention of the drafters in providing the term "arrangement". It was first introduced in the RSNT at the fourth session in 1976. According to the RSNT, States concerned shall take into account the delimitation principle in paragraph 1 in making the "provisional arrangement". It is to be recalled that the term "provisional arrangement" in the RSNT came up to replace the reference to the median or equidistance line, when there were worries of compulsory judicial settlement on delimitation being deleted. From the drafting history, it is clear that "provisional arrangements" in the RSNT were intended to take the form of provisional boundaries, because the provisions of paragraph 3 mentioned paragraph 1 of the same article which provided for rules on delimitation of boundaries.

By way of contrast, the proposal by Morocco in 1978 mentioned "provisional arrangements, regarding the activities in the *bona fide* disputed area". The joint proposal of India, Iraq and Morocco in 1979 mentioned "arrangements, whether of mutual restraint or mutual accommodation". The Chairman's proposal at the eighth session in 1979 viewed the provisional arrangement in connection with an obligation to refrain from aggravating the situation. The Chairman's proposal at the resumed eighth session in 1979 mentioned "provisional arrangement of a practical nature" and, along with that, it provided an obligation not to jeopardise or hamper the reaching of a final agreement. It appears, therefore, that in the course of negotiation since 1978, the provisional arrangement concept was envisaged as something related to activities in the overlapping area.

From the above review of the drafting history of Articles 74(3)/83(3) a number of conclusions can be reached. First, from the review of the provisions of the RSNT, it can be said that an arrangement could be a provisional line, because it said that paragraph 1 on the principle of delimitation is to be taken into account in making provisional arrangements. However, as the reference to paragraph 1 was subsequently dropped, the provisional arrangement need not necessarily be a provisional boundary.

Second, from the proposal by Morocco and the joint proposal by Morocco, India and Iraq, we can say that the provisional arrangement is about promoting or restricting activities in the disputed area. The Chairman's proposal at the eighth session and resumed session supports this observation. It is to be noted that the term "of a practical nature" was inserted, when the reference to paragraph 1 was dropped. This suggests that the objective of provisional arrangements as now provided in the LOS Convention is mainly for promoting and regulating activities in the overlapping zones.

3.2.3. *Provisional Arrangements and Treaties*

Note that at the beginning of Articles 74(3)/83(3), there is the phrase "pending agreement as provided for in paragraph 1", and in the last part of the same paragraph, is located the phrase "provisional arrangements". We can thus see that the drafters used the term "arrangement" in distinction with the term "agreement". The term "agreement" is used for the indication of a document containing a final delimitation, whereas the term "arrangement" is used to indicate a set of provisional rules regarding activities in a disputed area. As the arrangement of a practical nature is not an agreement on final delimitation, the arrangement can be less formal than the agreement on final delimitation. Or alternately, the arrangement can be informal, i.e. intended not to be legally binding. The arrangement can be in the form of a *Note Verbale*, Exchange of Notes, Agreed Minute, Memorandum of Understanding, Arrangement between fisheries departments of the two countries, or *Modus Vivendi*.[86] It was pointed out that Memorandum of Understanding (MOU) is commonly used for informal agreements.[87] Some States might prefer MOUs to formal agreements for provisional arrangements because these have some advantages in several aspects: no need to publish them as these are not treaties; no need for elaborate final clauses or the formalities surrounding treaty-making; easy amendment; and, no need to be submitted for an approval of the parliament.[88]

However, States are not hindered from choosing the term "agreement" if they would like to do so. States can also adopt a mutual policy of restraint as a provisional arrangement through diplomatic consultation without signing a formal document. Therefore the arrangement can even be in oral form. As there could be various possible forms of provisional arrangements, the legal nature of a provisional arrangement would be different from case to case. The decisive factor which determines the legal nature of an arrangement is the intention of the parties. In this regard, the following statement of the ICJ in the *South West African* Case is worthy of quotation:

> Terminology is not a determinant factor as to the character of an international agreement or undertaking. In the practice of States and of international organisations and in the jurisprudence of international courts, there exist a great variety of usage; there are many different types of acts to which the character of treaty stipulations has been attached . . .[89]

However, if the arrangement of a practical nature is in written form and parties intended to be legally bound whatever the nomenclature, it is a treaty under the terms of the "Vienna Convention on the Law of Treaties". Therefore, the Vienna Convention will apply. McNair defined the term treaty as "a written agreement by which two or more States or international organisations creates or intend to create a relation between themselves operating within the sphere of international law".[90]

Article 2 of the Vienna Convention stipulates that:

> "treaty" means an international agreement concluded between States in written form and governed by international law, whether embodied in a single instrument or in two or more related instruments and whatever its particular designation . . .

Even if an arrangement is in oral form, it can be a treaty under customary international law, as there is no rule of international law which prohibits oral agreement. Consequently, many of the principles in the Vienna

Convention will be applicable to an agreement in oral form. Following this, Lord McNair wrote that: "It cannot be asserted that international law regards writing as essential to the creation of an interstate agreement, though it is rare to find an oral agreement between States".[91] Aust wrote that: "For reasons of clarity and simplicity oral agreements were excluded from the Convention. But this does not affect their legal force, or the application to them of any of the rules in the Convention to which they would be subject under international law independently of the Convention, such as customary international law".[92]

What seems important in deciding the legal nature of a provisional arrangement is the intention of the parties expressed in the document.[93] In this regard the definition of international agreement by the American Law Institute deserves quotation. It is defined thus:

> "international agreement" means an agreement between two or more States or international organisations ***that is intended to be legally binding*** and is governed by international law. [emphasis added][94]

If it is a mere undertaking which is not a treaty whether oral or written, relevant general principles of international law, for example the principle of good faith and some other principles embodied in the Vienna Convention will be applied.[95]

If a provisional arrangement is a treaty such as the arrangement between the United Kingdom and Iceland in the *Fisheries Jurisdiction* Case, then the Vienna Convention would apply.[96] However, one significant difference between provisional arrangements and agreements on final delimitation might be that provisional arrangements are limited *ratione temporis*, whereas boundary agreements are permanent. However, if delimitation is eventually not agreed upon, a provisional arrangement may in fact become a perpetual one. In such a case, whether or when a party may terminate the arrangement would become a critical issue. Now suppose that there is no termination clause in an arrangement, then can it be regarded as containing implicitly the right of any party to discard the treaty unilaterally? In this regard, there is some uncertainty. For example, before the adoption of the Vienna Convention on the Law of the Treaties, it was once written in Oppenheim's *International Law* that "all such treaties as are either not expressly concluded for ever, or are apparently not intended to set up an everlasting condition of things" "can nevertheless be resolved after notice by one of the contracting parties".[97] But later, the assertion was withdrawn noting that "the position in customary law is less clear" and that "the general rule [in Article 56 of the Vienna Convention] is that in the absence of provision in the treaty, a party may not withdraw from it [treaty] or denounce it"[98] The ILC, when drafting the provisions on the Law of Treaties, noted that there were two opposite views on this issue. One was that an individual party may denounce or withdraw from a treaty only when such denunciation or withdrawal is provided for in the treaty or consented to by all the other parties; the other view was that a right of denunciation or withdrawal may properly be implied under certain conditions in some types of treaties.[99] After long debates, the following provision was finally adopted:

> A treaty which contains no provision regarding its termination and which does not provide for denunciation or withdrawal is not subject to denunciation or withdrawal unless it is established that the parties intended to admit the possibility of denunciation or withdrawal.[100]

Under the above provision, it is an important question whether or not the parties intended to admit the possibility of denunciation or withdrawal when there is no termination clause. With regards to this, it has been pointed out that the intention of the parties is essentially a question of fact to be determined not merely by reference to the character of the treaty but by reference to all the circumstances of the case.[101] Therefore it is inappropriate to say that a right of denunciation is implied in a arrangement without examining all the circumstances of the case.[102]

However, it can be assumed that the right of denunciation is implied at least after the final delimitation has taken place if it is clear from the circumstances that the arrangement was intended to be applied only pending ultimate delimitation even if there is no termination clause in a provisional arrangement. Here, note that a Party that wishes to terminate a provisional arrangement which does not contain a termination clause, should give no less than twelve months notice of its intention to denounce the arrangement.[103]

Here, a distinction can be drawn between a provisional arrangement and the provisional application of a treaty, which is provided for in Article 25 of the Vienna Convention. The provisional application of a treaty presupposes the existence of a formal treaty which has been concluded but which is yet to enter into force, whereas a provisional arrangement does not necessarily presupposes the existence of such a formal treaty. For instance, the United Nations Fish Stocks Agreement makes a distinction between "provisional arrangements" pending agreement on conservation and management measures and "provisional application" of the Agreement by a State or entity before the entry into the force of the Agreement.[104] However, an agreement for provisional application of a treaty which is not yet in force can be a "provisional arrangement" pending ultimate delimitation. For instance, the United States and Cuba agreed through Exchange of Notes to apply on a provisional basis a maritime boundary treaty between them signed in 1977, because the United States Senate has not yet given its consent for the ratification.[105] In this case the agreement between the United States and Cuba to provisionally apply the maritime boundary treaty is a provisional arrangement in the form of Exchange of Notes.

3.3. *Requirements for an Arrangement to be of a Practical Nature*

As the drafting history suggests, the phrase "practical nature" seems to mean that the arrangements are to provide practical solutions to actual problems regarding the use of an area and not to touch upon either the delimitation issue itself or the territorial questions underlying this issue.[106] This means that the provisional agreement itself or development activities under such a regime cannot influence the final delimitation. As the arrangement of a practical nature purports only to solve issues such as utilisation of living or non-living resources of the disputed areas, neither side may gain benefit on its position on the final delimitation or "sovereignty issues". In short, it would be "no loss, no gain" in terms of the final delimitation or sovereignty issues. Note that this element is specifically addressed in the final sentence: "Such arrangement shall be without prejudice to the final delimitation".

If the proposition that an arrangement of a practical nature is without prejudice to the final delimitation is accepted, then the proposition that an arrangement which is not practical could affect the final delimitation can be put forward. Although it was seen that an arrangement of a practical nature is to provide practical guidance on the activities in a disputed area and is not to touch upon either the delimitation issue, it would still be a difficult task to decide whether or not an arrangement is of a practical nature.

Unfortunately, no criterion is provided for deciding whether or not an arrangement is of a practical nature. However, it is to be recalled that intentions of Parties reflected in the provisions of a treaty are very important in deciding the status of the treaty.[107] Similarly, it appears that the intention of parties to provisional arrangements is most important in deciding whether an arrangement is intended to be of a practical nature or not, and thus not to create any legal basis for either party to reinforce its claim over the area in dispute. Then the next question is: how can a State make it clear that an arrangement is intended to be of a practical nature and thus shall be without prejudice to its position on ultimate delimitation? With regards to this, the British Institute of International and Comparative Law suggested the use of a "without prejudice clause" in its Model Agreement for States for Joint Development. The proposed clause reads:

> (1) Nothing contained in this Agreement shall be interpreted as a renunciation of any right or claim relating to the Zone by either State party or a recognition of or support for the other State party's position with regard to any right or claim to the Zone.
> (2) No act or activities taking place as a consequence of this Agreement or its operation shall constitute a basis for asserting, supporting or denying the position of either State with regard to rights or claims over the Zone.[108]

Article 4 of the Antarctic Treaty is a good example of the use in practice of a "without prejudice clause". It provides that:

> 1. Nothing contained in the present treaty shall be interpreted as:
> (a) a renunciation by any Contracting Party of previously asserted rights of or claims to territorial sovereignty in Antarctica;
> (b) a renunciation or diminution by any Contracting Party of any basis of claim to territorial sovereignty in Antarctica which it may have whether as a result of its activities or those of its nationals in Antarctica, or otherwise;
> (c) prejudicing the position of any Contracting Party as regards its recognition or non-recognition of any other State's right of or claim or basis of claim to territorial sovereignty in Antarctica
> 2. No acts or activities taking place while the present treaty is in force shall constitute a basis for asserting, supporting or denying a claim to territorial sovereignty in Antarctica or create any rights of sovereignty in Antarctica. No new claim, or enlargement of an existing claim, to territorial sovereignty in Antarctica shall be asserted while the present treaty is in force.

It has been pointed out that without the "without prejudice clause" the conclusion of the Antarctic Treaty would not have been possible.[109]

An important question arises here. What if there is no "without prejudice clause" in a provisional arrangement? Does the absence of a disclaimer clause then make room for a Party trying to reinforce *vis-à-vis* the other party its position on delimitation? It appears that the insertion of such a provision is an important factor in deciding whether an arrangement is of a practical nature or not. Regarding this, there is an argument that the disclaimer clause is needed as proposed by the British Institute of International and Comparative Law. Dr. Gao spelled out that:

> The without-prejudice clause is necessary for preserving or reserving each party's position and claim to sovereign rights over the disputed area. Only in such a way can an impartial and fair joint regime be arranged and realised in a win-win or, at least, a no loss/no gain basis.[110]

In the *Fisheries Jurisdiction* Case between the United Kingdom and Iceland the ICJ noted the presence of a "without prejudice clause" in the provisional arrangement between them when it examined the nature of the arrangement in relation to its possible impact on the proceeding. It said that:

> The interim agreement contained no express reference to the present proceedings before the Court nor any reference to any waiver by the United Kingdom or by Iceland, of any claims in respect of the matters in dispute. On the contrary, it was ***an interim agreement***, that it related to fisheries in the disputed area, that it was concluded ***pending a settlement of the substantive dispute***, and that it was ***without prejudice*** to the legal position or rights of either Government in relation to the substantive dispute. [emphasis added][111]

The British Institute of International and Comparative Law emphasised, in its commentary on the model agreement for joint development, the importance of such a clause. It said that:

> The Working Group believes that a forthright reservation or preservation of the claims of each of the parties to sovereign rights in respect of the Zone coupled with a shelving of delimitation throughout the duration of the Agreement was required. This could be coupled with an obligation upon the parties to resume negotiations on delimitation on the approaching expiry of the Agreement.[112]

The British Institute identified four elements to ensure that an arrangement would not prejudice the final delimitation. First, the use of a specific "without prejudice clause"; second, the insertion of provisions on shelving of delimitation; third, the presence of a duration clause; and last, provisions for carrying out negotiations on delimitation. The four elements seem important for ensuring that an arrangement is of a practical nature. However, it is unreasonable to say that all four elements are required always to ensure that an arrangement would not prejudice the final delimitation. Situations will differ from case to case.

With regard to the need for insertion of a "without prejudice clause", it can be argued that there is no longer the need in a provisional arrangement if the two Parties to the arrangement are also Parties to the LOS Convention, because there is a general "without prejudice clause" in the final sentence of Articles 74(3)/83(3): "Such arrangement shall be without prejudice to the final delimitation". Therefore, simply referring to the provisions of Articles 74(3)/83(3) in an arrangement would be sufficient for preserving position on final delimitation of each Party to an arrangement, if the Parties to the arrangement are also Parties to the LOS Convention.

3.4. *Obligations of Coastal States before Delimitation*

3.4.1. *Are there Any Obligations under Customary International Law?*

The LOS Convention imposes obligations on the States concerned (i) to make every effort to enter into provisional arrangements, and (ii) not to jeopardise or hamper the reaching of a final agreement. Are these obligations found also in customary international law? As we have discussed, the obligation to negotiate for the conclusion of provisional arrangements was not known to the 1958 Geneva Convention regime and what the ICJ mentioned in the *North Sea Continental Shelf* Cases was the obligation to negotiate for delimitation. Therefore, it seems inappropriate to argue that this is the obligation also under customary international law. What about the obligation "not to jeopardise or hamper the reaching of the final agreement"? Is there any customary international law in this regard? On this very point Churchill and Lowe stated that: "Arguably such a rule exists also in customary international law".[113] It seems important to take opportunities to look into relevant case laws and customary law in examining the obligation "not to jeopardise or hamper the reaching of the final agreement".

3.4.2. *Obligation to Negotiate*

According to the first sentence of paragraph 3 of Articles 74/83, the interested States "in a spirit of understanding and co-operation, shall make every effort to enter into provisional arrangements of a practical nature". In other words, States are under a heavy obligation to negotiate in good faith to conclude arrangements of a practical nature pending delimitation.

Undoubtedly, negotiation is one of the principal means for resolving disputes between States. Article 33(1) of the UN Charter, which provides various means of solving international disputes, places negotiation first before enquiry, mediation, conciliation, arbitration, and judicial settlement.[114] It has been observed that negotiation is employed more frequently than all other methods put together for settlement of international disputes, and it is also a technique for preventing international disputes from arising.[115]

As was seen earlier, under the 1958 Geneva Convention on the Continental Shelf there was no obligation upon the parties to carry out negotiations for the conclusion of a provisional arrangement. Furthermore, there is no evidence that States are obliged to negotiate for provisional arrangements under customary law pending maritime delimitation. In this regard, it is to be noted that there is no specific provision in the LOS Convention that requires States to negotiate for the delimitation of a maritime boundary. What Articles 74(1)/83(1) of the Convention require is that the delimitation of maritime boundaries "shall be effected by agreement". In the *North Sea Continental Shelf* Cases the ICJ relied on the "very general precepts of justice and good faith", and "equitable principles" to State "an obligation to enter into negotiations with a view to arriving at an agreement", but it did not say that the parties are under an obligation to negotiate for a provisional arrangement.[116] The obligation to negotiate for conclusion of a provisional arrangement is thus newly adopted by the LOS Convention but it is hard to say it is also a rule of customary law.

Now let us turn to the meaning of the phrase "shall make every effort". This phrase is not framed as a recommendation, but rather as an obligatory clause which binds parties to the LOS Convention. So if a State does not make every effort to establish a provisional arrangement it would constitute a breach of international law. However, this wording does not require a State to agree upon a particular provisional regime. It is of importance to stress that it does not oblige a State to agree on a provisional agreement. In examining the nature of an obligation to negotiate, Lord McNair said:

> It is, however, necessary, to distinguish between a true obligation to enter into a later treaty and an obligation merely to embark upon negotiations for a later treaty and to carry them on in good faith and with a genuine desire for their success. Less happily in our opinion, the term *pactum de contrahendo*[117] is applied to an obligation assumed by two or more parties to negotiate in the future with a view to the conclusion of a treaty. This is a valid obligation upon the parties to negotiate in good faith, and a refusal to do so amounts to a breach of the obligation. But the obligation is not the same as an obligation to conclude a treaty or to accede to an existing treaty . . .[118]

The obligation here is that the States concerned have to make "every effort" to conclude a provisional arrangement in good faith. However, the phrase "every effort" leaves some room for interpretation by the States concerned or instead by any dispute settlement body. According to the article, such negotiations should be carried on "in a spirit of understanding and co-operation", which in conjunction with the phrase "shall make every effort" appears to emphasise that the negotiation should be conducted "in good faith". Even if there is no mention of "good faith" in Articles 73(3)/84(3), good faith should be exercised in carrying out an obligation to negotiate. In this regard, Bin Cheng States that:

> A party which is bound by a *pactum de contrahendo* to negotiate a certain treaty is responsible for the consequences of its acts of bad faith. It may be said that in such cases good faith consists in a sincere and honest desire, as evidence by genuine efforts, to fulfil the substance of the mutual agreement.[119]

Regarding the obligation to negotiate in good faith, the ICJ referred to this as "a special application of a principle underlying all international relations".[120] In the *North Sea Continental Shelf* Cases, the ICJ held that:

> . . . the parties are under an obligation to enter into negotiations with a view to arriving at an agreement and not merely to go through a formal process of negotiation as a sort of prior condition for the automatic application of a certain method of delimitation in the absence of agreement; they are under an obligation so to conduct themselves that the negotiations are meaningful, which will not be the case when one of them insisted upon its own position without contemplating any modification of it.[121]

In the *Lac Lanoux* Case where the arbitral tribunal dealt with the issue of prior consultation between a upstream State (France) and a downstream State (Spain), the tribunal pointed out the essence of the negotiation thus:

> . . . the upstream State has, according to the rules of good faith, the obligation to take into consideration the different interests at stake, to strive to give them all satisfaction compatible with the pursuit of its interests, and to demonstrate that, on this subject, it has a real solicitude to reconcile the interests of the other riparian with its own.[122]

Another important point is, that States concerned should not engage in extreme claims when negotiating in good faith. It should be recalled that good faith implies the observance by the parties of a certain standard of fair dealing, and it requires that one party should be able to place confidence in the words of the other.[123] If a State makes extreme claims which lack legal grounds or reasonable proportionality, the extreme claims would make disputes more serious, and the opponent could argue that such claims constitute an abuse of right prescribed in Article 300 of the LOS Convention or in general international law. An abuse of rights can be defined as "any fictitious exercise of a right for the purpose of evading either a rule of law or a contractual obligation".[124] The prohibition of abuse of rights is derived from the principle of good faith since the latter requires every right to be exercised honestly and loyally.[125]

The area where negotiations for the establishment of provisional measures is needed will be the area where neighbouring States lay overlapping claims on reasonable ground. Such claims can be formally expressed by announcements or be implied by, for example, protests against other States and granting of fishery licenses to its own nationals or concessions to foreign oil companies. When carrying out such activities, States must act in good faith bearing in mind that their activities should have a reasonable foundation under current international law. In this context, it is to be recalled that the phrase "zones of overlapping boundaries claimed *bona fide*" was suggested by Ireland during UNCLOS III.

But the zones, on which States concerned lay overlapping claims, are not always clearly defined. In the *Aegean Sea* Case, Turkey made claims over a specific area contending that this area, in which Greece had already granted concession to a foreign company, belonged to it.[126] In the *Tunisia/Libya Continental Shelf* Case, the concessions granted by the two States overlapped in only a small area, whereas their claims to the continental shelf overlapped in large parts of the area.[127] The ambiguity of the extent of overlapping areas can also be caused when each State's position on applicable rules are not clearly established. Such was the *Anglo-French Continental Shelf Arbitration*, where the UK government argued for the application of Article 6 of the Geneva Convention on the continental shelf whilst the French government argued that rules of customary law as stated by ICJ in the *North Sea Continental Shelf* Cases were applicable.[128] When the geographical scope of overlapping claims is not clear ascertaining such scope is also a subject of negotiation.

The obligation to negotiate for a provisional arrangement appertains to all cases where a relevant maritime boundary has not been delimited between countries with adjacent or opposite coasts. This obligation to negotiate in good faith for provisional measures exists until the final delimitation is agreed upon. Thus this obligation can arise before the negotiation on boundary delimitation begins or after the negotiation has been thwarted.

Since an object of a provisional arrangement is to preserve the rights of parties concerned, it can be said that this obligation arises when a State raises the necessity of it. However, generally speaking, this kind of obligation occurs under situations in which overlapping claims over a certain area are revealed. The obligation to negotiate for a provisional arrangement ends as the agreement on final delimitation is reached.

3.4.3. *Obligation of Mutual Restraint*

According to paragraph 3 of Articles 74/83 of the LOS Convention, States are legally obliged not to jeopardise or hamper the reaching of a final agreement on delimitation. In other words, States have duties to maintain the *status quo* as to issues of final delimitation. The duty to maintain the *status quo* is understood as one aspect of the principle of good faith.[129] In this regard, Bin Cheng observed that:

> . . . whenever the parties have agreed to await a final decision concerning a certain matter, or are under an obligation to do so – a decision depending either upon the parties themselves or upon an independent third party – the principle of good faith obliges them to maintain the existing situation as far as possible so that the final decision, if taken on the basis of the *status quo*, would not be prejudiced in its effects by a unilateral act of one of the parties during the inevitable lapse of time.[130]

In a similar manner, the Permanent Court of International Justice (the forerunner of the ICJ) stated: "[T]he parties to a case must abstain from any measure capable of exercising a prejudicial effect in regard to the execution of the decision . . .".[131]

The necessity of mutual restraint for maintaining the *status quo* can be better understood in the light of the legal nature of interim measures or provisional measures, which are intended to preserve the rights of parties concerned before the final solution of disputes is made. Provisional measures of Article 41 of the Statute of the ICJ or Article 290 of the LOS Convention are all of the same purpose, and the provisional arrangements in Articles 74(3)/84(3) are intended to serve a similar purpose as mentioned earlier.

Article 18 of the Vienna Convention on the Law of Treaties is analogous in this respect. This article prescribes that a State is obliged after signing a treaty to refrain form acts that would defeat its object and purpose. Originally, it was proposed that this obligation is applied to the negotiating process as well, but the proposal was eventually abandoned.[132]

It is noteworthy that Sir Humphrey Waldock, who actively participated in the codification process of the Vienna Convention, revealed his opinion that "if a State exhausts, while carrying on negotiations, the natural resources of the area pending territorial waters delimitation, it would constitute a breach of such obligation".[133]

This obligation of mutual restraint is temporary since it arises when overlapping claims are made and it is terminated when the final delimitation is achieved. The issue of nature and scope of the mutual restraint obligation has arisen in a serious and practical manner between Denmark and Sweden. After the delimitation negotiations on the continental shelf around Kattegatt had been broken off in 1979 with Sweden, the Danish government permitted a Danish company to engage in exploratory drilling in the disputed sea and the subsequent drilling sparked off diplomatic protests from the Swedish government.[134] In a similar context, a diplomatic row arose in the Baltic sea in the middle of the 1980s between Sweden and the Soviet Union, when Soviet patrol boats inspected the fishing vessels of Sweden and certain third States in disputed areas.[135] Were these activities by Denmark and the Soviet Union prohibited by the rule of "mutual restraint"?

Since paragraph 3 of Article 74/83 does not list which particular activities are prohibited under this obligation, we have to examine what the activities are that endanger or hinder the final delimitation. It cannot be said that all the activities conducted in the overlapping area are prohibited.

The consideration of the ICJ in the *Aegean Sea Continental Shelf* Case between Turkey and Greece sheds some light on this issue. In this dispute which was caused in part by some exploration activities in the disputed area, the court said that:

> whereas no complaint has been made that this form of seismic exploration involves any risk of physical damage to the seabed or subsoil or to their natural resources; whereas the continued seismic exploration activities undertaken by Turkey are all of the transitory character just described, and do not involve the establishment of installations on or above the seabed of the continental shelf; and . . . no suggestion has been made that Turkey has embarked upon any operations involving the actual appropriation or other use of the natural resources of the areas of the continental shelf which are in dispute.[136]

In the 1972 *Fisheries Jurisdiction* Case between Iceland and the United Kingdom the ICJ ordered as an interim measure that the UK may annually catch the average annual catch of 1960-1969 within the disputed Ice-landic 50 N.M. fisheries zone.[137] In the same order, the ICJ ordered that; ". . . pending the Court's final decision in the proceedings, Iceland should refrain from taking any measures to enforce the Regulations of 14 July (on 50 N.M. exclusive fisheries zone) against United Kingdom vessels . . .".[138]

From the opinion of the ICJ in the above two cases, it appears that seismic exploration which does not involve physical damage to the seabed or subsoil or to their natural resources and some traditional fishing activities, would not be considered to be an irreparable prejudice to States concerned, while activities such as actual appropriation of petroleum, and enforcement action would be seem prejudicial to the rights of the other coastal State in the disputed area.

Although the ICJ decisions in the above two cases were made for the purpose of considering interim measures under its Statute, these will be of significance in considering what activities would be prohibited in the disputed areas under the obligation of mutual restraint.

The coastal States in dispute can of course agree on some specific measures of mutual restraint applicable between them. For instance, from 1977 to 1984, Canada and the United States applied this principle of mutual restraint in relation to the Gulf of Maine and Georges Bank. Article IX of the 1977 fishing agreement between them provided that:

> In the boundary regions, ***the following principles shall be applied as interim measures of mutual restraint*** pending the resolution of questions pertaining to the delimitation of areas subject to the respective fisheries jurisdiction of each party:
>
> 1. As between the parties, enforcement shall be conducted by the flag State.
> 2. Neither party shall authorise fishing by vessels of third parties in the boundary region
> 3. Either party may enforce against third parties in the boundary region. [emphasis added][139]

Having thus explored the nature of the obligation of mutual restraint, it would be appropriate to turn to the question of whether the obligation is also a rule of customary international law. As we have seen, there are various evidences which support an affirmative answer to the question. Firstly, under the principle of good faith and the duty to maintain the *status quo* which is derived from the principle of good faith, activities which might be prejudicial to the final delimitation are not allowed. Secondly, a consideration of the case law of the ICJ supports the conclusion that certain activities are not allowed in the disputed area which are not of a transitory character, for example drilling or setting up of a permanent installation. Thirdly, there is some consistent State practice to support this view. In this regard, Churchill and Ulfstein pointed to some examples: Tunisia and Libya halted offshore oil and gas operation in the disputed area while their case was being dealt with by the ICJ; the United States has not given hydrocarbon licensing on part of the continental shelf on Georges Bank, which was also claimed by Canada; Greece and Turkey agreed in 1976 to abstain from any initiative which might prejudice their negotiation; etc.[140]

Churchill and Ulfstein have stated that ". . . there probably is a rule of customary international law which is closely mirrored in the Law of the Sea Convention, (and) which prohibits unilateral oil and gas activities on disputed continental shelf . . .". Lagoni, expressed a broadly similar view.[141] Note, however, that this does not mean that all activities are not allowed in the disputed area under the rule of mutual restraint because it cannot be said that all activities in the disputed areas are prejudicial to the final delimitation. The problem here is that neither the LOS Convention nor case law provides an exhaustive list of activities which are prohibited in the disputed areas under the obligation of mutual restraint, even if it is clear that some activities would not be allowed. It is obvious that in considering what activities are prohibited in the disputed areas the aforementioned consideration of the ICJ in the *Aegean Sea Continental Shelf* Case and the *Fisheries Jurisdiction* Cases should be fully taken account of.

3.5. *Provisional Arrangements and Interests of Third Parties*

If a provisional arrangement is a treaty, it cannot impose, just as any other normal treaty, any right or duty upon third parties. Because a treaty is *res inter alios acta* (i.e., a matter which in law exclusively concerns others) as regards a third party. With regards to this, the Vienna Convention provides in Article 34 that: "a treaty does not create either obligations or rights for a third State without its consent". This complies with a maxim of Roman law, *pacta tertiis nec nocent nec prosunt* (i.e., contracts do not impose any burden, nor confer any benefits, upon third parties). The reasons for the rule embodied in Article 34 lies in the fundamental principle of the sovereignty and independence of States.[142] Aust wrote that:

> Similar rules apply in laws of contract, but the rule in the Convention rests firmly on the sovereignty and independence of States. Thus a treaty, whether bilateral or multilateral, cannot impose an obligation on a third State, nor modify in any way the legal rights of a third State without its consent.[143]

If the rule *pacta tertiis nec nocent nec prosunt* in the law of treaties is based upon the sovereignty and independence of States, then it would be unreasonable to argue that the rule applies only to formal written treaties. It was observed that even before the adoption of the Vienna Convention, "[b]oth legal principle and common sense are in favour of the rule *Pacta tertiis nec nocent nec prosunt*".[144] As the ILC observed, there appears to be "almost universal agreement" on this rule.[145] More positively, Oppenheim mentioned that: "[t]he general rule is well established that there is no need to cite authority for it".[146] Therefore, this rule would apply to a provisional arrangement regardless of whether or not the arrangement is a treaty under the terms of the Vienna Convention, if the arrangement is *pacta*.

Similarly, it can be said that under customary international law, coastal States should not conclude a provisional arrangement in such a manner as to impair the rights of the third parties, as the Roman maxim States "*sic utere tuo, ut alienum non laedas*" (i.e., do not exercise your rights in a manner as to impair others' rights).

However, it is observed that many treaties are incidentally unfavourable to non-parties.[147] Rights of a third party can be knowingly disregarded. The issue of remedies then become important. Lord MacNair has stated that:

> A State which learns that a treaty concluded between two other States has for its object or probable consequence the impairment of its rights . . . is entitled at once to lodge a diplomatic protest with those parties and to apply to International Court of Justice, or any other tribunal . . .[148]

The impairment of third parties' interests can take place in two situations. One is the case where a third State claims that it is also a coastal State and thus has the right to participate in a provisional arrangement between the other States, but these States refuse to allow the third State to participate in the arrangement. The other case is related to a provisional arrangement's possible interference with navigation and other rights and freedoms which a third party is entitled to enjoy under the LOS Convention, even if the third party is not a relevant coastal State in the area.

In regard to the former, it can be said that a provisional agreement is effective only within the geographical scope of explicit reference. This point seems important where a maritime area is disputed by more than two States. There are a number of maritime areas in the world where more than two States lay claims.[149] Malta claimed its rights to intervene in the *Tunisia/Libya Continental Shelf* Case in the ICJ.[150] Similarly, Italy claimed its right to intervene in the *Libya/Malta Continental Shelf* Case.[151] The Court, however, rejected the requests of interventions, indicating that its decisions are binding only upon the Parties to disputes, but it was careful not to delimit an area which might appertain to a third Party. The ICJ held that:

> The decision of the Court will . . . have binding force between the Parties, but not against third States. If therefore the decision is to be stated in absolute terms, in the sense of permitting the delimitation of the areas of

> shelf which "appertain" to the Parties, as distinct from the areas to which one of the Parties has shown a better title than the other, but which might nevertheless prove to "appertain" to a third State if the Court had jurisdiction to inquire into the entitlement of that third State, the decision must be limited to a geographical area in which no such claims exist.[152]

In the *Anglo-French Continental Shelf Arbitration* the tribunal was careful about possible impacts of the delimitation between UK and France on a future delimitation between UK and Republic of Ireland, and also made it clear that its decision would be *res inter alios acta* for any third States.[153] The arbitration tribunal in the *St. Pierre and Miquelon* Case between Canada and France, rejected the parties request to delimit beyond 200 N.M. because that would affect the third party, i.e. the International Seabed Authority.[154] In the *Yemen-Eritrea Arbitration* on maritime delimitation in the Red Sea, the arbitral tribunal was careful not to delimit the area which might involve third parties in the area, i.e., Saudi Arabia to the north and Djibouti to the south, and said that: "The Tribunal believes that these terminal points are well short of where the boundary line might be disputed by any third States".[155] From the observation made so far, it seems necessary for States to agree on the geographical scope of provisional arrangements in such a way as to avoid including an area which might fall into a zone of a third party.

With regard to the provisional arrangement's possible interference with navigation and other rights and freedom of the third party, the "due regard" rule should be recalled. Article 56(2) of the LOS Convention stipulates that:

> In exercising its rights and performing its duties under this Convention in the exclusive economic zone, the coastal State shall have due regard to the rights and duties of other States and shall act in a manner compatible with the provisions of this Convention.

A number of provisional arrangements have included provisions which require due regard for the rights of a third party. For instance, Article 27 of the joint development agreement of continental shelf between Korea and Japan provides that:

> Exploration and exploitation of natural resources in the Joint Development Zone shall be carried out in such a manner that other legitimate activities in the Joint Development Zone and its superjacent waters such as navigation and fisheries will not be unduly affected.[156]

Similarly, the model agreement for the joint development of the continental shelf formulated by the British Institute of International and Comparative

Law contains provisions for the protection of the rights of third parties. Article 24 reads as follows:

> (1) States Parties shall exercise their rights under this Agreement in such a manner as not to interfere unjustifiably with the rights and freedoms of other States in respect of the Zone as provided under generally accepted principles of international law.
> (2) In the event that any third party claims rights inconsistent with those of the States Parties under this Agreement then the States Parties shall consult through appropriate channels with a view to co-ordinating a response.

The first paragraph is concerned with the provisional arrangement's possible interference with navigation and other rights and freedom of the third party. The second paragraph is a procedural one to deal with a third party's claims. As this simply opens up the possibility of shaping a common stance between the parties to the agreement toward a competing third party claim, it will not be sufficient to protect alleged third party rights in the area of a provisional arrangement. The best way to protect such interests will be for the parties to the arrangement to have consultations with the third party. Alternatively, it might be desirable, if possible, for the parties to define the zone of a provisional arrangement so as to exclude the areas that might be also claimed by a third State. However, by way of contrast, third parties' claims can be easily disregarded.[157] Furthermore, there appears to be a tendency of the parties to such an arrangement, to co-operate against a possible third party claim. For example, Article 6 of the joint development agreement of 1974 between Sudan and Saudi Arabia States that the parties have equal sovereign rights in their joint development zone, a so-called "Common Zone", and the parties will protect their sovereign rights and defend them against third parties.[158] Such a defensive provision is also found in Article 5 of the joint development agreement of 1965 between Kuwait and Saudi Arabia. It provides, that if one of the parties cedes or alienates to any other States, such equal rights as are included in the Agreement the other Party is thereby relieved of any obligations under the Agreement.[159]

When a third party becomes aware that a treaty concluded between other States impairs its rights then it can make diplomatic protests, or commence proceeding before an international court or tribunal if it can establish jurisdiction. [160]

4. PROTECTION OF THE MARINE ENVIRONMENT IN DISPUTED AREAS

The coastal State has a broad jurisdiction with regard to the protection and preservation of the marine environment of its EEZ.[161] The coastal State has legislative and enforcement power in its EEZ to deal with the dumping of waste, other forms of pollution from vessels and pollution from sea-bed activities.[162] The coastal State also has the right to permit, regulate, and control the dumping of wastes in its continental shelf.[163] Here an important question arises as to whether the coastal State can effectively exercise its rights, with regard to the protection of the marine environment when there is no boundary with their neighbouring coastal States. Note that coastal States have competence to enforce its national laws and regulations and the applicable international rules and standards, regarding dumping within its exclusive economic zone or continental shelf. There is a possibility that a coastal State might attempt to stop and detain a vessel of the other coastal State in the event of a dumping violation in a disputed area relying on its power as a coastal State.[164] Thus it appears that the neighbouring coastal States need a provisional arrangement in order to avoid jurisdictional conflicts. On what circumstances, can jurisdictional conflicts happen?

First, let us begin with possible problems regarding dumping. Suppose that vessel X obtained an authorisation from coastal State A to dump waste into the disputed area and the other coastal State B objected to the dumping. Or suppose that vessel X did not obtain any authorisation either from State A or State B, and both State A and State B attempt to exercise enforcement jurisdiction against vessel X. In the former situation, it seems that State B has the right to oppose dumping on the basis of its assessment of the impact of dumping on the marine environment because State B is also a coastal State or affected State at the least. We can see here that there is potential conflict between the two coastal States regarding the marine environment in the disputed areas. The latter situation does not seem different from the situation of concurrent jurisdictions over a criminal matter and there is no clear rule to give an absolute priority either to State A or to State B in such situations.[165] Thus, a conflict can arise between the two coastal States with regard to jurisdiction over dumping in the maritime areas where there is no boundary in place.

Let us turn to the issue of possible concurrent jurisdiction between the two coastal States regarding pollution from the ship. Careful reading of Article 220 (Enforcement by coastal State) and Article 228 (suspension and restrictions on institution of proceeding) discloses that coastal State enforcement with regard to pollution from vessels is very restricted. According to Article 220, the coastal State, in normal circumstances, cannot exercise its enforcement jurisdiction against the polluting vessel even if the vessel is in passage in its exclusive economic zone. In a normal situation, when the vessel is in passage, the coastal State can only request the vessel to give some information regarding its identity and planned journey.[166] Normally, the coastal State can exercise its enforcement jurisdiction against the polluting vessel only when the vessel is voluntarily within a port or an off-shore terminal of the coastal State.[167] Therefore in normal situations, conflict of jurisdiction would not arise because the ship cannot be at the ports of the two States at the same time. The competing jurisdiction, however, can occur when the delinquent vessel is registered in coastal State A, and coastal State B attempts to exercise its enforcement jurisdiction against the vessel because it was found in a port belong to State B after polluting in the disputed area. In this situation, State A can successfully assert its jurisdiction against the vessel as a flag State of the vessel because a flag State's jurisdiction will normally override the coastal State's jurisdiction according to the LOS Convention.[168] Thus it can be said that the concurrent jurisdiction would not occur in the sea in a normal situation between the two coastal States regarding the protection of the marine environment in the disputed areas.[169]

However, there seems to be potential for jurisdictional conflicts between the two coastal States with regard to the pollution from ships in the disputed areas, when the situation is not normal, i.e., where there is "major damage or threat of major damage" to the environment of the coastal State or where there is a "substantial discharge causing or threatening significant pollution". Note that in the latter case, the coastal State can undertake physical inspection of the vessel in certain situations even if the vessel is in passage, and in the former case the coastal State can institute proceedings, including detention of the vessel even if the vessel is in passage.[170] We can see here that that the problem of concurrent jurisdictions could arise at sea when the two coastal States attempts to undertake a physical inspection of the vessel in passage in the disputed areas on the basis of "substantial discharge" or when the two coastal States attempt to exercise their enforcement jurisdiction against the vessel as it passes through the disputed area of the exclusive economic zone on the basis of "major damage". Or it can be the case, where a coastal State argues that there is no major damage and thus the other coastal States has no power to seize the vessel. Although it can be assumed that "substantial discharge" means lesser pollution than the major damage, the LOS Convention does not provide for the criteria for "major damage" or "substantial discharge".[171] It means that there can arise different views on these criteria.[172]

The above analysis shows that it is highly desirable for the contending neighbouring coastal States to have arrangements on the issue of enforcement jurisdiction in the disputed areas regarding the marine environment for the purpose of prevention of jurisdictional conflicts. It is to be recalled as seen in Chapter 3 that UK and Denmark (Faroe Islands) agreed to refrain from exercising jurisdiction regarding environmental issues against vessels of each country, in the Special Zone, without agreement of each other.[173]

5. MARINE SCIENTIFIC RESEARCH IN DISPUTED WATERS

Marine scientific research has contributed to the development of human knowledge of nature.[174] It has been crucial for the rational management of the oceans and its resources, and there is a pressing need for marine scientific research for the preservation of the marine environment.[175] However, the problem is that a marine scientific vessel today cannot travel the oceans with the same freedom from regulations as the *Beagle* did in the nineteenth century. Notice that the coastal States now have "the right to regulate, authorise and conduct marine scientific research in their exclusive economic zone and on their continental shelf",[176] although all States, in principle, have rights to conduct marine scientific research in the exclusive economic zone and on the continental shelf of other States subject to jurisdiction of the coastal States under the LOS Convention.[177]

It is not intended here to examine the legal characteristics of the freedom of the marine scientific research and jurisdiction of the coastal States. The main question here is whether it is advisable for two coastal States in a disputed area to enter into provisional arrangements of a practical nature in order to prevent jurisdictional conflicts on marine scientific research and to ensure the freedom of marine scientific research by third States.

In order to come up with an answer to the main question just raised, we need to consider these specific questions: (i) can coastal State A conduct marine scientific research without the consent of the other coastal State B in the disputed areas?; (ii) can the marine scientific research in the disputed area conducted by State A, without any protests from State B, reinforce its claims to the part of the disputed area or natural resources in the disputed area?; and (iii) can a third State conduct marine scientific research in the disputed area without consent from any of two coastal States?

Having thus raised these complex legal questions, we now face the reality that the LOS Convention is silent on these questions. Also academic literature on this topic is hard to find. However, it will be worthwhile to seek the answers to these questions because these are of some practical importance. After considering these three questions, I will consider whether it is advisable to have provisional arrangements on marine scientific research in the disputed areas.

5.1. *Can a Coastal State Conduct Marine Scientific Research in the Disputed Areas Without Consent from the Other Coastal State?*

As the provisions of the LOS Convention on the marine scientific research mainly deal with the rights and obligations between the coastal State and the researching State and thus do not deal with the legal relations between the two coastal States in the disputed areas, these provisions are not of direct relevance in considering the question whether coastal State A can conduct marine scientific research in the Disputed Areas without consent from the other coastal State B.

It is to be noted that, as will be seen in Chapter 4, the Chinese exploration activities in the East China Sea where there is no boundary sparked off diplomatic rows with Japan as it regarded these exploration activities by the Chinese vessels on the Japanese side of the hypothetical equidistance line as an infringement on its rights.[178] In this situation, the question necessarily arises as to whether all scientific research or explorations by a coastal State are not permitted at all in the disputed areas without consent from the other coastal State?

Interestingly, this question was raised in the *Aegean Sea Continental Shelf* Case. In this case, Greece requested the International Court of Justice to direct the two coastal States to refrain from all exploration activities or any scientific research with respect to the continental shelf areas in dispute, unless with the consent of each other pending the final judgment of the Court.[179] Greece argued that all unilateral exploration activities and scientific research in the disputed area are not allowed because these might prejudice the execution of any judicial decision.[180]

If we read Greece's argument carefully we can find that Greece tacitly presupposes a genuine coastal State which has the sole exclusive sovereign rights in the disputed area, when it argues that "the exclusivity of knowledge" of the coastal State on its continental shelf should be protected. Along this line of reasoning, Greece regarded any damages to exclusivity of knowledge as irreparable damage to the rights of the coastal State, which should be protected by the ICJ until it declared that Greece is the coastal State, i.e., the genuine coastal State.[181] Thus, Greece argued that, "Turkey's grants of exploration licenses and exploration activities must tend to anticipate the judgement of the Court, and that breach of the right of a coastal State to ***exclusivity of knowledge of its continental shelf*** constitute irreparable prejudice (emphasis added)".[182] Greece went on to argue that:

> . . . Turkey's seismic exploration threatens in particular to destroy the exclusivity of the rights claimed by Greece to acquire information concerning the availability, extent and location of the natural resources of the areas . . .[183]

However, the Court did not agree on the concept of "exclusivity of knowledge" of the genuine coastal State, but looked into the physical nature of the activities undertaken by the other possible coastal State. For the Court, seismic exploration was tolerable in the disputed area because it is merely "of the transitory character" which does not involve "any risk of physical damage to the seabed or subsoil or to their natural resources". The Court held that:

> the seismic exploration undertaken by Turkey, of which Greece complains, is carried out by a vessel traversing the surface of the high seas and causing small explosions to occur at intervals under water; whereas the purpose of these explosion is to send waves through the seabed so as to obtain information regarding the geophysical structure of the earth beneath it; whereas no complaint has been that this form of seismic exploration involves any risk of physical damage to the seabed or to their natural resources; whereas the continental seismic exploration activities undertaken by Turkey are all of the transitory character just described, and do not involve the establishment of installations on or above the seabed of the continental shelf; and whereas no suggestion has been made that Turkey has embarked upon any operations involving the actual appropriation or other use of the natural resources of the areas of the continental shelf which are in dispute . . .[184]

With regard to the alleged damage to the exclusivity of rights to the information on the natural resources, the Court held that:

> Whereas, in the present instance, the alleged breach by Turkey of the exclusivity of the right claimed by Greece to acquire information concerning the natural resources of area of continental shelf, if it were established, is one that might be capable of reparation by appropriate means, and whereas it follows that the Court is unable to find in that alleged

> breach of Greece's rights such a risk of irreparable prejudice to rights in issue . . .[185]

From the passage above, we find that the Court looked into the nature of the exploration, to decide whether any unilateral exploration by one coastal State is prejudicial to the rights of the other coastal State. If Turkey had undertaken any actual drilling into the disputed continental shelf or actual exploitation of oil or gas in the disputed area then the decision of the ICJ would have been different. Therefore, it can be said that a coastal State can conduct marine scientific research in the disputed area without consent from the other coastal State as long as the scientific research is of the "transitory character" which does not involve "any risk of physical damage to the seabed or subsoil or to their natural resources".

As the ICJ looked into the nature of marine scientific research and made a distinction between the transitory character and non-transitory character of the exploration, it might be appropriate to see whether there is such a classification of marine scientific research in the provisions of the Geneva Convention on the continental shelf and the LOS Convention.

We can see that the Geneva Convention makes a distinction between "purely scientific research into the physical or biological characteristics of the continental shelf" and other research and it provides that "the coastal State shall not normally withhold its consent" with regard to such "purely scientific research".[186] Although the Geneva Convention distinguished between "purely scientific research" and other scientific research, it did not indicate what this other research, which is not purely scientific, is. The other scientific research which is not purely scientific appears to be that research conducted for commercial or military purpose and thus can be referred to as applied scientific research.[187]

The reason why the Geneva Convention made a distinction between purely scientific research and other research, appears to be that there was a general view that the enhancement of knowledge about the ocean is for the benefit of mankind as a whole. When François, the Special *Rapporteur*, reported on the issue of scientific research in the eighth session of the International Law Commission (ILC) in 1956, he introduced the resolution of the International Council of Scientific Union (ICSU) on this issue, which asserts that "fundamental research by any nation carried out with the intention of open publication is in the interests of all" and the draft articles should be amended so as to ensure "such fundamental research at sea may proceed without vexatious obstruction".[188] Referring to the resolution of the ICSU, the Special *Repporteur* stated his opinion that the coastal State will not have the right to prohibit such purely scientific research.[189]

In the resolution by the ICSU, we can see an illustration of purely scientific research. In the resolution, this is "fundamental research in the geophysics, submarine geology, and marine biology of the sea-bed and subsoil of the continental shelf".[190] Alfred H.A. Soons, having examined the drafting history of the provisions, pointed out that:

> The reference to purely scientific research was intended to make it clear that the provision was not intended to cover exploration activities, i.e., the collection of data with a view to exploitation. Purely scientific research can yield results which are useful from the point of view of exploitation, but the research is conducted without paying attention to such practical applications of the results. The reference to "physical or biological characteristics of the continental shelf" was also intended to emphasise that the research covered by the provisions excludes the collection of data with a view of exploitation.[191]

Let us turn to the LOS Convention to see whether there is a similar distinction in the marine scientific research. The LOS Convention adopts basically the same regime on the regulation of marine scientific research on the continental shelf and in the exclusive economic zone as that of the Geneva Convention. Under the LOS Convention, all scientific research on the continental shelf and in the exclusive zone should be conducted with the consent of the coastal State.[192] The LOS Convention, like the Geneva Convention, makes a distinction between two different kinds of marine scientific research; one which is carried out "exclusively for peaceful purposes and in order to increase scientific knowledge of the marine environment for the benefits of all mankind" and the other scientific research.[193] The discretion of the coastal States with regard to the former type of marine scientific research is restricted in the sense that the coastal States shall, in normal circumstances, grant consent for marine scientific research of that type, whereas the coastal States have discretion to withhold its consent with regard to the latter type of the marine scientific research.[194] Here we can associate the former type of the marine scientific research under the LOS Convention with the purely scientific research in the Geneva Convention.[195]

The LOS Convention, unlike the Geneva Convention, elaborates what the other scientific researches are, for which the coastal States can withhold their consent. These other scientific researches are those which (a) are of direct significance for the exploration and exploitation of natural resources, (b) involve drilling into the continental shelf, the use of explosives or the introduction of harmful substances into the marine environment, and (c) involve the construction, operation or use of artificial islands, installations and structures.[196]

Having thus examined the provisions of the Geneva Convention and the LOS Convention regarding the distinction between purely scientific research and other research in the two conventions, an important question arises here as to whether the distinction between pure scientific research and other research adopted in the two conventions, have any relationship with the distinction between the marine scientific research of transitional character and of non-transitional character adopted by the ICJ in the *Aegean Sea Continental Shelf* Case. If there is some relationship between the two classifications, the provisions of Article 245 of the LOS Convention can also be illustrative of what research activities by a coastal State A in the disputed areas are prohibited without consent from the other coastal State B.

In fact, we can find a significant overlap between two ways of classification by the ICJ and the LOS Convention. We can see that paragraph 5 of Article 246 of the LOS Convention lists several forms of research for which the coastal State has discretion to withhold its consent. Furthermore, we find there in the paragraph the very illustrations of explorations of a non-transitory character which were mentioned by the ICJ in the *Aegean Sea Continental Shelf* Case.[197] Note that according to the ICJ, scientific research is not of a transitory character if the research involves either (i) "any risk of physical damage to the seabed or subsoil or to their natural resources", (ii) "the establishment on or above the seabed of the continental shelf, or (iii) "the actual appropriation or other use of the natural resources of the areas of the continental shelf which are in dispute".[198] Note that these kinds of marine scientific research are provided for in paragraph 5 of Article 246 of the LOS Convention, for which the coastal State can withhold its consent.[199]

Here, we can see that the two concepts of the applied scientific research in the LOS Convention and "exploration of transitory character" appear to be almost the same in reality.[200] Thus it can be presumed that a coastal State would be ordered by the ICJ to cease the marine scientific research if it involved the activities provided in paragraph 5 of Article 246 of the LOS Convention in the disputed areas without consent from the other coastal State.

By the same token, it can be said that the coastal State would not be ordered to cease the purely scientific research conducted "exclusively for peaceful purposes and in order to increase scientific knowledge of the marine environment for the benefits of all mankind", which does not involve any of the activities provided for in paragraph 5 of Article 246 of the LOS Convention.

However, it is not to be forgotten that even if one contending coastal State can carry out some form of research activity in the disputed area, it is nevertheless expected to follow other rules applicable to scientific research, such as general principles for the conduct of marine scientific research provided in Article 240 or the duty to publish and disseminate the information and knowledge acquired from the research as provided in Article 244 of the LOS Convention.

From the examination above, we can reasonably infer that some types of marine scientific research can be conducted by a coastal State in the disputed areas without consent from the other coastal State. However, the problem is that there might be a situation where the coastal States have different views on which research is permissible in the disputed areas without the consent from the other contending coastal State. Or it might be the case where one of the coastal States argues that any marine scientific research in disputed areas without its permission is prejudicial to its rights. Therefore, we can see that there is a clear need for neighbouring littoral States to talk in order to have a common understanding as to which research can be conducted in the disputed areas without the consent from the other coastal State and to enter into arrangements for marine scientific research in the disputed areas. It can be seen in Chapter 3 that there are several instances where arrangements are in place in the disputed areas. These are the joint regime area between Colombia and Jamaica, the common zone between Sudan and Saudi Arabia and the common scientific and fishing zone between the Dominican Republic and Colombia.[201]

5.2. *Does the Marine Scientific Research Conducted by State A, Without Any Protests, Reinforce Its Claims to the Part of the Disputed Area or Natural Resources in the Disputed Area?*

The question just raised seems very important. When there is a doubt as to whether the marine scientific research conducted by State A, without any protests from State B might reinforce A's claims to the part of the disputed area or natural resources in the disputed area, then State B is tempted to intervene in the research activities by State A, and State A, in turn, is tempted to carry out its research activities as much as possible in the hope to reinforce its position on the delimitation.

Canada and the United States conducted petroleum exploration activities on the continental shelf in the Gulf of Maine, including those locations on Georges Bank to which both Parties laid claims.[202] The question in the *Gulf of Maine* Case was as to whether unilateral geographical limits for exploration activities by one coastal State can turn into a valid international boundary on the basis of acuiescence by the other coastal State.[203]

In 1964 Canada began to issue, on its own side of what it regarded as the median line dividing Georges Bank, permits for the exclusive exploitation of hydrocarbons, and from 1964 onwards seismic research was carried out under the authority of Canada.[204] Canada alleged that it was known to the United States authorities that Canada had issued such permits but the United States had not protested to Canada with regard to such issuance of permits, refraining from authorising any explorations in Canada's claimed area in the Georges Bank.[205] Canada went on to argue that this was evidence of acquiescence regarding a median line as a boundary between the two countries.[206] The United States challenged the existence of evidence for such acquiescence, arguing that Canada never issued an official proclamation or any other publication for the purpose of making its claims known internationally and therefore it could not agree on the existence of such a median line claimed by Canada.[207] The United States went on to argue that, on the contrary, Canada had not even taken an official stand in the Truman Proclamation and its possible implications for the continental shelf in the Georges Bank area, which was included in its entirety therein.[208]

In this situation, the Court repudiated the claims by Canada pointing out that there were "uncertainties" and "a fair degree of inconsistency" in the attitude of the United States. It said that:

> In the view of the Chamber, it may be correct that the attitude of the United States on maritime boundaries with its Canadian neighbour, until the end of the 1960s, revealed uncertainties and a fair degree of inconsistency. Notwithstanding this, the facts advanced by Canada do not warrant the conclusion that the United States Government thereby recognised the median line once and for all as a boundary between the respective jurisdictions over the continental shelf; nor do they warrant the conclusion that mere failure to react to the issue of Canada exploration permits, from 1964 until the aide-memoir of 5 November 1969, legally debarred the United States from continuing to claim a boundary following the North East Channel, or even including all the areas south west of the adjusted perpendicular.[209]

Although, the ICJ rejected the Canadian claims, the decision of the ICJ might have been different if there had been some degree of consistency in the United States attitude. Having read the judgment, we can see that some degree of consistency by one coastal State in regard to the permits of exploration and exploitation by the other coastal State might lead to acquiescence of the *de facto* or *de jure* maritime boundary.

However, unlike the Geneva Convention, the LOS Convention provides a set of general provisions applicable to marine scientific research in section 1 of Part XIII and in Article 241 which contains an important provision relevant in our discussion. It provides that:

> Marine scientific research activities shall not constitute the legal basis for any claim to any part of the maritime environment or its resources.

Can this provision be applicable to the claims by the coastal States laying claims to the parts or whole of the disputed areas and trying to reinforce its positions on the basis of scientific research undertaken by itself in the disputed areas? There is no direct answer to it in the LOS Convention. However, it can be reasonably assumed that if marine scientific research cannot constitute a legal basis for the marine environment or to the resources, then it can be said *a priori*, marine scientific research cannot constitute a claim to part of the disputed area itself.

Although the observation made above appears to have reasonable grounds, it seems desirable to make it clear in the provisional arrangements that marine scientific research in the disputed areas cannot constitute a claim to part of the disputed area itself.

5.3. *Can a Third State Conduct Marine Scientific Research in a Disputed Area Without Consent from Any of Two Coastal States?*

If a third State wishes to conduct marine scientific research in a disputed area, it would be faced with a number of questions, all connected to each other. Firstly, which coastal State it has to apply to for the consent; secondly, whether it has to apply to both coastal States; and finally whether it can conduct marine scientific research in the disputed areas without consent from the any of the coastal States. The coastal States in the disputed area would also not be free from uncertainty regarding its jurisdiction with regard to marine scientific research by a third State in the disputed area. In this uncertainty, the coastal State in the disputed area might wish to refrain from giving consent to the marine scientific research in the disputed area lest its consent should aggravate the dispute.[210] Or the coastal State might give its consent for marine scientific research by a third State in the disputed area on its own assumption that the disputed area is not disputed and it has rights to the area.

Can a third State conduct a marine scientific research in the disputed area without having consent from any of the two contending States? The answer seems to be no. It seems unreasonable to assume that third States can undertake marine scientific research without consent from the coastal States, simply because the area of research is disputed. It should be recalled that the disputed area is not the high sea but part of the exclusive economic zone/continental shelf which is to be delimited, and "the coastal States have the rights to regulate, authorise and conduct marine scientific research in their exclusive economic zone and on their continental shelf".[211] Rather, it might be the case where the two contending coastal States ask the researching State to get consent from each of them. Therefore the situation can arise where the researching State needs to get consent from both coastal States.

It is not clear whether the researching State is required to get plural consent from all the contending coastal States in order to conduct marine scientific research. It is probable that the marine scientific research by a third State in a disputed area will be faced with protest by a coastal State whose consent it does not have and thus the marine scientific research will be interrupted. It means that the freedom of marine scientific research by third States is in fact significantly restricted in the disputed area. Therefore, the littoral States in the disputed areas need to negotiate for the guarantee of the freedom of the marine scientific research in the disputed areas, and it can be done by inserting in the provisional arrangements some relevant provisions on the research by third parties in the disputed areas.

6. OBSERVATIONS

In disputed areas, unilateral boundaries cannot be valid solutions to the overlapping rights and jurisdictions under the provisions of the LOS Convention.

Even if there is a critical view on the use of Article 74(3) and 83(3) of the LOS Convention, these provisions are the result of long negotiations and contain substantial rules applicable to disputed areas. These are (i) an obligation to negotiate for conclusion of provisional arrangement of a practical nature, and (ii) an obligation of mutual restraint not to jeopardise or hamper the reaching of the final agreement on delimitation.

Provisional arrangements share some common aspects with interim measures ordered by domestic or international courts pending their final judgments on the merits. Although the term "arrangement" is used frequently to indicate documents less formal than treaties, an arrangement can take a formal treaty form and can be a treaty under the Vienna Convention on the Law of Treaties. Through such arrangements, coastal States can promote some exploitation activities in disputed areas, while preserving their positions on the final delimitation. Therefore, the insertion of a "without prejudice clause" is regarded by some jurists as important to preserve the positions of coastal States.

With regard to the obligation of mutual restraint, it can be safely argued that there is a customary law rule. However, it does not necessarily follow that all activities are prohibited in disputed areas. It remains to be seen if or when a black list of prohibited activities in disputed areas can be completed through development of case law and State practice.

We saw that there is a possibility of jurisdictional conflicts between coastal States in the disputed area with regard to dumping, and pollution from ships when these are a suspected substantial discharge or a major damage to the environment. This means that it is advisable to enter into a provisional arrangement between the coastal States in the disputed areas for co-operation in the exercise of enforcement jurisdiction and avoidance of jurisdictional conflicts, with regard to the protection of the marine environment.

With regard to the marine scientific research in the disputed areas, there is also a possibility of jurisdictional conflict between coastal States in the disputed area. However, it seems unreasonable to argue that all scientific research or exploration by one coastal State should be prohibited without the consent of the other coastal State. Alternatively, it would also be unreasonable to suggest that all kinds of activities of scientific research or exploration can be conducted by one coastal State without the consent of the other coastal State. It is to be recalled that in the *Aegean Sea Continental Shelf* Case, the Court made the distinction between exploration of a transitory character and that of a non-transitory character.[212] Even though the same distinction by the Court was not found in the provisions of the LOS Convention, they are, in fact, very similar.

As to the question of whether the marine scientific research or permit by one coastal State can reinforce its claims to part of a disputed area or natural resources, some caution should be taken because the ICJ considered whether there was consistency in the reaction of the United States to the permits issued by Canada. It is not plain whether Article 241 in the LOS Convention, which provides that "marine scientific research activities shall not constitute the legal basis for any claim to any part of the marine environment or its resources", can also be applicable to the two contending coastal States which lay claims, not to the environment or resources, but to the area itself. However, it seems safe to assume that if marine scientific research cannot constitute a legal basis for the marine environment or to the resources then, *a priori*, marine scientific research cannot constitute a claim to part of the disputed area under the LOS Convention.

It seems unreasonable to assume that third States can undertake marine scientific research without consent from the coastal States simply because the area of research is disputed. It should be recalled that the disputed area is not part of the high seas but part of the exclusive economic zone/continental shelf, which is to be delimited. The two contending coastal States might ask the researching State to get consent from them. Therefore the situation can arise where the researching State needs to get consent from both coastal States. It means that the freedom of the marine scientific research by third States in the disputed areas can be seriously restricted and thus there is a need for an arrangement for co-operation between the coastal States, for safeguarding the right of the third States to the freedom of marine scientific research.

NOTES

1. Churchill and Lowe, *The Law of the Sea*, Manchester University Press (1999), p. 198.

2. *Anglo-Norwegian Fisheries* Case, 1951 *ICJ Reports*, p. 132.

3. 1969 *ICJ Reports*, para. 85.

4. As the mention of the equidistance/median line in each case was made in a different context, it is difficult to draw a conclusion as to the customary status of the median line as a principle of delimitation. Some States used the term equidistance/median line as their preferred method in delimitation agreement. Other States mention the equidistance/median line not as a boundary but as a limit (entitlement) of maritime zone when the area is less than 400 miles in width and make clear that the equidistance/median line is subject to a delimitation by agreement. For the domestic laws of each State in the world on EEZ and the continental shelf, see R.W. Smith, *Exclusive Economic Zone Claims: An Analysis and Primary Document*, Nijhoff Publisher (1986), pp. 41-498: The countries which mention equidistance line in their legislation are as follows: Barbados (equidistance line by agreement), Comoros (equidistance line save as otherwise specially agreed), Dijibouti (delimitation by agreement and no beyond a median line pending agreement), Equatorial Guinea (shall not extend beyond the equidistant median line except where otherwise provided in international treaties), Fiji (median line where no other line is for the time being specified), Grenada (median line where there is no boundary), India (delimitation by agreement and equidistance line pending agreement and unless any other provisional arrangement), Indonesia (delimitation by agreement and

equidistance line pending agreement), Japan (median line or boundary by agreement), Kiribati (by international agreement or the award of any international body or median line), North Korea, Morocco (delimitation by equitable principle and the outer limit shall not extend beyond a median line), Mozambique (whenever there is no boundary, the boundary shall not extend beyond the equidistance line), New Zealand (median line subject to international agreement, arbitral award, or judgement), Nigeria (median line subject to the provision of any treaty), Norway (not beyond the median line), Oman (median line), Qatar (median line in the absence of any particular agreement), Saint Lucia (equidistance line pending delimitation agreement), Solomon Islands (median line when no other line is for the time being specified), South Korea (delimitation by agreement and shall not exercise its right beyond the median line unless otherwise provided in international treaties), Spain (median line subject to international treaties), Tuvalu (median line subject to agreement), United Arab Emirates (agreement and median line), Western Samoa (median line subject to international agreement, and award or judgement).

5. For example, in the East China Sea where there is no boundary between China and Japan, Japan regards the equidistance line pending delimitation of boundaries whereas China does not. As China conducted exploration activities in the waters in the Japanese side of the hypothetical equidistance line, a dispute arose between the two countries. See "3.3. Delimitation in the East China Sea" in Chapter Four.

6. R.W. Smith, *Exclusive Economic Zone Claims: An Analysis and Primary Document*, Nijhoff Publisher (1986), pp. 41-498.

7. See, 1953 *ILC Yearbook*, Vol. II, p. 216; see also E.D. Brown, *International Law of the Sea*, Dartmouth Publishing Company (1994), p. 162.

8. D.W. Bowett, *The Law of the Sea*, Manchester University Press (1967), p. 40.

9. 1969 *ICJ Reports*, para. 34.

10. 1969 *ICJ Reports*, para. 35.

11. 1969, *ICJ Reports*, para. 85(a).

12. E.D. Brown, *op. cit.* (1994), p. 159.

13. Bin Cheng, *General Principles of Law as Applied by International Courts and Tribunals*, Stevens and Sons (1953), pp. 25-26.

14. Bin Cheng, *ibid.*, p. 24.

15. P.W. Birnie and A.E. Boyle, *International Law and The Environment*, Oxford University Press (1992), p. 89.

16. P.W. Birnie and A.E. Boyle, *idem.*

17. Article 197 of the LOS Convention.

18. Article 31 of the Vienna Convention on the Law of Treaties provides that: "A

treaty shall be interpreted in good faith in accordance with the ordinary meaning to be given the terms of the treaty in their context and in the light of its object and purpose".

19. For the literal or logical interpretation of treaties, see G. Schwarzenberger, *A Manual of International Law* (1967), p. 165.

20. D.P. O'Connell, *International Law* (1970), vol. 1, p. 263.

21. Article 32 of the Vienna Convention on the Law of Treaty provides that: "Recourse may be had to supplementary means of interpretation, including the preparatory work of the treaty and the circumstances of its conclusion, in order to confirm the meaning resulting from the application of article 31, or to determine the meaning when the interpretation according to article 31 (a) leave the meaning ambiguous or obscure; or (b) leads to a result which is manifestly absurd or unreasonable".

22. M.H. Nordquist *et al.* eds., *United Nations Convention on the Law of the Sea 1982: A Commentary*, Vol. 2, p. 801, Martinus Nijhoff Publishers (1993).

23. See, M.H. Nordquist and Choon-ho Park eds., *Reports of the United States Delegation to the Third United Nations Conference on the Law of the Sea*, The Law of the Sea Institute (1983).

24. R. Lagoni, "Interim Measures Pending Maritime Delimitation", 78 *A.J.I.L.* (1984), p. 350.

25. M.H. Nordquist *et al.* eds, *op. cit.* (1993), vol. 2, p. 805; A/CONF.62/ C.2/L.14 (1974), *Official Records of the UNCLOS III*, vol. 3, p. 190.

26. A/CONF.62/C.2/L.14 (1974), Official Records of the UNCLOS III, vol. 3, p. 190.

27. The Greek proposal read:

> 1. Where the coasts of two or more States are adjacent or opposite to each other, the delimitation of the continental shelf boundaries shall be determined by agreement among themselves.
>
> 2. Failing such agreement, no State is entitled to extend its rights over the continental shelf beyond the median line every point of which is equidistant from the nearest points of the baselines, continental or insular, from which the breadth of the continental shelf of each of the two States concerned.; A/CONF.62/C.2/L.32 (1974), Official Records of the UNCLOS III, vol. 3, p. 210.

28. A/CONF.62/C.2/L.31/Rev.1 (1974), Official Records of the UNCLOS III, vol. 3, p. 211 (Japan).

29. A/CONF.62/C.2/L.43 (1974), Official Records of the UNCLOS III, vol. 3, p. 220 (Ireland).

30. A/CONF.62/WP.8/Part II (ISNT, 1975), Article 61, Official Records of the UNCLOS III, vol. 4, p. 151 (Chairman, Second Committee).

31. B.H. Oxman, "The Third United Nations Conference on the Law of the Sea: The 1976 New York Sessions", 71 *A.J.I.L.* (1977), p. 267; The document presented at Caracas (the second session in 1974) by an informal group suggested three basic options regarding the compulsory judicial settlement. First, an effective dispute settlement system must apply to all disputes relating to the interpretation and application of the LOS Convention. Second, the dispute settlement machinery should have no jurisdiction over specified categories of the Convention. The third option contained an "opt-out" system, which would allow States to exclude at least from compulsory binding procedures. In 1975, the ISNT adopted the third option whereby disputes, *inter alia*, concerning maritime boundary delimitation can be opted out. Even if the issue on dispute settlement was classified as one of the hard issues and dealt by the Negotiation Group 7(NG7) established in 1978, the basic structure of ISNT was maintained through UNCLOS III; see M.H. Nordquist *et al.* eds., *op. cit.* (1993), vol. 5, pp. 87-145; Article 298 of the LOS Convention; A.O. Adede. *The System for Settlement of Disputes under United Nations Convention on the Law of the Sea*, Martinus Nijhoff Publishers (1987), pp. 56-61.

32. A/CONF.62/WP.8/Rev.1 (1976), Article 18, para. 2(a), *Official Records of UNCLOS IIIV*, Vol. V., p. 190 (President).

33. A/CONF.62/WP.8/REV.1/Part II (ISNT, 1976), Article 62, *Official Records of UNCLOS III*, Vol. V, p. 151 (Chairman, Second Committee); Article 71 of the RSNT, which dealt with delimitation of continental shelf, had the same provision in its paragraph 3 as in paragraph 3 of article 62 of RSNT.

34. Nordquist *et al.* eds, *op. cit.* (1993), vol. 2, p. 810.

35. Spain (1977, mimeo.), Article 62; reproduced by Platzöder, *Third United Nations Conference on the Law of the Sea: Documents*, vol. IV, p. 467.

36. A/CONF.62/WP.10 (ICNT, 1977), Article 83, Official Records of the UNCLOS III, vol. 8, p. 17.

37. M.H. Nordquist and Choon-ho Park eds., *op. cit.* (1993), pp. 211-212; M.H. Nordquist *et al.* ed., *op. cit.* (1993), Vol. 2, p. 812.

38. NG7/2 (1978, mimeo.), Article 83 (Bahamas, Barbados, Canada, Colombia, Cyprus, Democratic Yemen, Denmark, Gambia, Greece, Italy, Japan, Kuwait, Malta, Norway, Spain, Sweden, United Arab Emirates, United Kingdom, Yugoslavia); reproduced by Platzöder, *op. cit.*, vol. IX p. 392.

39. NG7/3 (1978, mimeo.), Article 83 Morocco; reproduced by Platzöder, *op. cit.*, vol. IX p. 394.

40. NG7/15 (1978, mimeo.), Article 83 (Papua New Guinea); reproduced by Platzöder, *op. cit.* vol. IX, p. 406.

41. M.H. Nordquist *et al.* eds., *op. cit.*, vol. 2, p. 967.

42. *Ibid.*, p. 968.

43. B.H. Oxman, "The Third United Nations Conference on the Law of the Sea: The Seventh Session (1978)", 72 *A.J.I.L.* (1978), p. 23.

44. B.H. Oxman, *idem.*

45. Nordquist *et al.* ed., *op. cit.*, vol. 2, p. 969.

46. M.H. Nordquist and Choon-ho Park, *op. cit.* (1993), p. 285.

47. Nordquist *et al.* eds., *op. cit.* no. 15, vol. 2, p. 970.

48. NG7/32 (1979, mimeo.) Article 83 (India, Iraq and Morocco); reproduced by Platzöder, *op. cit.*, vol. IX, p. 453.

49. NG7/39 (1979, mimeo.), Articles 74/83 (Chairman, NG7); reproduced by Platzöder, *op. cit.*, vol. IX, p. 459.

50. A/CONF.62/WP.10/Rev.1 (ICNT/Rev.1, 1979, mimeo.), Articles 74/83; reproduced by Platzöder, *op. cit.* no. 29 vol. I, pp. 375-423.

51. A/CONF.62/91 1979, *Reports* to the plenary Conference, NG7/45 Chairman, NG7, *Official Records of the UNCLOS III*, vol. 12, pp. 71-101.

52. NG7/2/Rev.1 (1980, mimeo.), Article 83 (Bahamas, Gambia, Greece, Guyana, Italy, Japan, Kuwait, Malta, Norway, Portugal, Spain, Sweden, United Arab Emirates, United Kingdom, Yugoslavia); reproduced by Platzöder, *op. cit.*, vol. IX, p. 393.

53. NG7/10/Rev.1 (1980, mimeo.), Article 83 (Algeria, Argentina, Bangladesh, Benin, Burundi, Congo, France, Iraq, Ireland, Ivory Coast, Kenya, Liberia, Libyan Arab Jamahiriya, Madagascar, Maldives, Mali, Mauritania, Morocco, Nicaragua, Pakistan, Papua New Guinea, Poland, Romania, Senegal, Syrian Arab Republic, Somalia, Turkey, Venezuela and Viet Nam); reproduced by Platzöder, *op. cit.*, vol. IX, p. 403.

54. A/CONF.62/WP.10/Rev.2 (ICNT/Rev.2, 1980, mimeo.), Article 83. Reproduced by Platzöder, *op. cit.*, vol. II, p. 179.

55. *Black's Law Dictionary*, West Publishing Co. (1990) p. 784.

56. D.W. Bowett, *The Law of International Institutions*, Stevens and Sons (1982), p. 40. The provisional measures within the scope of Article 40 of the Charter have as their subject matter the suspension of hostilities, troop withdrawal, and the conclusion of or adherence to a truce; see Bruno Simma ed., *The Charter of the United Nations: A Commentary*, Oxford University Press (1995), pp. 618-620.

57. *Fisheries Jurisdiction* Cases, 1972 *ICJ Reports*, para. 22.

58. J.G. Merrills, "Interim Measures of Protection in the Recent Jurisprudence of the International Court of Justice", 44 *I.C.L.Q.* (1995), p. 100.

59. S. Rosenne, *The Law and Practice of the International Court*, A.W. Sijthoff-Leyden (1965), p. 427.

60. The UN Fish Stocks Agreement of 1995 expands the legal base for provisional measures by allowing a court or tribunal to prescribe provisional measures to prevent damage to the fish stocks in question; Article 31 of the UN Convention on Straddling Fish Stocks and Highly Migratory Fish Stocks; A.E. Boyle, "Problems of Compulsory Jurisdiction and the Settlement of Disputes Relating to Straddling Fish Stocks", 14 *I.J.M.C.L.*, (1999), pp. 23-24.

61. In the *M/V Saiga* Case *no. 1*, the Tribunal made a judgement on 4 December 1997 which included an order that Guinea should promptly release the M/V Saiga and its crew from detention, upon the posting of a bond of $ 400,000. However, without accepting the guarantee of a bank, which Guinea submitted to St. Vincent and Grenadines as a bond, Guinea proceeded criminal proceedings against the Captain of the M/V Saiga. On 22 December 1997 St. Vincent instituted a new proceeding against Guinea under Part XV of the LOS Convention, alleging, *inter alia*, that Guinea had failed to comply promptly with the release of the M/V Saiga. And on 13 January 1998, St. Vincent asked the ITLOS to order a provisional measure requiring, *inter alia*, to release the M/V Saiga. The Tribunal ordered on 11 March 1998 provisional measures on the basis of paragraph 1 of Article 290 of the LOS Convention thereby requiring Guinea to comply without delay with the judgement of the Tribunal on 4 December 1997.

62. For the analysis of the *M/V Saiga* Case, see E.D. Brown "The M/V Saiga Case on prompt release of detained vessels: the first Judgement of the International Tribunal for the Law of the Sea", 22 *M.P.* (1998), pp. 307-326; Vaughan Lowe, "The M/V Saiga: The First Case in the International Tribunal For the Law of the Sea", 48 *I.C.L.Q.*, (1999), pp. 186-199; Louise De La Fayette, "International Tribunal for the Law of the Sea: The M/V Saiga (No. 2) Case" 49 *I.C.L.Q.* (2000), pp. 467-475.

63. ITLOS, *The M/V Saiga* Case *No. 2*, 1998 Order, para. 42-44.

64. *Ibid.*

65. For the analysis of the cases see R.R. Churchill, "International Tribunal for the Law of the Sea and the *Southern Bluefin Tuna* Cases (*New Zealand* v. *Japan*: *Australia* v. *Japan*): Order for Provisional Measures of 27 August 1999", 49 *I.C.L.Q.* (2000), pp. 979-990.

66. ITLOS, The *Southern Bluefin Tuna* Cases, 1999 Order, para. 90 (1).

67. ITLOS, The *Southern Bluefin Tuna* Cases, 1999 Order, paras. 69, 70, 80.

68. R.R. Churchill, *op. cit.* (2000/ *I.C.L.Q.*), pp. 979-980.

69. A.O. Adede, *op. cit.* (1987), p. 281.

70. McNair, *The Law of the Treaties*, Oxford University Press (1961), p. 20.

71. D.P. O'Connell, *International Law*, Stevens and Sons (1970), p. 201.

72. W.C. Gilmore, *Combating International Drugs Trafficking: The 1988 United Nations Convention Against Illicit Traffic in Narcotic Drugs and Psychotropic Substances*, Commonwealth Secretariat (1991), p. 35.

73. John Siddle, "Anglo-American Co-operation in the Suppression of Drug Smuggling", 31 *I.C.L.Q.* (1982), p. 738.

74. W.C. Gilmore, *op. cit.* (1991), p. 52.

75. A. Aust, "The Theory and Practice of Informal International Instruments", 35 *I.C.L.Q.* (1986), pp. 797-799. An Agreement which is not intended to be legally binding is sometime called "non-binding agreement": Manfred Lachs, "Non-binding Agreements", in R. Bernhardt (ed.), *Encyclopaedia of International Law*, vol. 7, pp. 353-371).

76. *22 Code of Federal Regulation*, Part 181; reproduced by A. Aust, *ibid.*, pp. 797-799.

77. N. Bentwich and A. Martin, *A commentary on the Charter of the United Nations*, Routledge and Kegan Paul L.T.D. (1950), pp. 109-110.

78. D.W. Bowett, *op. cit.* (1982), p. 166.

79. The full title is "Agreement for the Implementation of the Previsions of the United Nations Convention on the Law of the Sea of 10 December 1982 relating to the Conservation and Management of Straddling Fish Stocks and Highly Migratory Fish Stocks". The Agreement was adopted on 4 August 1995 and opened for signature on 4 December 1995. It remained open for signature until 4 December 1996 and it will enter into force 30 days after the deposit of the thirtieth instrument of ratification or accession: text and introduction is available at <http://www.un.org/Depts/>. See also D.H. Anderson, "The Straddling Fish Stocks Agreement of 1995-an initial assessment", 45 *I.C.L.Q.* (1996), pp. 463-475: Peter G. Davis and Cathrrine Redgwell, "The International Legal Regulation of Straddling Fish Stocks", 74 *B.Y.I.L.* (1997), pp. 265-272.

80. Article 1 of the UN Fish Stocks Agreement.

81. 1974 *ICJ Reports*, para. 33.

82. 1974 *ICJ Reports*, *idem.*

83. 1974 *ICJ Reports*, *idem.*

84. 1974 *ICJ Reports*, para. 35.

85. 1974 *ICJ Reports*, para. 39.

86. David Colson used the term "Modus Vivendi" in explaining the arrangement between Norway and the Soviet Union on the Barents Sea; see Charney and L. Alexander, *International Maritime Boundary* (Martinus Nijhoff Publishers, (1993):

Lord McNair explained that "Modus Vivendi" is frequently used to indicate a temporary arrangement prolong the duration of an agreement from time to time; McNair, *op. cit.* (1961), p. 24: Anthony Aust wrote that: "The term is used more to describe a treaty which is intended to be temporary. It is used for MOUs, particularly on fisheries matters;" Anthony Aust, *Modern Treaty Law and Practice*, Cambridge University Press (2000), p. 25.

87. Anthony Aust, *ibid.*, p. 18.

88. Anthony Aust, *ibid.*, pp. 34-39. For the dangers in using MOSs, see Anthony Aust, *ibid.*, pp. 39-41.

89. *South West Africa* Cases (Preliminary Objection), 1961 *ICJ Reports*, p. 31.

90. McNair, *op. cit.* (1961), p. 15.

91. McNair, *ibid.*, p. 7.

92. Anthony Aust, *Modern Treaty Law and Practice*, Cambridge University Press (2000), p. 7.

93. The International Law Commission observed that "the restriction of the use of the term 'treaty' in the draft articles to international agreements expressed in writing is not intended to deny the legal force of oral agreements under international law or to imply that some of the principles contained in later parts of the Commission's draft articles on the law of treaties may not have relevance in regard to an oral agreement." (1966 *ILC Yearbook*, Vol. II, p. 189.)

94. American Law Institute, *Restatement of the Law: The Foreign Relations Law of the United States* (Third Edition, 1986), vol. 1, p. 149.

95. I. Brownlie stated that "a good number of articles are essentially declaratory of existing law"; Brownlie, *Principles of International Law* (1990), p. 604. Shabtai Rosenne examined the codification process of rules on good faith in the Vienna Convention and showed that the references to good faith in the Convention are reflection of customary law; see S. Rosenne, *Development in the Law of Treaties 1945-1986*, Cambridge University Press (1989), pp. 135-180.

96. Note that the UK and Iceland used an Exchange of Notes which was a treaty for provisional arrangement: see "3.2.1. Practice in Use of the Term 'Arrangement'".

97. L. Lauterpacht, ed., *Oppenheim's International Law*, Longman (1955), p. 938.

98. R. Jennings and A. Watts, ed., *Oppenheim's International Law* (9th ed.), Longman (1992), p. 1299: See also, Anthony Aust, *Modern Treaty Law and Practice*, Cambridge University Press (2000), pp. 233-234.

99. 1966 *I.L.C. Yearbook* 1966 Vol. II, pp. 250-251.

100. Article 56 (1) of the Vienna Convention on the Law of Treaties.

101. 1966 *ILC Yearbook*, Vol. II, p. 251.

102. However, it has been pointed out that without setting out the categories of treaties said to be terminable at will by their nature, the formulation of Article 56(1) of the Vienna Convention is a total failure, and thus its meaning will be forever debated by all the parties to any dispute; see K. Widdows, "The Unilateral Denunciation of Treaties Containing No Denunciation", 53 *B.Y.I.L.* (1982), p. 113.

103. Paragraph 2 of article 56 of the Vienna Convention on the Law of Treaties.

104. The Agreement's full title is "Agreement for the Implementation of the Provisions of the United Nations Convention on the Law of the Sea of 10 December 1982 Relating to the Conservation and Management of Straddling Fish Stocks and Highly Migratory Fish Stocks". This Agreement is adopted on 4 August 1995 and not in force yet. See D.H. Anderson, "The Straddling Stocks Agreement of 1995-An Initial Assessment", 45 *I.C.L.Q.* (1996), pp. 463-475.

105. See Robert W. Smith, "Report Number 1-4: Cuba-United States", *International Maritime Boundary*, pp. 419-421.

106. R. Lagoni, "Interim Measures Pending Maritime Delimitation", 78 *A.J.I.L.* (1984), p. 358.

107. Anthony Aust, *op. cit.* (2000), p. 20. He wrote that ". . . the name does not, itself, determine the status of the instrument; what is decisive is whether the negotiating States intended the instrument to be (or not to be) legally binding".

108. Fox *et al.* eds., *Joint Development of Offshore Oil and Gas*, vol. 2, British Institute of International and Comparative Law (1990), p. 6.

109. J.G. Merrills, *International Dispute Settlement*, Cambridge University Press (1998), p. 14.

110. Zhiguo Gao, "The Legal Concept and Aspects of Joint Development in International Law", *Ocean Yearbook* 13, The University of Chicago Press, p. 120.

111. 1974 *ICJ Reports*, para. 37.

112. H. Fox *et al.* ed., *Joint Development of Oil and Gas: A Model Agreement for States for Joint Development with Explanatory Commentary*, British Institute of International Law and Comparative Law (1989), p. 378.

113. Churchill and Lowe, *op. cit.* (1999), p. 192.

114. Article 33 (1) of the U.N. Charter provides that:

> The parties to any dispute, the continuation of which is likely to endanger the maintenance of international peace and security, shall, first of all, seek a solution by negotiation, enquiry, mediation, conciliation, arbitration, judicial settlement, resort to regional agencies or arrangements, or other peaceful means of their own choice.

115. J.G. Merrills, *op. cit.* (1998), pp. 2-3.

116. 1969 *ICJ Reports*, para. 85(a).

117. The term *pactum de negotiando* is used along with the term *pactum de contrahendo* in a similar context. According to a view the legally binding obligations resulting from the *pactum de contrahendo* are more extensive than those flowing from the *pactum de negotiando*. However, there is the other convincing view that there is no relevant distinction between the two *pacta* in the legal quality of the obligations; see Ulrich Beyrlin, "Pactum de Cotrahendo, Pactum de Negotiando", in R. Bernhardt (ed.), *Encyclopaedia of International Law*, vol. 7, p. 376. For *Pactum de Contrahendo* see also, Anthony Aust, *op. cit.* (2000), p. 25.

118. McNair, *op. cit.* (1961), p. 20.

119. Bin Cheng, *General Principles of Law as Applied by International Courts and Tribunals*, Stevens and Sons (1953), p. 118.

120. *North Sea Continental Shelf* Cases, 1969 *ICJ Reports*, para. 86.

121. ICJ, *ibid.*, para. 85(a).

122. 24 *I.L.R.* (1957), p. 119. Much of the award is reproduced with explanation in Whiteman, *Digest of International Law*, US Department of State Publication (1970), Vol. 3. pp. 1066-1073.

123. Bin Cheng, *op. cit.* (1953), pp. 107-117.

124. Bin Cheng, *idem.*

125. Bin Cheng, *ibid.*, p. 123.

126. 1976 *ICJ Reports* at 6, para. 15.

127. 1982 *ICJ Reports* at 42, para. 34.

128. M. Nordquist, S.H. Lay and K.R. Simmonds eds, *New Directions in the Law of the Sea*, Oceana Publications (1980), p. 29; see also A.E. Boyle, "The Law of Treaties and the Anglo-French Continental Shelf, 29 *I.C.L.Q.* (1980), pp. 498-508.

129. Bin Cheng, *op. cit.* (1953), p. 140.

130. Bin Cheng, *ibid.*, p. 141.

131. The *Electricity Company of Sofia and Bulgaria* Case, 1939 PCIJ *Series* No. 79, p. 199.

132. *UN Doc.A/CN.4/SER.A/1966/Add.1.*

133. *United Nations Conference on the Law of the Treaties, Official Records of the First Session*, UN Doc. A/Con. 39/11 para. 26 (1968).

134. Rainer Lagoni, *op. cit.* (1984/*A.J.I.L.*), p. 364; *Frankfurter Allgemeine Zeitung*, 21 July 1983, p. 5.

135. Alex G. Oude Elferink, *The Law of Maritime Boundary Delimitation: A* Case *Study of the Russian Federation*, Martinus Nijhoff Publishers (1994), pp. 209-210.

136. *Aegean Sea Continental Shelf* Case, 1976 *ICJ Reports*, para. 30.

137. *Fisheries Jurisdiction* Case, 1972 *ICJ Reports*, paras. 25, 26, 43.

138. 1974 *ICJ Reports*, para. 33.

139. Reproduce by David A. Colson, "Report Number 1-3: Canada-United States (Gulf of Maine)", *International Maritime Boundaries*, pp. 402-403.

140. R. Churchill and G. Ulfstein, *Marine Management in the Disputed Areas: The* Case *of the Barents Sea*, Routledge (1992), p. 87.

141. R. Lagoni, "Interim Measure pending Maritime Delimitation Agreement", 78 *A.J.I.L.* (1984), p. 366.

142. Malcolm. N. Shaw, *International Law*, Cambridge University Press (1994), p. 579.

143. Anthony Aust, *op. cit.* (2000), p. 207.

144. McNair, *The Law of Treaties*, Oxford University Press (1961), p. 309.

145. *ILC Yearbook* (1966), p. 226.

146. R. Jennings and A. Watts, ed., *Oppenheim's International Law* (9th ed.), Longman (1992) Vol. 1, p. 1261.

147. Christine Chinkin, *Third Parties in International Law*, Oxford University Press (1993), p. 71.

148. McNair, *op. cit.* (1961), p. 321.

149. A number of maritime areas in the world have more than two relevant coastal States. For example, in the East China Sea, Korea, Japan, China and Taiwan lay claims to overlapping areas of the sea. A similar situation has arisen in the North Sea, where three agreements UK-Netherlands, UK-Denmark, and the Netherlands-Denmark, met an equidistant tri-point, without regard to the interests of Germany in the area. Following the *North Sea Continental Shelf Cases*, the agreements between the Netherlands, Germany and Denmark have concluded and the agreement between the UK and the Netherlands and the UK and Denmark had to be modified to open up the centre of the North Sea for Germany. Sometimes tripartite negotiations have achieved tri-junction. The examples are, *inter alia*, the agreement among Indonesia, Malaysia and Thailand on the delimitation of northern continental shelf in the Strait of Malacca (1971), the agreement among India, Maldives, and Sri Lanka on the Gulf of Manaar in the Indian Ocean(1976), and the agreement among India, Indonesia and Thailand on the Andaman Sea (1978). Bilateral agreements can also establish tri-junction. For example, Sweden and the USSR agreed maritime boundary in the Baltic in 1988 and, subsequently, in 1989

Poland and Sweden agreed on the maritime boundary using the terminus agreed in the Sweden and the USSR agreement. In the *Libya/Malta Continental Shelf Case*, the ICJ modified the boundary of the continental shelf to take into account the Italian claim; for explanations of some of the cases, see D. Colson, "The Legal Regime of Maritime Boundary Agreements", *International Maritime Boundary*, pp. 61-62; for the agreement between Poland, Sweden and the USSR, see Alex G. Oude Elferink, *The Law of Maritime Boundary Delimitation: A* Case *Study of the Russian Federation*, Martinus Nijhoff Publishers (1994), p. 193 and pp. 212-213.

150. 1981 *ICJ Reports*, paras. 28, 33, 35.

151. 1985 *ICJ Reports*, paras. 23.43.

152. The *Libya/Malta Continental Shelf* Case; 1985 *ICJ Reports*, para. 21.

153. 18 *R.I.A.A.*, The *Anglo-French arbitration*, para. 28: the award is reproduced at 18 *I.L.M.* (1979), pp. 379-494.

154. 31 *I.L.M.* (1992) p. 645; see Alan E. Boyle, "Dispute Settlement and the Law of the Sea Convention: Problems of Fragmentation and Jurisdiction", 46 *I.C.L.Q.* (1997), p. 45.

155. 1999 *Yemen-Eritrea Arbitration Award*, para. 164. In this case, Saudi Arabia sent a letter to the tribunal expressing its positions in this regards but Djibouti did not intervene the proceeding and thus the tribunal considered the Djibouti position *proprio motu*.

156. Similar provisions are found, for example, in Article 6 of the Convention between France and Spain Agreement on the Delimitation of the Two States in the Bay of Biscay, and in Article 15 of the Agreement between Sudan and Saudi Arabia Relating to the Joint Exploration and Exploitation of the Natural Resources of the Sea-Bed and Sub-Soil in a Defined Area of the Two Countries in the Red Sea. See for the France and Spain Agreement, *New Directions in the Law of the Sea*, vol. V, p. 251; for the Sudan and Saudi Arabia, see *New Directions in the Law of the Sea*, vol. V, p. 393.

157. When Korea and Japan negotiated in the early 1970s the agreement on the joint development of continental shelf China's participation was not made partly due to the fact that neither Korea nor Japan had diplomatic relations with China. There have been strongly worded protests from Beijing in this regard. Later in the 1990s similar situations have arisen among the three States. When China and Japan concluded a provisional fisheries agreement in 1997, Korea protested through a diplomatic channel that the agreement infringed upon the areas to which Korea has an entitlement of EEZ. Again in 1998 when Korea and Japan concluded a provisional fisheries agreement, China made its position clear through a diplomatic channel that China would not recognise the joint fishing zone between Korea and Japan established in the East China Sea as it intruded to the area where China has a lawful claims.

158. Agreement between Sudan and Saudi Arabia Relating to the Joint Exploration and Exploitation of the Natural Resources of the Sea-Bed and Sub-Soil of the Red Sea in a Defined Area of the Two Countries in the Red Sea, 16 May 1974; see *New Directions in the Law of the Sea*, vol. V, p. 393.

159. "Kuwait and Saudi Arabia Agreement Relating to the Partition of the Neutral Zone", 7 July 1965. 4 *I.L.M.* (1965), p. 1134.

160. McNair, *op. cit.* (1961), p. 93.

161. Paragraph 1(b) (iii) of Article 56 of the LOS Convention provides that: "1. In the exclusive economic zone, the coastal State has: . . . (b) jurisdiction as provided for the relevant provisions of this Convention with regard to: . . . (iii) the protection and preservation of the marine environment".

162. Article 210(5), 216; Article 211(5)-(6), 220, 234; and Article 208, 214.

163. Besides dumping, other provisions are not found in the LOS Convention that manifestly deal with the continental shelf coastal State's jurisdiction with regard to pollution. Paragraph 5 of Article 210 provides that: "Dumping within the territorial sea and the exclusive economic zone or onto the continental shelf shall not be carried out without the express prior approval of the coastal State, which has the right to permit, regulate and control such dumping after due consideration of the matter with other States which by reason of their geographical situation may be adversely affected thereby". In relation to enforcement on the dumping on the continental shelf, the LOS Convention provides that coastal States can enforce its relevant law and regulation with regard to dumping within its territorial sea or its exclusive economic zone or onto its continental shelf: Article 216(1) of the LOS Convention provides that "1. Laws and regulations adopted in accordance with this Convention and applicable international rules and standards established through competence international organisations or diplomatic conference for the prevention, reduction and control of pollution of the marine environment by dumping shall be enforced: (a) by the coastal States with regard to dumping within its territorial sea or its exclusive economic zone or onto its continental shelf".

164. Kari Hakapää wrote that: "In the event of a dumping violation in the exclusive economic zone coastal authorities may stop the vessel and, if necessary, detain it for the purpose of instituting legal proceedings at a coastal port": Kari Hakapää, *Marine Pollution in International Law*, Suomalinen Tiedeakatemia (1981), pp. 24-248.

165. D.W. Bowett, "Jurisdiction: Changing Patterns of Authority over Activities and Resources", 52 *B.Y.I.L.* (1983), p. 14. He wrote that: "However trite it may seem, the question of the purpose and utility of the rules or principles of jurisdiction, summarised above, needs to be answered. The answer cannot be that these rules serve to indicate the State which ought to exercise jurisdiction. For, generally speaking, situations of concurrent jurisdiction are normally enough, so the propriety of a given State exercising jurisdiction is rarely raised in an absolute sense . . .".

166. Article 220(5) of the LOS Convention.

167. Article 220(1) of the LOS Convention.

168. Article 228 of the LOS Convention. A similar situation actually arose between Korea and Japan regarding the Korean cargo vessel "Sunny Rose" in 1997. On 14 December 1997, the Japanese law enforcement vessel "Yanaka" approached the Sunny Rose at the 50 miles away from the Japanese Hamada Port on charge of discharge of 311 litres of bilge when it was in passage, and then took the Sunny Rose to the Hamada Port after the officers of the Yanaka failed to board the Sunny Rose due to the bad weather. In the Hamada Port, the Japanese officers boarded the vessel, inspected the vessel and took evidence from the vessel. On the following day, the Sunny Rose and its crews were released from detention after they were formally charged and paid the bond of one million Yen to the Japanese authorities. Then the Korean government then pointed out to the Japanese government that the discharge of 311 litres of oil could not be seen as "major damage" and thus the Japanese government infringed upon the free passage of the Sunny Rose. Then Korean government initiated its own criminal proceedings against the Sunny Rose and asked Japan to forsake its proceeding against the Sunny Rose. Later the Japanese government ceased its proceeding after the Korean court imposed fines on the Sunny Rose. Source: Korean Foreign Ministry, *Legal Examination of the Sunny Rose* Case (written in Korean for internal use), March 1998.

169. However, a dispute can arise between two coastal States when coastal State A tries to enforce its law on Ship X of coastal State B when the vessel is in State A's port with regard to a discharge in the disputed areas, if State B argues that the discharge in question has occurred in State B's waters and not in State A's waters and thus State A does not have jurisdiction over the vessel X. However, this kind of dispute would not occur at sea between the two law enforcement vessels of the two coastal States, but would occur between the foreign ministries of the two countries.

170. Article 220(5) of the LOS Convention provides that: "Where there are clear grounds for believing that a vessel navigating in the exclusive economic zone or the territorial sea of a State has, in the exclusive economic zone, committed a violation referred to in paragraph 3 resulting in a substantial discharge causing or threatening significant pollution of the marine environment, that State may undertake physical inspection of the vessel for matters relating to the violation if the vessel has refused to give information or if the information supplied by the vessel is manifestly at variance with the evident factual situation and if the circumstances of the case justify such inspection".

171. Kari Hakapää, *op. cit.*, pp. 244-245.

172. According to a commentary, the term "major damage" was intended to indicate incidents like the Amoco Cadiz incident: see Myron H. Nordquist, eds., *United Nations on the Law of the Sea 1982 A Commentary, Martinus Nijhoff Publishers* (1991), vol. IV, p. 301.

173. See “3.1.6. United Kingdom and Denmark” in Chapter Three.

174. John A. Knauss, *op. cit.*, 1 *O.D.I.L.* (1973), p. 94.

175. Alfred H.A. Soons, “The International Legal Regime of Marine Scientific Research”, 24 *N.I.L.R.* (1977), p. 401.

176. Article 246 of the LOS Convention. The LOS Convention uses the term “jurisdiction” in relation to marine scientific research. But the 1958 Geneva Convention on the Shelf did not use the term “jurisdiction” in relation to marine scientific research. The term “jurisdiction” was used in the Geneva Convention only once in relation to installation and other devices for exploration and exploitation; see articles 55(1)(b) and 246 of the LOS Convention and article 5 of the Geneva Convention. For the development of restriction on the marine scientific research, see John A. Knauss, “Development of the Freedom of Scientific Research Issue of the Third Law of the Sea Conference”, 1 *O.D.I.L.* (1973), p. 94.

177. Article 238 of the LOS Convention.

178. *Sankei Shinbun* (A Daily News Paper published in Japanese) of 29 August 2000.

179. The Aegean Sea Continental Shelf Case, 1976 *ICJ Reports*, para. 2. A question can arise with regard to the relationship between the marine scientific research and exploration. In this regard, it is understood the term the marine scientific research used in the LOS Convention includes the concept of exploration: see P. Birnie “The Legal Framework Relating to Ocean Resources and Their Development, Management and Protection: Implications for Marine Scientific Research (provisional version)”, a paper delivered at the International Conference on Oceanography, held in Lisbon, 14-19 November 1994, p. 12.

180. 1976 *ICJ Reports*, paras. 2 and 17.

181. 1976 *ICJ Reports*, paras. 17 and 26.

182. 1976 *ICJ Reports*, para. 17.

183. 1976 *ICJ Reports*, para. 26.

184. *Ibid.*

185. 1976 *ICJ Reports*, para. 33.

186. Article 5(8) of the Geneva Convention provides that:

> The consent of the coastal State shall be obtained in respect of any research concerning the continental shelf and undertaken there. Nevertheless, the coastal State shall not normally withhold its consent if the request is submitted by a qualified institution with a view to purely scientific research into the physical or biological characteristics of the continental shelf, subject to the proviso that the coastal State shall have the right, if it so desires, to participate or to be represented in the research, and that in any event the results shall be published.

187. E.D. Brown, *The International Law of the Sea*, p. 419.

188. For the drafting history of the article 5 of the Geneva Convention, see Alfred H.A. Soons, "The International Legal Regime of Marine Scientific Research", 24 *N.I.L.R.* (1977), pp. 412-429; United Nations, *ILC Yearbook* 1956, vol. II, p. 10.

189. United Nations, *ILC Yearbook* 1956 vol. II, p. 11.

190. Paragraph 5 of the Resolution by the International Council of Scientific Unions in April 1954: reproduced in the *ILC Yearbook* 1956, vol. II, p. 10.

191. Alfred H.A. Soons, "The International Legal Regime of Marine Scientific Research", 24 *N.I.L.R.* (1977), p. 429.

192. Paragraph 3 of Article 246 of the LOS Convention provides that:

> Marine scientific research in the exclusive economic zone and on the continental shelf shall be conducted with the consent of the coastal State.

193. Paragraph 3 and 5 of Article 246 of the LOS Convention.

194. *Ibid.*

195. Interestingly, Churchill and Lowe used the term "pure research" for the former and "applied research" for the latter when they explained the provisions on the marine scientific research of the LOS Convention: R.R. Churchill and Lowe, p. 405.

196. Paragraph 5 of Article 246 of the LOS Convention. According to the paragraph, the coastal States also have discretion to withhold their consent to the conduct of a marine scientific research when the project contains information communicated pursuant to Article 248 regarding the nature and objectives of the project which is inaccurate or if the researching State or competent international organisation has outstanding obligations to the coastal State from a prior research project.

197. Compare paragraph 5(b) and (c) of Article 246 of the LOS Convention with the paragraph 30 of the Order of the ICJ in the *Aegean Sea Continental Shelf* Case.

198. Quoted from the 1976 *ICJ Reports*, para. 30.

199. The elements of Paragraph 5 of Article 246 were introduced in 1976 by Informal Group of Juridical Expert: Myron H. Nordquist *et al.* eds., *United Nations Convention on the Law of the Sea 1982 A Commentary*, Martinus Nijhoff Publishers (1990), Vol. IV, p. 505.

200. However, it is not easy to draw a line between pure and applied research in actual circumstances. For example, research on the geological structure of the continental shelf can turn out to be of important value for the location of the gas and oil because the presence of oil and gas is closely related to the geological structure of the continental shelf: Alfred H.A. Soons, *Marine Scientific Research and the Law of the Sea*, Kluwer Law and Taxation Publishers (1982), pp. 30-33. Therefore, in our discussion, a premise should be made that in order to qualify as

a purely scientific research it should not involve the activities provided for in Paragraph 5 of Article 246.

201. It is also to be noted that CIESM (International Commission for the Scientific Exploration of the Mediterranean Sea) was established as early as in 1919 and the text of the new Statute of the CIESMwas adopted in 1967. The CIESM is an inter-governmental body whose members are 20 coastal States and 3 non-coastal States: see Tullio Scovazzi, "Reason for International Co-operation in the Mediterranean", in Tullio Scovazzi, ed., *op. cit.* (1999), p. 59.

202. *The Gulf of Maine* Case, 1984 *ICJ Reports*, paras. 60-78.

203. 1984 *ICJ Reports*, paras. 128-129.

204. 1984 *ICJ Reports*, para. 131.

205. 1984 *ICJ Reports*, paras. 131 and 135.

206. 1984 *ICJ Reports*, para. 128.

207. 1984 *ICJ Reports*, para. 134.

208. *Ibid.*

209. 1984 *ICJ Reports*, para. 138.

210. According to Warren S. Wooster, 7 percent of the US requests to conduct marine scientific research in areas under foreign jurisdiction were denied between 1972 and 1978, and disputed jurisdiction was one of the reasons given for denials of consent to the US: see Warren S. Wooster "Research in Troubled Waters: US Research Vessel Clearance Experience, 1972-1978", 9 *O.D.I.L.* (1981), pp. 219-231.

211. Paragraph 1 of Article 246 of the LOS Convention.

212. 1974 *ICJ Reports*, para. 30.

CHAPTER THREE

Practice on Provisional Arrangements in Disputed Areas

1. POSSIBLE PROVISIONAL ARRANGEMENTS IN DISPUTED AREAS

As discussed in Chapter 2, Articles 74(3)/83(3) aim to promote the adoption of certain interim measures pending maritime delimitation. They oblige the States concerned to negotiate in good faith for provisional arrangements. An important question then arises: what provisional arrangements are possible?

As the drafting history examined in the previous chapter demonstrates, such provisional arrangements are intended mainly to encourage utilisation of undelimited areas of the sea. However, the States concerned are not barred from prohibiting specific development activities, or choosing a total moratorium of exploitation in the overlapping areas. Similarly, coastal States can agree to halt development activities on non-renewable resources, such as petroleum or natural gas, and concur in exploiting only fisheries.

Suppose, you are a policymaker considering a provisional arrangement in a disputed area. A number of choices should be made. The choices would be, *inter alia*, as follows:

- Do you prefer to set up a joint exploitation zone?
- If you are going to set up a joint exploitation zone, do you wish to clearly define the joint zone by co-ordinates or to leave the precise geographical limits of the zone vague?
- Are you going to deal with all resources whether living or non-living, or certain resources only, i.e., fisheries resources or oil and gas only, or both of them?
- Even if you wish to have an arrangement that covers only fisheries resources, there is still a further choice: are you going to deal with all

fish stocks in the disputed area or only limited number of specific fish stocks?
- Do you want effective management or loose management of the resources in the disputed area?
- Would you prefer a provisional line rather than a zone?
- Are you going to tacitly recognise a *de facto* boundary, if there is any, which is in place with your neighbouring State?
- Do you want a powerful joint committee to deal with the management issues of a joint zone or weak joint committee?
- Do you wish to deal with the issues of marine scientific research and the marine environment as well?

Many policy options are available to a policymaker from various combinations of the choices mentioned above. How can a policymaker decide among them? There may be policymakers in other States who have already thought through these choices. It might, therefore, be very useful to find out, or try to figure out the reasons why States choose different options, by carefully examining instances of relevant provisional arrangements. By doing so, we can see what the advantages of one option over the others are.

Generally speaking, joint exploitation zones either of the continental shelf or for fishing purposes are by far the most widely used form of interim measure in practice. Indeed, joint development zones and joint fishing zones had been in use in a number of areas throughout the world even before the adoption of the LOS Convention. However, it does not mean that all joint development zones have the same characteristics. There are variations in accordance with the combinations of different choices.

The basic ideas of joint development zones and joint fishing zones are the same and can be found in the judgement of the ICJ in the *North Sea Continental Shelf* Cases. The Court mentioned the possibility of "a regime of joint jurisdiction, use or exploitation for the zones of overlap or any part of them". It said that:

> if the delimitation leaves to the Parties areas that overlap, these are to be divided between them in agreed proportions or failing agreement, equally, unless they decide on a regime of joint jurisdiction, use, or exploitation for the zone of overlap or any part of them.[1]

In the same case, Judge Jessup elaborated on this issue in his separate opinion. He pointed out that: "Even if it is not yet considered to reveal an emerging rule of international law, the principle of co-operation may at least be regarded as an elaboration of the factors to be taken into account in the negotiations now to be undertaken by the Parties".[2]

Sharing the concepts in the *North Sea Continental Shelf* Cases, States have developed variations of joint exploitation zones to meet their specific needs and policy objectives. However, other forms of co-operative arrangements which are not based upon joint zones but upon provisional lines or upon a *de facto* boundary, have also been utilised. We will also examine instances of such arrangements and relevant issues.

In analysing instances of provisional arrangements in this chapter, the main focus will be on the following:

– background: the reasons why Parties chose a joint development zone and/or a joint fishing zone, and why different options were chosen from place to place.
– the method in shaping the joint development zone or joint fishing zone;
– forms of the instruments for arrangements and provisions on future delimitation, without-prejudice, and duration and termination of the arrangements; and
– main co-operative mechanism, effectiveness of arrangements, and power of joint committees.

2. POLICY OPTION 1: JOINT DEVELOPMENT ZONES

The British Institute of International and Comparative Law defined a joint development zone as "an agreement between two States to develop so as to share jointly in agreed proportion by inter-State co-operation and national measures the offshore oil and gas in a designated zone of the seabed and subsoil of the continental shelf to which both or either of the participating States are entitled in international law".[3] Lagoni identifies four essential elements in the joint development zones: first, designation of a special zone; second, the specification of the resources to which it applies; third, the determination of the jurisdiction and the laws under which the operation shall be carried out; and finally, the terms and conditions of exploration.[4] Churchill has identified three types of the joint development in the context of the continental shelf: (i) joint development as an alternative to a boundary, (ii) joint development as an additional element in a boundary settlement, and (iii) joint regulation of the exploitation of an oil or gas field straddling a boundary.[5]

Here a distinction should be drawn between the concept of condominium and joint development. The latter is concerned with "sovereign rights" with regard to economic uses of natural resources, whereas the former is a case of "sovereignty" which is jointly exercised by two or more

States on a basis of equality.[6] On the concept of condominium, Oppenheim remarked: "In this case a piece of territory consisting of land or water is under the Joint Tenancy of two or more States, these several States exercising sovereignty jointly over it, and over the individuals living thereon".[7]

It appears that the need for a joint development zone arises when there is a possibility of the existence of oil and gas deposits and coastal States wish to exploit the resources before they reach an agreement on a maritime boundary. It can be imagined without difficulty that no international oil company would be willing to invest a huge amount of money in a disputed area, when there is no clearly defined joint development zone with a detailed provisions on exploitation, if there are no boundaries. It is to be recalled that Greece asked the ICJ to order Turkey to halt seismic explorations in the disputed area in the Aegean Sea. This illustrates that a huge amount of money spent for explorations in a disputed area would turn into a huge waste when a competent judicial institution orders the cessation of the exploration activities. We will examine six instances of joint development zones which purported to be provisional arrangements, pending maritime delimitation.

2.1. *Kuwait-Saudi Arabia*

Background

A joint development zone in the Persian Gulf between Kuwait and Saudi Arabia developed from the "Neutral Zone" established by the Al Uqair Protocol of 1922 in their land border areas where a land boundary line could not be agreed upon. This provided, that ". . . the Government of Najd [Saudi Arabia] and Kuwait will share equal rights [in the Neutral Zone] until through the good offices of Great Britain a further agreement is made . . . concerning it".[8] The Neutral Zone covered a land area of 5,700 km^2. Through the history of exploration, common and co-ordinated modes of exploitation have been established in the Neutral Zone. Similar practices of joint exploration and exploitation also have been established in offshore areas of the Neutral Zone.

In 1965 an international boundary line was drawn between the countries in the Neutral Zone.[9] The boundary line extends from land to the territorial sea. However, the boundary does not affect the nature of the Neutral Zone and its offshore areas in terms of exploitation of natural resources. With respect to the submerged area beyond the territorial sea, the two countries agreed to exercise their equal rights by means of shared exploration unless the two countries would agree otherwise.[10]

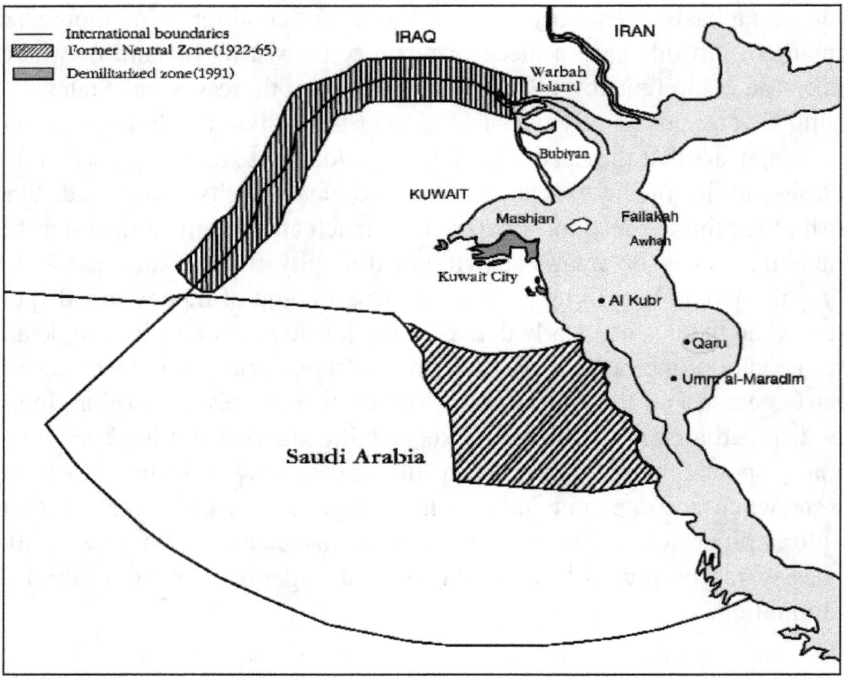

Map 1: Neutral Zone between Kuwait and Saudi Arabia

(Source: M. Miyoshi, *The Joint Development of Offshore Oil and Gas in relation to Maritime Boundary Delimitation*, IBRU Maritime Briefing 1999)

Methods of Shaping the JDZ

As the background of the establishment of the Neutral Zone shows, the Neutral Zone in the land territory corresponded to the area where the two countries could not agree on land boundaries. The submersed maritime zone pertaining to the Neutral Zone of the land also became a joint development zone later on. However, the seaward limit of the joint continental shelf seems obscure because there is no maritime boundary between Kuwait and Iran until now.[11]

Operational Mechanism and the Effectiveness of the Arrangement

The Protocol of 1922 does not provide a detailed mechanism to facilitate joint development. What is clear from the Protocol is that the two States have equal shares in joint and undivided property in the Neutral Zone. Each government granted concessions to their own concessionaires but let their concessionaires co-operate with each other. For example, the Pacific Western

Oil Corporation (subsequently Getty Oil Company), which is the concessionaire of Saudi Arabia, entered into an agreement on joint drilling with American Independent Oil Company (Aminoil) in 1956. The two companies signed a joint operating agreement in 1960. The joint drilling and joint operation of the two concessionaires were under the supervision of a Joint Operating Committee. Later, in 1965, the two governments established a Joint Permanent Committee.[12] It is noteworthy that of the seven jointly developed fields in the Persian Gulf, four, including the largest, are found in the Neutral Zone.[13]

Delimitation Issue

As there are two islands in dispute between Kuwait and Saudi Arabia, the delimitation of maritime boundaries has proved to be difficult. The two islands are Qarah and Umm al-Maradim. These have been under the Kuwaiti control and claimed by Saudi Arabia. In 1961 Kuwait offered to Saudi Arabia a half share of any income accruing from future exploitation in return for the Saudi Arabian acknowledgement of the Kuwaiti sovereignty over the islands. However, Saudi Arabia has yet to make a decision on the offer.[14]

2.2. *South Korea-Japan*

South Korea and Japan concluded an agreement for establishing a joint development zone in the East China Sea in 1974. This agreement will be examined thoroughly in Chapter 5.

2.3. *Malaysia-Thailand*

Background for Joint Development

In 1972 Malaysia and Thailand agreed on a lateral shelf boundary which is about 29 N.M. long in the Gulf of Thailand in 1972. However, the two countries could not agree on a boundary further seaward because of a serious disagreement between them with regard to the effect of the Thai rock, "Ko Losin" on the delimitation. Ko Losin is about 1.5 metres high at high tide and located some 39 N.M. north (i.e., Thailand side) of what would have been the agreed equidistant lateral boundary if the effect of the island had been disregarded.[15] Thailand argued for the full effect of Ko Losin, whereas Malaysia denied any effect of Ko Losin.

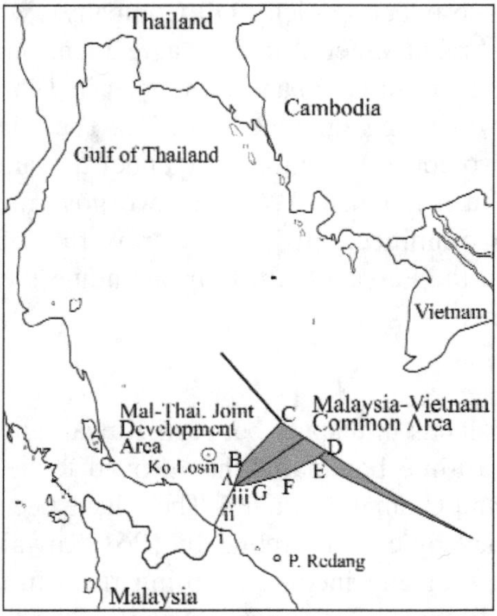

Map 2: Joint Development Zone between Malaysia and Thailand and Malaysia and Vietnam

(Source: M. Miyoshi, *The Joint Development of Offshore Oil and Gas in Relation to Maritime Boundary Delimitation*, IBRU Maritime Briefing 1999)

Method of Shaping the Joint Development Zone

As an interim measure, the Parties agreed to establish a joint development zone which is of a roughly triangular-shape covering 5,439 km^2 in area in 1979: the outer limits of the joint development zone were made by using the lines which were claimed by the two countries.[16]

Forms and Provisions

The above arrangement is in the form of an annexed Memorandum of Understanding to the main Memorandum of Understanding which provides for a partial continental shelf boundary which was constructed by connecting point i, point ii, and point iii. The two countries agreed to continue to seek to delimit the remaining part of the continental shelf by negotiations or such other peaceful means as may be agreed by them.[17] There is no provision on *ratione temporis* in the Memorandum. However, a Joint Authority is to be established for a period of 50 years.[18] If both States arrive at a satisfactory solution on the delimitation within 50 years, then the Joint Authority is to be wound up. If, however, no satisfactory solution were to be found within 50 years, then the current arrangement would continue.[19]

Operational Mechanism and Effectiveness of the Arrangement

An important feature of the provisional arrangement is that it has a powerful Joint Authority. The Joint Authority assumes "all rights and responsibilities on behalf of both Parties for the exploration and exploitation of the non-living resources of the sea-bed and subsoil" in the joint development area.[20] In 1990, the two countries agreed on a production sharing system. The production sharing system provides terms and conditions for exploitation. Among others, the conditions State that: the duration of the contract shall not exceed 35 years; the payment of 10% of gross production of petroleum shall be made by the contractor to the Joint Authority as royalty; 50% of gross production shall be taken by the contractor for the recovery of costs; and the remainder of gross production shall be profit and divided equally between the Joint Authority and the contractor.[21]

2.4. *Malaysia-Vietnam*

Background

In the Gulf of Thailand, Vietnam and Malaysia lie opposite to each other and have overlapping claims on the continental shelf. Vietnam wanted the area to be delimited by an equidistant line between the coasts of the mainland of the two countries, ignoring the Malaysian island of Redang and the Vietnamese island of Hon Khoai altogether.[22] The Malaysian claim on the continental shelf made in 1979, however, was based upon an equidistant line constructed between the Vietnamese mainland and Malaysian Islands.[23]

Method of Shaping the Joint Development Zone

The joint development, the so-called "Defined Area", is a long sword-shape. In shaping the Defined Area the overlapping claims of both countries have been accommodated. Points A and F are based on Malaysia's claim. They are equidistant points constructed by giving full weight to Malaysian island of Redang while disregarding Vietnamese offshore islands. Point E is the terminal point of the 1969 Indonesia-Malaysia continental shelf boundary. It is the equidistant point from Indonesia, Malaysia and Vietnam constructed by taking account of all the offshore features in the area. Points C and D are derived from Vietnamese claims. These are equidistant points between the mainland of the two countries.[24]

Forms and Provisions

Malaysia and Vietnam concluded a Memorandum of Understanding for the exploration and exploitation of petroleum in the Gulf of Thailand in 1992.

In the Memorandum, they noted that "it is in the best interests of both countries, pending delimitation of their continental shelves located off the north-east coast of West Malaysia and off the south-west coast of Vietnam, to enter into an interim arrangement for the purpose of exploring and exploiting petroleum in the seabed in the overlapping area".[25] The Memorandum of Understanding has a "without prejudice" clause in Article 4. It provides that, "Nothing in this Memorandum of Understanding shall be interpreted so as to in any way prejudice the position and claim of either party in the relation to and over the Defined Area". The Memorandum of Understanding shall come into force on a date to be specified by an exchange of diplomatic Notes and shall continue to be in force for a period to be specified by an exchange of Diplomatic Notes between the two parties.[26]

Operational Mechanism and the Effectiveness of the Arrangement
Under the arrangement, all costs incurred and benefits derived from exploration within the Defined Area are to be shared equally by the Parties. In the event that a petroleum field is found in the Defined Area or partly outside the Defined Area, the Parties should seek to agree upon terms for the exploration and exploitation of the field.[27]

2.5. *Australia-Indonesia*

Background
Australia and Indonesia have settled their boundary delimitation stage by stage. In 1971 the two countries delimited their continental shelf boundaries in part of the Arafura Sea (Point A1-Point A12). Subsequently in 1972, the two countries agreed on the boundary of the remaining part of the continental shelf in the Arafura Sea (Point A13-Point A16) and in the offshore areas of West Timor (Point A17-Point A25). However, the area off East Timor, which was then a colony of Portugal, was left undelimited and was called the "Timor Gap".[28] After East Timor was annexed by Indonesia in 1975, various rounds of negotiation have been held for delimitation in the area of the Timor Gap.[29] Australia argued in favour of, and Indonesia argued against, the principle of natural prolongation for the purpose of the delimitation in this area. In 1985 when the ICJ adopted the distance criterion as the basis of title to the continental shelf in the *Libya/Malta Continental Shelf* Case, Australia's position was weakened.[30]

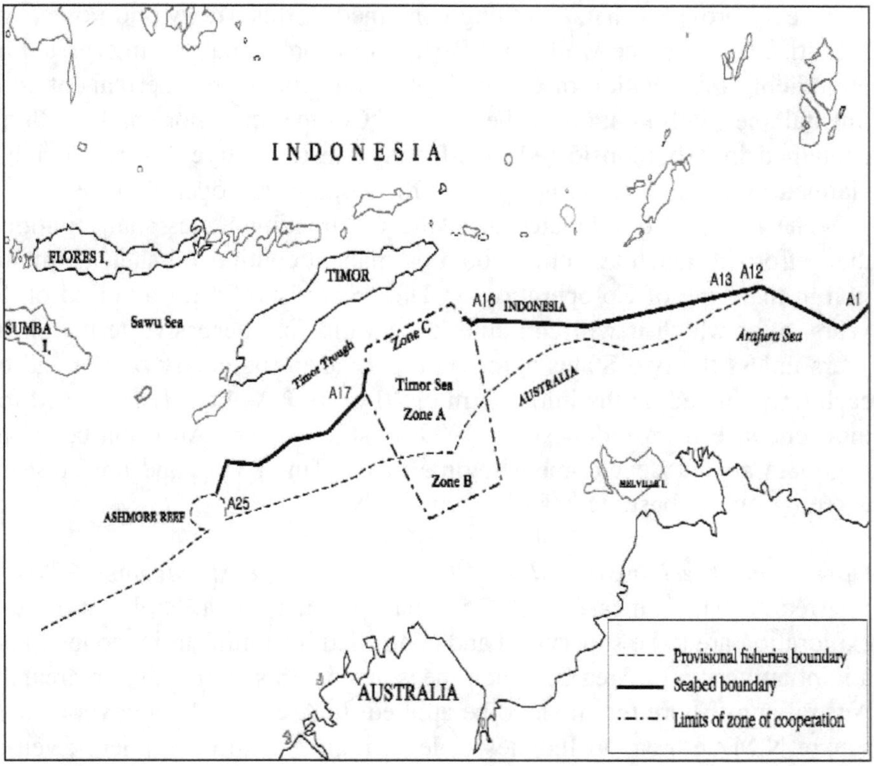

Map 3: Maritime Boundaries and Joint Development between Australia and Indonesia (Source: Australian Government's paper)

Method of Shaping the Joint Development Zone

In shaping the zone of co-operation, the Australian position for a natural prolongation line along the Timor Trough and the Indonesian position for a median line were both accommodated. The zone is divided into three areas; Area C, Area A and Area B from north to south. The northern limit of the zone is drawn along the bathymetric axis of the Timor Trough and its southern boundary is drawn along the 200 N.M. line from the Island of Timor. The median line is adopted to divide Area A and Area B in the zone. Area A and Area C is divided by a modified line of 1,500 metres isobath.[31]

Forms and Provisions

The provisional arrangement was signed in December 1989 in the form of a formal Treaty.[32] It is entitled "Treaty on the Zone of Co-operation in an Area between the Indonesian Province of East Timor and Northern Australia".

The treaty provides that, "Nothing contained in this Treaty and no acts or activities taking place while this Treaty is in force, shall be interpreted as prejudicing the position of either Contracting States on a permanent continental shelf delimitation in the Zone of Co-operation nor shall anything contained in it be considered as affecting the respective sovereign rights claimed by each Contracting State in the Zone of Co-operation".[33]

The treaty also stipulates that "the Contracting States shall continue their efforts to reach agreement on a permanent continental shelf delimitation in the Zone of Co-operation".[34] This treaty is valid for a period of 40 years, after which it will continue to be valid for successive terms of 20 years unless the two States agree on a permanent boundary by the end of each term, including the initial term of 40 years.[35] As East Timor voted for independence from Indonesia in 1999, East Timor and Australia began to negotiate a joint development regime in the Timor Gap and are close to agreeing on the basic tenets of a new treaty.[36]

Operational Mechanism and the Effectiveness of the Arrangement

In Area A, which measures 9,375 square N.M. in area, exploitation and exploration are to be supervised and controlled by a ministerial council and a joint authority. In Area B, which measures 5,178 square N.M. in area, an Australian offshore regime is to be applied. In Area C, which covers 1,576 square N.M. in area, an Indonesian legal regime is to be applied. Even if Area C and B are under the jurisdiction of Indonesia and Australia respectively, each country is to pay the other country certain tax revenues derived from exploitation taking place in those areas.[37] On 14 March 1997 the two countries signed the Treaty Establishing an Exclusive Economic Zone Boundary and Certain Seabed Boundary. However, the 1997 Treaty does not affect the legal regime of the Treaty in the Zone of Co-operation.[38]

2.6. *United Kingdom-Argentina in the South West Atlantic*

Background

Although the British government has exercised effective control over the Falkland Islands in the South West Atlantic since the 1830s, the Argentine government's claim to the Islands has been maintained to the present. It is not intended here to examine the sovereignty issue over the Falkland Islands/Islas Malvinas.[39]

The UK government claimed, as early as in 1950, the continental shelf around the Falkland Islands. According to the Order in Council of that year, the continental shelf of the Falkland Islands was generally up to the 100-fathom isobath from the Islands.[40] Note here that the Order of 1950 was concerned only with the continental shelf and thus the extent of fisheries jurisdiction around the Falkland Islands remained the three-mile territorial sea until 1986.[41] In 1986 the UK government established a Falkland Islands Interim Conservation and Management Zone, which extended from a central point in the Islands up to 150 N.M. seawards, but modified in the south-west to reduce the area of potential overlap with Argentina's then 200 N.M. territorial sea.[42] We will discuss this zone in detail at a later stage, when we discuss fisheries issues in this area. Also in 1986 the British government proclaimed that the Falkland Islands continental shelf extended to 200 N.M. from the coasts or to such other limit as is prescribed by the rules of international law.[43] On 22 November 1991 the Governor of the Falkland Islands issued a Proclamation thereby bringing the outer limits of the Falklands continental shelf further out to the same limits as the outer limits of the Falkland fisheries zone, which had been extended up to 200 N.M. from the baselines of the Islands in 1990.[44] Along with the Proclamation dated 22 November 1991, the Legislative Council of the Colony of the Falkland Islands passed a Continental Shelf Ordinance to provide an interim framework for preliminary exploration of the continental shelf.[45]

The Falkland Islands Governor's Proclamation of 22 November 1991 brought forth an immediate protest from the Argentine government.[46] The two governments began to explore the possibilities of accommodating the positions of both sides over the continental shelf. Note here that a co-operation scheme for fisheries management in the region took place in 1990.[47] We will discuss this at a later stage. After many years of negotiation, the two governments issued the "Joint Declaration on Co-operation over Offshore Activities in the South West Atlantic" on 27 September 1995.[48] They agreed in the Declaration to establish a joint development zone, a so-called "Special Area" to the south west of the Falkland Islands.

Method of Shaping the Joint Development Zone

Even though the Argentine government claims, in principle, sovereignty over the whole of the Falkland Islands, the Special Area, whose co-ordinates are shown in the Annex to the Joint Declaration, is limited to an area situated to the south west of the Islands. The joint development zone, which covers 20,000 km^2 in area, is constructed by lines connecting 12 points.[49] However, the possibility of establishing additional joint development zones is not ruled out. Importantly, in this context, paragraph 4 (b) refers to "additional trenches either within the sedimentary structure referred to in the Annex or in a further area to be agreed by the Governments on the recommendation of the Commission".

Forms and Provisions

The joint development scheme between the United Kingdom and Argentine Republic was agreed in the form of a Joint Declaration signed by the British and Argentine Foreign Ministers. Paragraph 1 thereof provides a "without-prejudice" clause, based upon the so-called "Madrid Formula".[50] It stipulates, *inter alia*, that, "nothing in the content of the Present Joint Declaration or of any similar subsequent joint statements shall be interpreted as a change in the position of the United Kingdom with regard to the sovereignty or territorial and maritime jurisdiction over the Falkland Islands, South Georgia and the South Sandwich Islands and the surrounding maritime areas".

Despite this Joint Declaration, uncertainty regarding exploration and exploitation activities on the continental shelf around Falkland Islands arose when the Argentine government issued a statement, on the same day the Joint Declaration was made, to the effect that it would introduce charges on companies operating in the continental shelf around the "Islas Malvinas".[51] However, responding to the Argentine statement, the British government made a Declaration, stating that: "HMG do not accept any Argentine claim to impose such charges on companies by reason of their activities on the continental shelf around the Falkland Islands under Falklands license".[52]

Operational Mechanism and the Effectiveness of the Arrangement

Exploration for and exploitation of hydrocarbon deposits by the oil and gas industry is to be carried out in accordance with sound commercial principles and good oil field practice.[53] Co-operation is to be furthered by means of the establishment of a Joint Commission. The Joint Commission has a mandate to promote exploration and exploitation of hydrocarbons in the maritime areas of the South-West Atlantic by way of submitting recommendations to the Parties on any related matter.[54] The first meeting of the Joint Commission was held in March 1996.[55] The Joint Committee meets at least twice a year and the Sub-Committee meets at more frequent intervals in order to co-ordinate activities in the Special Area.[56] The Sub-Committee's discussion aims to complete its preparations for a licensing round in the Special Co-operation Area as soon as possible.[57]

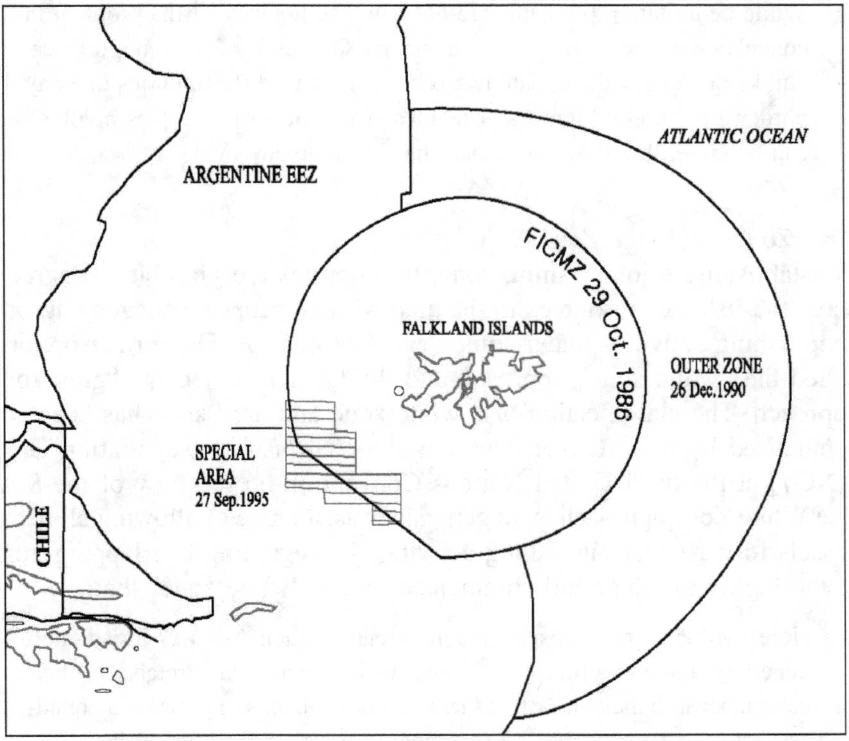

Map 4: Joint Development Zone between UK and Argentina

(Source: M. Miyoshi, *The Joint Development of Offshore Oil and Gas in Relation to Maritime Boundary Delimitation*, IBRU Maritime Briefing 1999)

3. POLICY OPTION 2: JOINT FISHING ZONES

The Need for Joint Fishing Zones

If an EEZ or an exclusive fisheries zone boundary were delimited between neighbouring countries there would be no need to establish a joint fishing zone. If there is such a boundary, a State can implement its own fisheries laws and regulations in its EEZ or exclusive fisheries zone by giving or denying permission to nationals and vessels of other countries to fish in the area within its side of the boundary and by exercising enforcement jurisdiction on illegal fishing activities. However, where no boundary has been agreed in an area of overlapping claims, then something needs to be done about fisheries. Usually, fisheries pose more urgent issues than mineral resources pending ultimate delimitation. In this regard, Oxman observed that:

> While delimitation problems are not new, the urgency of the problem in connection with extensions of fisheries jurisdiction is of some importance. Unlike oil development, fishing has been conducted for centuries in many border areas under high sea principles. The number of peoples involved can be large; the basic investment has been already made . . .[58]

Grey Zones or White Zones?

In establishing a joint fishing zone, two options are possible: to agree to leave the fisheries resources in the area without proper management, or to keep fishing activities under some degree of control. The former option is called the "white zone" approach and the latter is called the "grey zone" approach. The classification of a white zone and grey zone has been conceptualised by Mr. Manner who was the chairman of Negotiation Group 7(NG7) at the third United Nations Convention on the Law of the Sea.[59] The White Zone approach is, in general terms, a mode of allowing all fishing vessels to participate in fishing activities by regarding overlapping zones as the high seas. Churchill commented on the "white zone" that:

> Here two or more States, although extending their fisheries jurisdiction generally, agree to refrain from doing so along particular stretches of their coasts, because they know that there will be problems in drawing a boundary between what would be overlapping areas of fisheries jurisdiction. This restraint means that there will be areas of high seas left ("white zone") where all States can fish under the freedom of the high seas, although under some regulation by an international fisheries commission.[60]

The Grey Zone approach, on the other hand, is a mode which purports to manage the living resources in the disputed area.[61] As Churchill has explained: "A Grey zone agreement is an agreement that regulates fisheries in all or part of a disputed area".[62] On the same zones, Lagoni has observed that: "These zones are designed to regulate fishing within the area of overlapping claims on a provisional basis and to exclude the fishing fleets of third States from the area".[63]

Precursor of Joint Fishing Zones: Common Oyster Fisheries Zone between the United Kingdom and France

We can find a precursor of joint fishing zones in the 1830s. At that time disputes between British and French fishermen became frequent and acrimonious in the English Channel and there were complaints from the English fishermen about the encroachments of the French fishermen.[64] However, French fishermen also complained about the English fishermen, particularly about their dredging for oysters off the French coast.[65] To solve the fisheries dispute the two governments concluded a Fisheries Convention in 1839.[66] The Convention defined the limits within which the fisheries were to be exclusively reserved to the fishermen of each Party. The limits for the exclusive fisheries were generally three miles from the low-water mark along the whole extent of the coasts of each country.[67] Article 3 of the Convention provided that:

> The oyster fishery outside the limits within which that fishery is exclusively reserved to French and British subjects respectively, as stipulated in the preceding Articles, shall be common to the subjects of both countries.

This Convention was the first to establish through an international agreement the three-mile fishery zone off the British coasts.[68] There were fishing regulations in the common fisheries zone prepared by a Mixed Commission and confirmed by the respective governments. The regulations were concerned with the numbering and lettering of fishing vessels and with restrictions on the fishing apparatus to be used.[69] However, it is to be noted that this Convention was binding only on the British and French subjects and thus the fishing regulations in the common fisheries zone and the three-mile fishery limits were not applicable to fishing vessels from third countries.[70]

Can Joint Fishing Zones Have Impact on the Sovereignty Issue over Disputed Islands?

A question arises as to whether a joint fishing zone can have any impact on issues of sovereignty disputes over islands which are geographically within the zone. In the *Minquires and Ecrehos* Case the French government tried to reinforce its position by using the Fisheries Convention of 1839. The French government asserted, and the UK government denied, that the Minquires and Ecrehos groups were included in the common fisheries zone. The French government went on to argue that the acts performed by each Party on the islets and rocks after the Fisheries Convention were not capable of being set up as manifestations of territorial sovereignty.[71] In this matter, the Court held that:

> The Court does not consider it necessary, for the purpose of deciding the present case, to determine whether the waters of the Ecrehos and Minquires groups are inside or outside the common fishery zone established by Article 3. Even if it be held that these groups lie within this common fishery zone, the Court cannot admit that such an agreed common fishery zone in these waters would involve a regime of common use of the land territory of the islets and rocks, since the Articles relied on refer to fishery

> only and not to any kind of use of land. Nor can the Court admit that such an agreed common fishery zone should necessarily have the effects of precluding the Parties from relying on subsequent acts involving a manifestation of sovereignty in respect of the islets.[72]

3.1. *Grey Zones*

Why do some States choose white zones rather than grey zones? Generally speaking, it appears to be the norm that those coastal States which wish to effectively manage fisheries resources in a disputed area, choose a grey zone rather than a white zone, since the former zone purports to provide an effective management for fisheries in the disputed areas. Then, a question arises as to why some States do not choose a comprehensive joint exploitation, but a grey zone only for the management of fisheries resources. The reason might be that in some areas fisheries pose a more urgent issue than oil and gas; that there is no commercially exploitable quantity of oil and gas in a disputed area; or coastal States wish to await a final delimitation of the continental shelf and exploit oil and gas after delimitation; or a coastal State is worried about the possible image of condominium when there is a joint development regime of oil and gas along with a joint fishing zone. We will soon see that these various reasons have been taken into consideration in practice.

3.1.1. *Soviet Union and Norway in the Barents Sea*

Background

The Barents Sea is an important world fishing ground. In recent years the total catch in the Barents Sea has been about 1-1.5 million tons a year, which is about 1.5-2.0 per cent of the world total.[73] The Barents Sea is also thought to have a potential for oil and gas resources. However, no commercial production activities have been conducted even though there have been some exploration activities.[74] In the Barents Sea, Russia (formerly the Soviet Union) and Norway are the only coastal States. Norway issued a Continental Shelf Proclamation in 1963, and the Soviet Union did so in 1968.[75] The two countries thereafter embarked upon the negotiations for the delimitation of the continental shelf. In the negotiations Norway argued for an equidistance line while the USSR put forward a sector line.[76] The overlapping area is some 175,000 km^2 in area. The negotiation expanded to the matter of delimitation of exclusive economic zones in 1977 as Norway proclaimed an economic zone and the USSR proclaimed a 200 exclusive

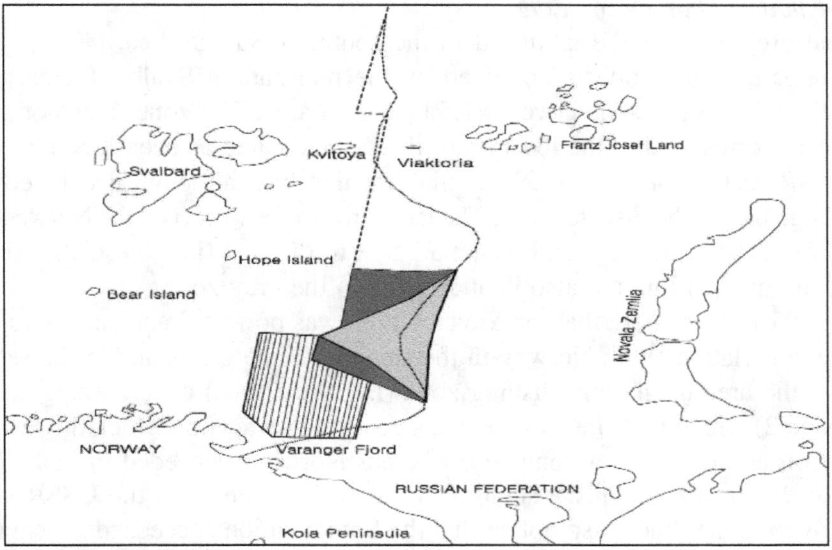

Equidistance line
Sector line
Area covered by the agreement Norway/Soviet Union 1978
Area beyond 200 miles from the baselines of both states
Area within 200 miles from the Norwegian baseline but beyond 200 miles from the baseline of the Russian Federation

Map 5: Grey Zone between Norway and Soviet Union

(Source: A.G. Oude Elferink, *The Law of Maritime Delimitation: A Case Study of the Russian Federation*)

fisheries zone in 1976, while the negotiations for the delimitation of the continental shelf were under way.[77]

While the negotiations were in stalemate, the urgent need to manage and conserve the fisheries resources in the Barents sea was recognised by the two countries. By contrast, there was no immediate urgency for the regulation of continental shelf resources as neither the Soviet Union nor Norway desired to begin early exploration for oil and gas in the disputed area.[78]

In particular, both countries shared a common concern about the impact of fishing by third States on fisheries resources in the disputed area, as well as possible jurisdictional conflicts between their enforcement authorities over fishing vessels from third countries.[79] The objective of prohibiting access to the disputed area by fishing vessels of third States was an important incentive in agreeing on a provisional arrangement.[80] These common and urgent concerns led the two countries to reach the so-called "Grey Zone Agreement" on 11 January 1978.

Method of Shaping the Zone

The grey zone was established in the southern Barents Sea, beneath the Svalbard zone which is governed by the multilateral Svalbard Treaty of 1920.[81] The grey zone covers 41,500 km^2 in area. The zone does not precisely correspond to the location of the disputed area as occasioned by the USSR sector line and the Norwegian median line, although it does cover a large part of the disputed area. The large area to the west (i.e., the Norwegian side) of the sector line and the small area to the east (i.e., the USSR side) of the median line are also included within the grey zone.

We can see here that the Soviet Union was powerful and successful in the negotiations with Norway in the shaping of the grey zone in the sense that the area of the joint fishing zone (i.e., grey zone) covers a large area (Area 2) which is on the Norwegian side of the sector line which the Soviet Union argued for as a boundary, whereas Norway succeeded in including into the grey zone only a small area (Area 3) which is on the USSR side of the median line. Also notice that the Soviet Union succeeded in leaving out some of the disputed areas (Area A, B, C) from the grey zone and making them subject to its exclusive jurisdiction (Area B and C), or to the high seas regime (Area A). Note that Area A is beyond the 200 mile line from the USSR coasts but within the 200 mile line from the Norwegian coasts.[82] Norway has a small area (Area D) to the north of the grey zone where it can exercise exclusive fisheries jurisdiction. But this area is small compared to areas A, B and C.[83]

Forms and Provisions

Several instruments together constitute the arrangement: the Agreement, an exchange of letters with an annex on the relevant co-ordinates, and a protocol containing substantive provisions and corresponding statements made by the two parties. The last of these purports to confirm the parties' intention to reach a final delimitation solution and the non-prejudicial character of the arrangement.[84] The "without prejudice" clause of the Agreement read:

> This Agreement does not prejudice the positions or views of either party with regard to the boundaries of the parties' areas of fisheries jurisdiction, or to the delimitation of the continental shelf and areas referred to in this Agreement.[85]

The arrangement was initially concluded for only one year, but it has been extended for annual periods in every subsequent year.[86]

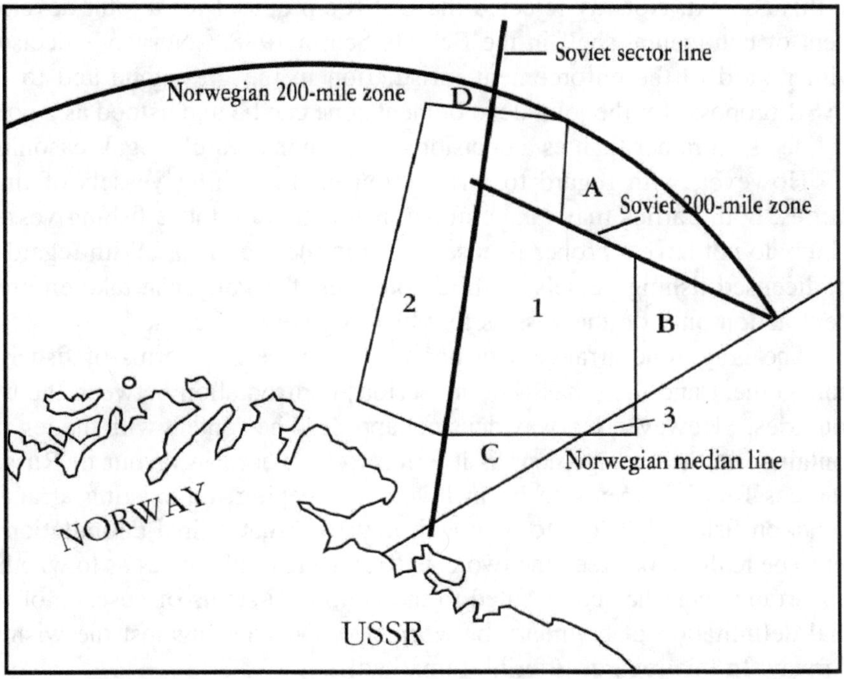

Map 6: Shaping of Grey Zone between Norway and Soviet Union

(Source: R. Churchill and G. Ulfstein, *Marine Management in the Disputed Areas: The Case of the Barents Sea*)

Operational Mechanism of the Arrangement

The Total Allowable Catch (TAC) is to be set by the joint Norwegian-Soviet fisheries commission and shall be utilised in essentially equal shares by the two States, while a part of the TAC can be allocated to third States after consultation between the parties. Both parties recognised the necessity of common regulations for fishing vessels from the two States and from third States. Either party can authorise fishing vessels of third States to have access to the grey zone, but in such a case the Party in question shall notify the other.[87] But one Party may not adopt unilateral regulations for vessels from the other Party or for vessels from third States licensed by the other party.

The enforcement mechanism is based on flag-State jurisdiction with each other: one Party cannot enforce its fishing regulation on fishing vessels of the other party. Flag-State jurisdiction in the grey zone was a crucial element for Norway, because Norway wanted that the grey zone would not be viewed as something like a condominium between Norway and the Soviet Union.

In this context, Norway rejected the USSR's proposal for a joint development of continental shelf in the Barents Sea in 1988.[88] Norway's decision with regard to the enforcement jurisdiction in the grey zone and to the USSR proposal for the joint development zone can be understood as a political decision rather than as a decision based upon a purely legal reasoning.

However, with regard to jurisdiction to the fishing vessels of third Parties, both Parties may take enforcement action on those fishing vessels which do not have a proper license to fish in the grey zone. With regard to the licensed fishing vessels of third countries, the Party can take enforcement action only on the vessels to which it gave the license.[89]

The grey zone arrangement has been a success in terms of fisheries management and there has been no serious confrontation between the two countries.[90] However, Norway does not appear to be content with the regime contained in the arrangement as it is felt to be biased in favour of Russia, whereas Russia appears to be in favour of keeping the existing arrangements on fisheries.[91] It is to be noted, however, that a final delimitation is yet to be realised between the two countries. Here doubt arises as to whether this arrangement has contributed to the continued status of absence of the final delimitation of boundary between two countries, against the wish of Norway. In this regard, Elferink remarked:

> . . . the Soviet Union may have preferred the absence of an agreement over a boundary line . . . Furthermore, the provisional arrangement on fisheries concluded in 1978 took away any pressure to reach a compromise, which otherwise might have arisen, had this area remained a white zone area to which the fishing vessels of third States would have had free access.[92]

3.1.2. *Sweden and Denmark*

Background

Sweden legislated on the continental shelf and on the territorial sea in 1966. Later in 1977 Sweden proclaimed its fisheries zone beyond 12 N.M.[93] Denmark took almost the same measures as were taken by Sweden.[94] Although they agreed to the principle of equidistance, the question as to how much effect their islands should have on the delimitation was difficult to solve.[95]

The Danish islands in question were Bornholm and Aertholmarna in the Baltic Sea and Laeso, Anholt, and Hesselo in the Kattegat.[96] These are for small islands except for Bornholm.[97]

Similarly, the small Swedish islands of Utklippan and Falsterborev also became involved in the issue of confrontation with regard to their effect.[98]

Each State argued for the full effect for its own islands but for reduced effect for the islands of the other State. In the case of Danish Bornholm the point of dispute did not lie in the size of the island but in its location, because it is a big island with a large population but located far away from the mainland of Denmark.

In the deadlock of negotiations on delimitation, Sweden and Denmark agreed in 1977 to jointly exercise fisheries jurisdiction in the area beyond 12 N.M. in the Kattegat. The Exchange of Notes of 29 December 1977 provides that:

> Pending the conclusion of such an agreement, . . . the area of the Kattegat situated outside the present 12-nautical-mile-fishery limits should be placed under joint Danish-Swedish fisheries jurisdiction. Detailed regulation concerning the exercise of such fisheries jurisdiction in respect of fishing by third countries in the area shall be established by agreement . . .[99]

It is regrettable that the detailed regulation of the arrangement between Sweden and Denmark has not been made known to the public.[100] It is also not clear what regime was in place in the area around Bornholm in the Baltic. Possibly however, it can be presumed from the Swedish decrees establishing the limits of the Swedish fishing zone that Sweden and Denmark agreed to jointly exercise fisheries in all of the disputed area between them, beyond 12 N.M. from the coasts.[101]

Termination of the arrangement and delimitation

This agreement was abrogated in 1984 after both parties assented to a boundary of the EFZ and the continental shelf.[102] In delimiting the boundary in the Kattegat, the three Danish islands (Laeso, Anholt, and Hesselo) were all given full effect.[103] However, Danish Falsterborev received a half effect.[104] The island of Bornholm was also given full effect, but Aertholmarna was given only limited effect mainly due to its small size.[105] The Swedish island of Utklippan, which is uninhabited and has an approximate size of 0.12 km^2 was similarly given only limited effect.[106]

3.1.3. *Denmark and Poland*

Poland extended its territorial sea to 12 N.M. in 1977, and in the same year extended its fisheries zone beyond 12 N.M.[107] Along with the extension of its fisheries zone, Poland also legislated on its continental shelf.[108] In negotiations on the maritime boundaries of the continental shelf and fishing zone between Denmark and Poland, the effect of the Danish island of Bornholm which is situated far away from the Danish mainland turned out to be a point of dispute. The two countries agreed in 1978 to exercise joint fisheries jurisdiction in the overlapping zone, with the exclusion of fishing vessels

Map 7: Delimitation and Islands in the Baltic Sea

(Source: E. Franckx, "Maritime Boundaries in the Baltic: Past, Present and Future", IBRU Maritime Briefing)

from third countries.[109] Although the detailed contents of the arrangement between Poland and Denmark are not available, there is a possibility that the grey zone between them is mainly to exclude fishing by fishing vessels from third countries while applying flag-State jurisdiction between the two Parties. The delimitation between Denmark and Poland is still awaited.

3.1.4. *Venezuela and Trinidad and Tobago*

After the Santo Domingo Declaration of 1972,[110] the need arose for maritime delimitation between Venezuela and Trinidad and Tobago. The two countries started negotiations for delimitation in 1973. While negotiations hardly progressed, the need to regulate the fishing activities arose because Trinidad fishermen were frequently found fishing in the waters off the Venezuelan coasts.[111]

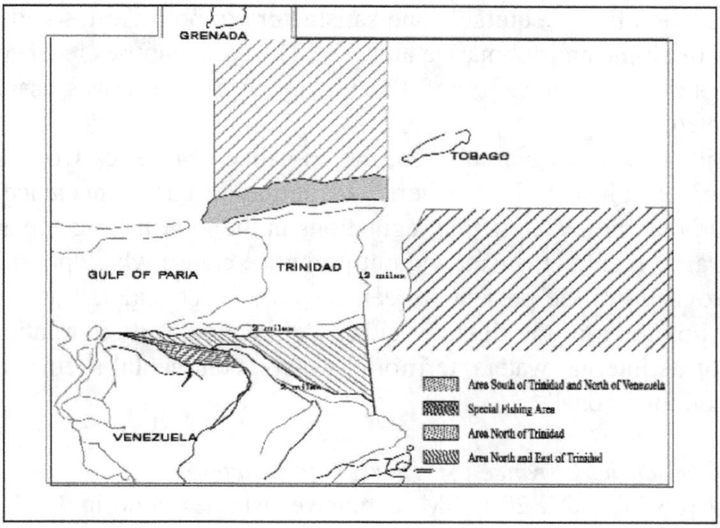

Map 8: Joint Fishing Zones between Trinidad and Tobago and Venezuela (Source: 2 I.J.E.C.E.L (1987))

In 1977, the two countries concluded a fisheries agreement whereby they established two grey zones – one was to the north and the other was to the south of Trinidad and Tobago.[112] In 1985 they revised the agreement thereby creating an additional joint fishing zone to the east of Trinidad.[113] Fishermen of both Parties are allowed, in principle, to fish within these joint fishing zones except within two miles from the coast of either country.[114] However, the legal quality of the three zones is not the same. In the area to the south of Trinidad, fishermen of both Parties have freedom of fishing access.[115] In the area to the north of Trinidad (the so-called "Area North of Trinidad") the government of Trinidad has "sovereignty" and "jurisdiction" and has the power to grant fishing permits to Venezuelan fishing vessels.[116] Also in the area to the east of Trinidad (the so-called "Area North and East of Trinidad"), Trinidad has "sovereignty" and "jurisdiction".[117] They also agreed to establish the so-called "Special Fishing Area" in the internal waters of the Venezuela. It is agreed that Venezuela should allow fishing access to the vessels of Trinidad in the Special Fishing Area although Trinidad has sovereignty over the area.[118] There are detailed provisions for regulation on fishing activities in these areas.

On the same day the Fishing Agreement was signed the two countries signed a Joint Declaration. Paragraph 1 of the Declaration stated that: "both Governments will make all necessary efforts to achieve as soon as possible a delimitation that is equitable and satisfactory to both Parties with respect to their maritime and submarine areas".[119] The two countries reached delimitation agreement in 1990 and the Fishing Agreement was assumed be terminated.[120]

A joint fisheries commission, the so-called "Fisheries Commission", was then established. The Fisheries Commission has competence to recommend the Parties on fishing regulations in the joint fishing zones.[121] The fishing agreement contains a without-prejudice clause which provides that: "No provision in the present Agreement is to be considered as a diminution or limitation of the rights of either Contracting Party in relation to the limits of its internal waters, territorial waters, continental shelf and exclusive economic zone".[122]

3.1.5. *Canada and France (St. Pierre and Miquelon)*

Canada proclaimed a 200 N.M. exclusive fisheries zone in 1977, and so did France in the same year. France argued that the boundary should be an equidistance line by giving full effect to the French islands of St. Pierre and Miquelon, whereas Canada maintained that the French maritime zone around St. Pierre and Miquelon should be limited to the 12 N.M. territorial sea.[123]

In March 1989 the two countries signed two agreements: one referred the boundary dispute to an *ad hoc* Court of *Arbitration*, and the other (a Proces-Verbal) provided fishing quotas for France for the period of 1989-1991 in the Gulf of St. Lawrence and also provided rules to govern activities in the disputed area.[124]

Regarding the disputed area, each country agreed: (i) to undertake not to change fundamentally the intensity, nature, or method of fishing for the specified fish stocks, (ii) to refrain from enforcement actions on vessels flying the flag of the other Party and from conducting any drilling activity or exploiting the mineral resources of the seabed and subsoil, and (iii) when inspecting fishing vessels flying its flag in the disputed zone, to take on board the inspection vessels an agent, designated by the other Party at the request of the other Party.[125]

Canada noted in its Note Verbal sent to France on the same day as the Proces-Verbal, that France intended to unilaterally set an annual fishing quota for French vessels in the disputed area. Canada maintained that such a French measure had no legal basis; the annual fishing quota for French vessels in the area including the disputed zone would be set by Canada; and "any fishing by French vessels in excess of the quota set by the Canadian authorities constitutes over-fishing which can be tolerated only in consideration of the present special circumstances".[126] In this regard, France argued that Canada had no jurisdiction to set the annual quota in the zone including the disputed area and that "France is legally at liberty to set its own quotas in the 'disputed zone' which constitutes its own exclusive economic zone over which it exercises sovereign rights as determined by international law".[127] Even though a heated debate arose on quota-setting powers between Canada and France, the actual conflicts at sea should not have arisen because they agreed to adopt flag-State jurisdiction in the disputed zone.

The disputed zone between Canada and France disappeared after an arbitration court rendered its award on 10 June 1992. The award gave France a 24 N.M. arc to the west and 12 N.M. arc to the east of the St. Pierre and Miquelon, and established for France a 10.5 N.M. wide corridor running from St. Pierre and Miquelon due south to the 200 N.M. limit.[128]

3.1.6. *United Kingdom and Denmark (Faroe Islands)*

Rockall, around which the UK government claimed extended maritime zones,[129] has been a point of dispute between the UK and other neighbouring countries – Ireland, Denmark and Iceland.[130] The United Kingdom proclaimed a continental shelf in 1964, and a 200 N.M. fisheries zone in 1976. The British government designated some 52,000 square miles of the continental shelf around Rockall in 1974 and used Rockall as a base point under its Fisheries Act 1976.[131]

Denmark proclaimed a 200 N.M. fisheries zone in 1976 in respect of the Faroe Islands, which overlapped with the UK fisheries zone around Rockall. In 1985 Denmark also designated the Faroe's continental shelf which again overlapped with the British designation of the continental shelf around Rockall.[132] It was notable that the UK government renounced the 200 N.M. fisheries zone around Rockall in 1997 when she joined the LOS Convention, declaring that Rockall is a rock under Article 121(3) of the LOS Convention.[133] This move has taken away one of the main obstacles in delimitation.

The United Kingdom and Denmark (Faroe Islands) concluded an Agreement on maritime delimitation on 15 May 1999, disregarding Rockall but using St. Kilda as a base-point, which lies 190 miles to the east (i.e., landwards) of Rockall. The Agreement establishes a continental shelf boundary between the United Kingdom and the Faroe Islands and also delimits some parts of the boundaries of the fisheries zone between them.[134] The Agreement also established a grey zone between them, the so-called "Special Area" where the boundaries of the exclusive economic zone were left out. The Special Area covers 8,000 km^2 in area.

The reason why the two countries adopted the grey zone was to maintain traditional fishing patterns in the area.[135] Interestingly, in the Agreement

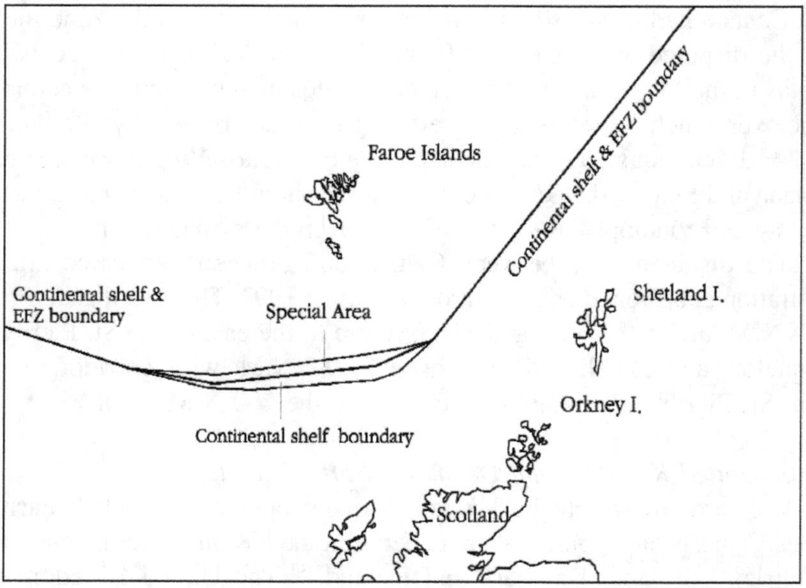

Map 9: Grey Zone between UK and Denmark (Faroe Islands)
(Source: UK government, *Command Paper* No. Cm 4373)

no detailed scheme for joint fisheries management is found. In the Special Area, each Party is to "apply the relevant rules and regulations applicable within its zone of fisheries jurisdiction concerning the management, including the issuing of fishing licenses, and conduct of fisheries".[136]

With regard to enforcement in the Special Area, each of the Parties is obliged to refrain from inspection and control of fishing vessels which operate in the Special Zone solely under licenses issued by the other Party, and also to refrain from any action that would disregard or infringe upon the exercise of fisheries jurisdiction by the other Party.[137] Note that there is no provision for establishing a joint commission in the Agreement. However, each Party may call for consultation, *inter alia*, on issues of the Special Area.[138]

As there is a continental shelf boundary passing through the Special Area, clarification was needed on the relationship between the continental shelf boundary and the Special Zone, which is a joint fishing zone. In this regard, it was provided that each Party, in exercising its continental shelf jurisdiction and rights in the Special Area, shall take due account of the interests of the other Party in maintaining its fishing possibilities.[139] With regard to jurisdiction for the protection of the marine environment in the

Special Area, the two countries agreed to co-operate with each other and refrain from exercising such jurisdiction unless agreed otherwise.[140]

3.1.7. *Joint Fishing Zone between South Korea and Japan of 1965*
South Korea and Japan concluded a fisheries agreement in 1965 when they normalised their diplomatic relations.[141] Under the agreement, each country recognised the rights of the other Party to establish a 12 N.M. exclusive fisheries zone "from the baseline of the coasts".[142] Beyond and adjacent to the Korean exclusive fisheries zone there was established a "joint fishing zone". The Agreement was terminated in January 1999 when a new fisheries agreement came into effect. I will examine the Fisheries Agreement of 1965 and new Fisheries Agreement of 1999 in detail in Chapter 5.

3.1.8. *New Fisheries Agreements in North East Asia*
In North East Asia, three Fisheries agreements came into being: the Fisheries Agreement between South Korea and Japan which entered into force in January 1999; the Fisheries Agreement between the People's Republic of China (China) and Japan which entered into force on 1 June 2000; and the Fisheries Agreement between South Korea and China which entered into force on 30 June 2001. The three agreements are all intended to deal with fisheries issues pending delimitation of boundaries of the exclusive economic zone by setting up joint fishing zones in the overlapping areas. The three agreements will be discussed in Chapter 5.

3.2. *Joint Fishing Zones Focusing on Specific Fish Stocks*

As we have seen in the previous analysis, the task of shaping joint fishing zones or joint development zones is a complicated one in an area where only a specific fish stock matters, and effective management can be achieved by setting up a management regime for specific fish stocks in a disputed area without establishing a joint fishing zone. However, even if this option is chosen, there is still a need to define the area of application of the management regime broadly, like for example, "Northern Pacific". Two instances of this approach can be found. Churchill coined the term "light grey zone" to denote this kind of fisheries arrangement.[143]

3.2.1. *Agreement between the US and Canada on the Enforcement Measures Regarding the Conservation of Halibut in the Northern Pacific and the Bering Sea*
Halibut fishing in the North Pacific has been regulated by a series of Conventions between the United States and Canada since 1923 with an

international commission established and reshaped by these Conventions.[144] The last Convention, namely the Convention between the United States and Canada for the Preservation of Halibut Fishery of the Northern Pacific Ocean and the Bering Sea of 1953, established the International Pacific Halibut Commission.[145] The Convention applied to "the territorial waters and the high seas off the western coasts of the United States and Canada, including the southern as well as the western coasts of Alaska".[146] Fishing vessels of the Parties engaged in fishing on the high seas of the Convention waters in violation of the Convention were subject to seizure by authorities of either Party irrespective of the flags they fly, although prosecution against seized vessels was to be conducted by authorities of the Party under which flag fishing vessels fly.[147]

After the United States and Canada proclaimed their respective 200 N.M. fisheries zone in 1977 they concluded a Protocol in 1979 amending the 1953 Convention on the Conservation of Halibut Fishery of the Northern Pacific and the Bering Sea.[148] The Protocol changed the definition of the Convention waters into "the waters off the western coasts of the United States and Canada including the southern as well as the western coasts of Alaska, ***within the respective maritime areas in which either Party exercises exclusive fisheries jurisdiction*** (emphasis added)".

The Protocol dealt with, *inter alia*, the issues of conservation of the halibut fisheries and enforcement measures in the waters where both States claimed exclusive fisheries zones.[149] In these areas of overlapping claims, the allocation for the halibut catch was divided into 40 per cent for the United States and 60 per cent for Canada and no quota was allocated for third States.[150] The Protocol provides in Article 2 that: "Each Party shall have the right to enforce this Convention in that portion of the Convention waters in which it exercises exclusive fisheries jurisdiction against nationals of fishing vessels of either Party or of third Parties". However, the two Parties noted that there were no boundaries of fisheries zones between them and thus Article 2 could not be applied until delimitation of boundaries. Therefore, they agreed as a provisional measure to apply flag-State jurisdiction between the Parties in overlapping areas, while agreeing that either Party may enforce the provisions of the Convention on fishing vessels from third Parties.[151] Note that the provisional arrangement does not apply to fishing other than halibut in the disputed area and thus there have been a number of disputes over such fishing.[152]

3.2.2. *Denmark (Greenland), Iceland and Norway (Jan Mayen) on Capelin Fishing*

In 1989 there was no maritime boundary in place either between Iceland and Greenland or between Greenland and Jan Mayen, while there was a boundary between Iceland and Jan Mayen. Under these circumstances Denmark (Greenland), Iceland and Norway (Jan Mayen) concluded a fisheries agreement on capelin fishing in the waters between Greenland, Iceland and the Norwegian island of Jan Mayen. The area of this agreement only has a conceptual scope, which is not specifically defined.[153] According to the Agreement, all parties shall try to reach an agreement on the Total Allowable Catch (TAC) of capelin each year. If an agreement is not achieved Iceland has the responsibility of deciding the TAC. Once the TAC is decided upon, Iceland receives 78 per cent of it and Denmark and Norway both get 11 per cent.[154] Vessels flying the flag of one Party are allowed to fish capelin in the waters even within 200 miles from the other Parties' coasts under the limits of their flag-States' fishing quota.[155]

3.3. *White Zones*

Usually, a white zone approach can be chosen when there is no urgent need to regulate the fishing activities or the States concerned expect that delimitation will soon take place. A white zone approach can also be chosen for the purpose of keeping traditional fishing patterns. Interestingly, however, a white zone approach can be taken even when there is an urgent need for effective management of fisheries resources. Several reasons can be given for the latter case. A coastal State might choose a white zone approach in expectation that an early agreement on delimitation would take place, because when there is effective management of fisheries through a grey zone approach then the urgent need for a final boundary would not be felt. By way of contrast, where there is no effective management of fisheries resources then there would arise an urgent need for a final boundary. There are choices: early delimitation at the cost of effective management, or effective management at the cost of an early final delimitation. For example, the Soviet Union concluded two fisheries agreements in the late 1970s: one is a grey zone agreement with Norway and the other is a white zone approach with Sweden. In 1987 the Soviet Union and Sweden reached an agreement on final delimitation, whereas an agreement on delimitation between the Soviet and Norway is still to be concluded. Perhaps delimitation is unlikely to take place as far as the effective resources management scheme is in place because there is no urgent need to delimit as far as fisheries resources are concerned.

3.3.1. *Sweden and the Soviet Union in the Baltic Sea*

Background

As mentioned earlier, the Soviet Union proclaimed a 200 miles exclusive fisheries zone in 1976, and Sweden extended its fisheries zone beyond 12 miles in 1977.[156] Even though Sweden and the Soviet Union held five rounds of talks on maritime delimitation on the basis of the equidistance principle from April to December 1977, they were not able to reach an agreement.[157] The main reason for this failure was the different positions adopted on the effect to be given to the Swedish islands of Gotland, and other small islands lying north-east of Gotland which were treated altogether with Gotland in the delimitation negotiations.[158] Sweden argued for full effect to be afforded to the islands and the Soviet Union denied that they should have any effect, except for a 12 N.M. territorial sea.[159] The resulting overlapping area was about 13,500 km^2 which was rich in fisheries resources. Sweden proposed a third-party settlement of the issue but the Soviet Union declined.[160] The Soviet Union suggested that a grey zone should be applied in the disputed area, but Sweden turned down the Soviet proposal because it feared that such an approach would weaken its position on the delimitation. Furthermore, for Sweden, the Gotland region was important for defence reasons.[161]

White Zone Arrangement

In 1977 both countries agreed to leave the disputed area as a white zone where neither country could take any measures which might prejudice the result of future negotiations on maritime delimitation.[162] The provisional arrangement was established by a protocol to an agreement on their mutual relations regarding fisheries of 22 December 1977.[163] The size of the overlapping area amounts to approximately 13,000 km^2.

Although neither of the States exerted jurisdiction over fisheries in this area fisheries resources were partly managed multilaterally by the "International Baltic Sea Fishery Commission (IBSFC)".[164] The IBSFC was established by the Convention on Fishing and Conservation of the Living Resources of the Baltic Sea and the Belt, the so-called "Gdansk Convention" which was concluded in 1973 by the Baltic States – Denmark, Finland, West Germany, East Germany, Sweden, Poland, the USSR.[165] The main objective of the Gdansk Convention is to promote co-operation among the Baltic States with a view to preserving and managing the living resources in the region.[166] The main competence of the IBSFC is to recommend conservation measures to the Parties. The IBSFC has a duty to collect and disseminate information on fisheries in the region.[167] However, it seems that management of the living resources in the white zone, which was treated as high seas between Sweden and the Soviet Union, was not effective under the loose multilateral Convention regime and thus over-fishing has occurred.[168] The over-fishing was the most important factor in leading the two countries to an agreement on the delimitation in 1988.[169]

Delimitation

The white zone arrangement between Sweden and the Soviet Union was terminated in 1988 when the two countries agreed on a single maritime boundary of the continental shelf, the exclusive fisheries zone (Sweden); and the exclusive economic zone (the Soviet Union), by giving 75 per cent effect to Gotland and other small associated Swedish islands.[170] The boundary line delimits the disputed area from 75 per cent to 25 per cent, leaving a greater share to Sweden. Perhaps as compensation for this favourable result for Sweden in terms of delimitation, the Soviet Union received favourable treatment in the field of fisheries. It is provided that:

> Within the framework of this Agreement and during a period of 20 years after the line of delimitation has been established, Sweden shall allow the Soviet Union a catch quota of 18,000 tonnes annually, 240 tonnes of which refer to salmon, in its part of the previous disputed area. During the same period, the Soviet Union shall allow Sweden a catch quota of 6,000 tons annually, 80 tons of which refer to salmon, in its part of the previously disputed area.[171]

The above provisions came to be applied between Sweden and Estonia, Latvia, Lithuania and the Russian Federation after the demise of the Soviet Union in 1991.[172]

3.3.2. *Sweden and Poland*

In the negotiations on the delimitation of the continental shelf and the fisheries zone, Poland and Sweden disagreed on the effect to be given to the Swedish island of Gotland. Poland objected to the position of Sweden that Gotland should be allowed full effect and argued for disregard of the island.[173] The position of Poland on the effect of Gotland appears to have been affected by two facts: first, the Soviet Union was also in dispute over the effect of Gotland with Sweden and thus it seemed better to wait and see the result of the negotiations between the USSR and Sweden; and second, if Poland had recognised the effect of Gotland, then it might have weakened its position *vis-à-vis* Denmark which argued for the full effect to the Danish island of Bornholm.[174]

The disputed area covered some 500 km^2. This area remained a white zone until 1989 when they agreed on a boundary of the continental shelf

Map 10: White Zone between Sweden and the Soviet Union

(Source: A.G. Oude Elferink, *The Law of Maritime Boundaries: A Case Study of the Russian Federation*)

and the fisheries zone one year after the USSR and Sweden concluded their maritime boundaries in 1988.[175] They agreed to give 75 per cent effect to Gotland, which was treated in the same way as between Sweden and the Soviet Union in 1988.[176]

3.3.3. *White Zones between Finland and the Soviet Union and Finland and Sweden*

Background

Finland laid a claim to the continental shelf in 1965, established an 8 N.M. fishing zone beyond and adjacent to its territorial sea in 1975 and extended its fishing zone beyond 12 N.M. only in the Gulf of Bothnia in 1977.[177] Finland extended its fishing zone beyond 12 N.M. from its territorial sea baselines in all the maritime areas along its coast in 1981, except the area of Bogskar.[178] The reason why Finland did not extend its fishing zone beyond 12 N.M. in the area of Bogskar was that the Soviet Union and Sweden challenged the full effect of Bogskar in the negotiations on maritime boundary between Finland and the Soviet Union and Finland and Sweden.[179] Note that Finland undertook negotiations on the maritime boundaries with its neighbouring countries before it took a legislative action on its fishing zone

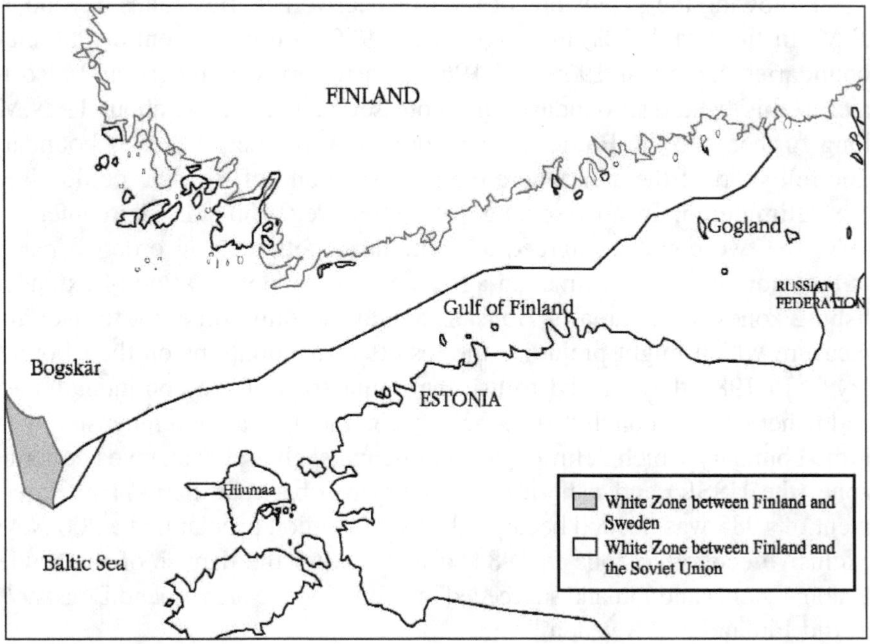

Map 11: White Zones between Finland and the Soviet Union; and Finland and Sweden (Source: A. G. Oude Elferink, *The Law of Maritime Boundaries: A Case Study of the Russian Federation*)

in 1981, because it wished to include precise co-ordinates of the outer limits of its fisheries zones in its law on the fisheries zone. It appears that the Soviet Union and Sweden challenged the effect of the Finish Bogskar in the delimitation relying on its small size and the distance between Bogskar and the Finish coasts.[180]

White Zones and the Delimitation Issue

The dispute regarding the effect of the Finnish Bogskar between Finland and the Soviet Union and Finland and Sweden led the countries concerned not to extend their fisheries jurisdiction to this disputed area, thus leaving the disputed areas as parts of the high seas.[181] Each of these white zones covers a relatively small area, which is about 500 km^2 or less. It will be convenient to briefly examine the delimitation issue between Finland and the Soviet Union and Finland and Sweden. Finland and the Soviet Union delimited their continental shelf boundaries in the agreements of 1965 and 1967.

Following the extensions of their respective fisheries zones beyond 12 N.M. in the late 1970s, they agreed in 1980 to use the continental shelf boundaries agreed in 1965 and 1967 as a fisheries boundary and also to add to this fisheries boundary one more segment which is about 19 N.M-long further into the Baltic Sea.[182] Notice that the new fisheries boundary line fell short of the area where the Finnish island of Bogskar could affect the delimitation. In an Exchange of Letters regarding the Agreement of 1980 the two countries agreed to continue negotiations in order to reach, in the shortest possible time, an agreement on the delimitation of extended fishing zones in the area of Bogskar and at the same time not to take any measure which might prejudice the results of negotiations on their boundary.[183] In 1985 they agreed to use the boundaries of the continental shelf and fisheries zone concluded so far between them, as a comprehensive maritime boundary which delimits the continental shelf and exclusive economic zone (the USSR) and fisheries zone (Finland) between them. This agreement of 1985 was needed because the Soviet Union proclaimed a 200 N.M. exclusive economic zone in 1984. In 1996, after the demise of the Soviet Union, Estonia and Finland succeeded in delimiting the area around Bogskar,[184] as did Finland and Sweden.[185]

Let us turn to the delimitation issue between Finland and Sweden. Finland and Sweden agreed on delimitation of the continental shelf in the Gulf of Bothnia in 1972, except in the area of Finnish Bogskar, since the effect of Bogskar was also the point of dispute between them.[186] In 1980 the two countries reached agreement on the boundary of fisheries zones in the Gulf of Bothnia before they took legislative action for extending their respective fisheries zone in 1981.[187] But in 1980 they also agreed not to extend their fisheries zones beyond 12 N.M. in the area around Bogskar as they could not agree on fisheries boundary between them in that area.

4. POLICY OPTION 3: FISHERIES ARRANGEMENT ON THE BASIS OF *DE FACTO* BOUNDARIES

Why do States choose this option? As there are only a few instances of this option in use, it is hard to generalise on this question. Yet, if we look at the Provisional Fisheries Agreement of 1977 between USSR and Japan, we realise that this option was the only possible solution. Suppose that there is a *de facto* boundary in place which is favourable to State A and its determination to keep the *de facto* boundary is very strong. Suppose also that there is an urgent need for State B to fish in the area beyond the *de facto* boundary. There is a need, then, for State B to agree on a fisheries agreement tacitly recognising the presence of the *de facto* boundary but at the same time saving its official position that it does not recognise the *de facto* boundary. This is the situation which actually arose between the Soviet Union and Japan.

4.1. *Fisheries Agreement between the USSR and Japan*

The USSR issued the Edict of the Presidium of the Supreme Soviet on the provisional establishment of a 200 miles fisheries zone in December 1976, and then defined the limits of its fisheries zone in a concrete manner through a Decision of the Council of Ministers in February 1977.[188] Japan also proclaimed a 200 miles fisheries zone in 1977 adopting the median line principle for boundaries but without such detailed provisions on the limits as was found in the Decision of the USSR Council of Ministers.[189]

The two countries never embarked upon negotiations for the delimitation of their maritime boundaries, apparently because of the territorial dispute over the four islands in the southernmost part of the Kuril islands.[190]

However, a fisheries treaty was urgently needed as Japanese fishermen had been heavily fishing in the waters which had been the high seas and then were claimed as part of the fisheries zone by the USSR. They started to negotiate in March 1977. The main obstacle was the definition of the geographical scope of the Soviet fisheries zone as the USSR insisted upon referring to, in the fisheries treaty, its Decision of the Council of Ministers of February 1977. Note that the Decision was formulated in such a way as to show that the disputed southern group of the Kuril Islands belongs to the USSR.[191]

In May 1977 they concluded the two provisional fisheries agreements – one was to give access to Japanese fishermen to the USSR exclusive fisheries zone and the other was to give access to USSR fishermen to the Japanese fisheries zone. In the former they agreed on a compromise formula on the description of the Soviet exclusive fisheries zone by avoiding mentioning the Decision of the Council of Ministers of February 1977 but by referring to "Article 6 of the Edict of the Presidium of the Supreme Soviet" and "decisions of the Soviet Government". The provisional fisheries agreements included a "without-prejudice clause", which provided that:

> Nothing in this Agreement shall be deemed to prejudice the positions or views of either Contracting Party in regarding any question of the law of the sea or any question pertaining to mutual relations.[192]

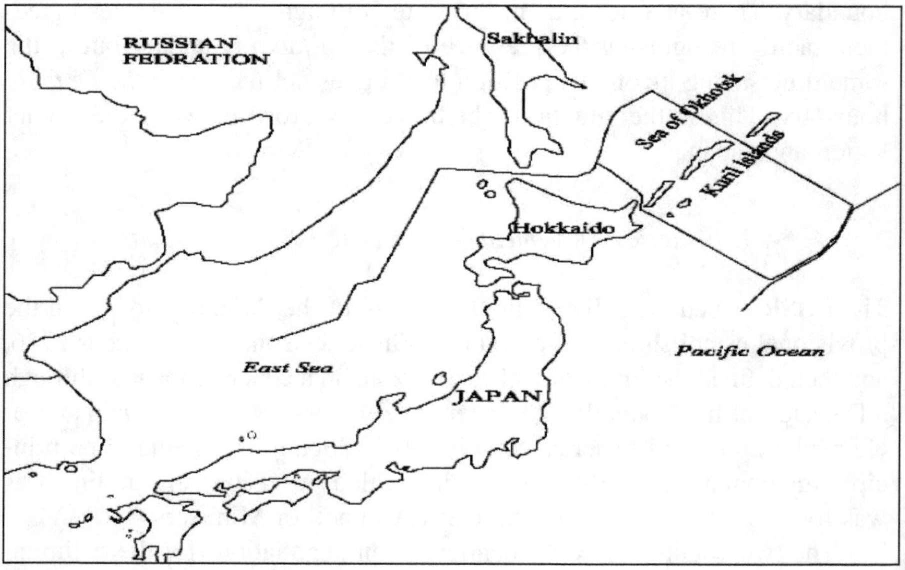

Map 12: Kuril Islands

(Source: A.G. Oude Elferink, *The Law of Maritime Boundary Delimitation: A Case Study of the Russian Federation*)

In implementing the agreements, Japan has acted upon the reality that the USSR exercised jurisdiction in the disputed islands, and recognised the equidistance line between the disputed islands and the Japanese islands as *de facto* maritime boundaries.[193] These provisional agreements were renewed every year until they were consolidated into one formal treaty in 1984, when the USSR proclaimed its 200 miles exclusive economic zone in accordance with the provisions of the LOS Convention which was adopted in 1982.[194] In 1997 Russia and Japan concluded a new fisheries agreement regarding the Japanese fishermen's access to the territorial waters of the disputed islands.[195]

4.2. *Co-operative Fisheries Arrangements around the Falkland Islands between the UK and Argentina*

The South-West Atlantic supports one of the richest fisheries resources in the world.[196] The major fish stock in the region is squid, which is exploited principally by fishing vessels from South Korea, Japan, Taiwan, Spain and Eastern Europe.[197]

In 1986 the British government established a fisheries zone (FICMZ: Falkland Islands Conservation and Management Zone).[198] It extended from a central point in the islands up to 150 N.M. seawards, but was modified in the South East. Why was 150 N.M. chosen rather than 200 N.M.? Symmons suggested three reasons: first, after the armed conflict in 1982, a 150 N.M. protection zone was already in place; second, most of the fishing was done within a 150 N.M. radius of the Falkland Islands; and third, the 150 N.M. might have been seen by the UK government as less likely to upset Argentina than would a full-200 N.M. zone.[199]

Although the 150 N.M. FICMZ achieved its immediate objective of bringing some order to fishing activities by fishing vessels from the Far East and Eastern Europe,[200] a need for more effective fisheries management in the region was felt by the British government.[201] There were two reason for this: First, many fish stocks straddle the FICMZ and the Argentine 200 N.M. zone; and second, a good deal of the fishing by foreign fishing vessels took place beyond the FICMZ but within the 200 N.M. of the Falkland Islands.[202] The two governments started negotiations on this issue after the Madrid Joint Statement of February 1990, whereby the British government and the Argentine government agreed to re-establish full diplomatic relations and set out a number of co-operative agendas.[203] After several meetings both governments agreed on the Joint Statement on the Co-operation of Fisheries on 28 November 1990.[204] The two governments re-affirmed the Madrid Formula that each State's position on the sovereignty issue is reserved and agreed to (i) establish a UK/Argentine "South Atlantic Fisheries Commission" (SAFC) to exchange information on fishing activities between 45° S and 60° S, and make recommendations relating to conservation of the fisheries resources in the area; and (ii) impose a temporary total ban on commercial fishing in those areas lying between the 150-mile FICMZ and a 200-mile line drawn from the Falkland coasts but excluding the area which would fall within the 200-mile EEZ of Argentina.[205]

In December 1990 the Falkland Islands government issued a proclamation establishing an Outer Fishery Conservation Zone (OFCZ), the location of which is the same as the zone where the total commercial fishing ban is applied.[206] When the British government in 1994 extended the OFCZ at its northern end to embrace a further 500 square miles of the ocean,[207] and also extended *ratione materiae* of the FICMZ and OFCZ so as to include the jurisdiction for the purpose of protection of the marine environment in the FICMZ and OFCZ, the Argentine government protested against the British action.[208] But notice that the arrangement is still in place. Under the arrangement, both governments made a commitment to maintain and conserve fish stocks in the South West Atlantic and three specific measures were agreed. These measures are for the stability of fish stocks, enhanced levels of co-operation and communication between the prospective governments and fishing departments, and co-ordinated policy to deter illegal fishing.[209] It is to be noted that the co-operative arrangement between the British government and the Argentine government could not be possible unless there was recognition by the Argentine side of the zones proclaimed by the British government around the Falkland Islands.

5. POLICY OPTION 4: COMPREHENSIVE JOINT EXPLOITATION ZONES

Theoretically speaking, comprehensive exploitation zones would be ideal in disputed areas because they can address all issues of exploitation in overlapping zones. Two coastal States can agree to exercise jurisdiction jointly with regard to marine scientific research and for the purpose of protection of the marine environment in a overlapping area. However, in practice the instances of comprehensive zones are very few. Why? It might be that policy makers tend to, by nature, seek specific solutions to specific problems. Or it could be that policy makers do not want to make problems by seeking solutions to issues which do not require solutions at present. It is to be recalled here that Norway refused the USSR proposal to set up another joint development zone along with an already existing joint fishing zone, and instead decided to await a final delimitation before exploiting oil and gas. The policy makers in Norway did not want a flavour of condominium felt in co-operation with the Soviet Union. From the following examination of four instances of a comprehensive resort to joint zones, we can see that there are different reasons for each.

5.1. *Iran-Sharjah (U.A.E.)*

Background

The reason for the joint development in the Persian Gulf between Iran and Sharjah lies in the fact that there is a long standing sovereignty dispute over the island of Abu Musa between the two countries.[210] Abu Musa lies approximately 31 N.M. from the nearest land tip of Sharjah. When the British government announced in 1968 its intention to withdraw from the region by the end of 1971, the sovereignty issue over the island of Abu Musa in the Gulf became a political issue between Iran and Sharjah.[211] Various discussions on the disputed island were held between Iran and Sharjah through mediation by the British government.[212] In November 1971 the Emirate of Sharjah accepted a British proposal to give Iran the rights to deploy its troops on the northern part of the island.[213] On 28 November Iranian troops landed and took up strategic positions and hoisted the Iranian flag. The Iranian troops' sudden landing on the disputed island and the subsequent political tension led to an arrangement on the utilisation of natural resources in the territorial waters of Abu Musa.[214]

Forms and Provisions

The main document, forming the basis of the arrangement, is a "Memorandum of Understanding" between Iran and Sharjah, which was announced on 29 November 1971, one day after the Iranian troops' landing on the island. However, there are various other documents relating to the arrangement.[215] A series of correspondence involving the ruler of Sharjah, the Iranian Ministry of Foreign Affairs, the British Foreign Office, the National Iranian Oil Company and Buttes Gas and Oil Co. are also elements of the arrangement.[216]

The first article of the Memorandum States that: "Neither Iran nor Sharjah will give up its claim to Abu Musa nor recognise the other's claim". However, the same article goes on to provide for Iran's full jurisdiction in the agreed areas occupied by Iranian troops and Sharjah's retention of full jurisdiction over the remaining part of the island.[217]

Operational Mechanism and the Effectiveness of the Arrangement

Although the territorial waters of the island are divided into two parts in accordance with the division of the land of the island between the two countries,[218] a special regime applies to the exploitation of oil and fisheries resources in the territorial waters of Abu Musa.

The arrangement includes a scheme for the sharing of revenue from the exploitation of oil in the territorial sea of the Island. The Buttes Gas and Oil Company is designated to carry out exploitation of oil in the territorial waters of the island. The company is to carry out its exploitation activities under the terms agreed with Sharjah and is to pay half of the oil revenues from the exploitation directly to Iran and the other half to Sharjah.[219]

The Memorandum has provisions on the fisheries in the territorial waters of Abu Musa. Article 5 of the Memorandum States that "The nationals of Iran and Sharjah shall have equal rights to fish in the territorial sea of Abu Musa".

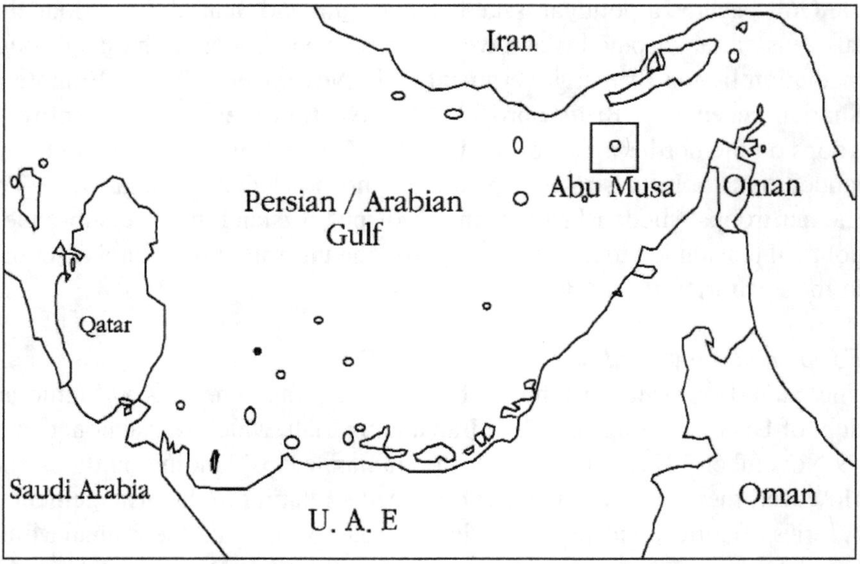

Map 13: Abu Musa

(Source: M. Miyoshi, "The Joint Development of Offshore Oil and Gas in Relation to Maritime Boundary Delimitation", IBRU Maritime Briefing)

5.2. *Colombia-Jamaica*

Background

In 1993 Colombia and Jamaica agreed on a part of their maritime boundary (point 1, 2, 3, 4, 5). But, since they were not able to draw a line further to the west of point 1, they agreed to establish a joint development and fishing zone, called the Joint Regime Area (the "JRA"), to the west of the boundary. The main reason why they could not reach an agreement on a boundary in this area is related to the presence of Serrandilla Bank and Low Cay (Bajo Nuevo or Los Bajos), because it is not settled as to whether the features belong to Colombia or to Honduras.[220] Jamaica is not a country which lays claims to the two features but Jamaican fishermen have been heavily fishing in the waters around the features. Therefore, it would have been better for Jamaica to keep its fishing access to the waters around the features through establishing a joint exploitation zone pending the settlement of territorial status of the two features and delimitation in the area between Colombia and Honduras. Colombia appears to have been tempted to make use of the establishment of the Joint Regime Area for reinforcing its claims to the features. In the legend of the map annexed to the treaty which establishes the Joint Regime Area, the understanding can be seen that the territorial waters around the features are regarded as Colombian, although there is no provision in the treaty itself on the status of the Bajo Neuvo or Serranilla Bank.[221]

Method of Shaping the Joint Development Zone

The JRA covers 4,500 square miles and is roughly a triangular-looking shape. However, upon closer examination, the JRA is found to be composed of lines connecting 11 points and two arcs. The circles of 12 N.M. radius around Serranilla Bank and Bajo Neuvo are excluded from the JRA, though Alice Shoal which belongs to Jamaica is enclosed in the JRA.[222]

Forms and Provisions

The provisions on the JRA are provided for in the maritime delimitation treaty between Jamaica and the Republic of Colombia.[223] The maritime boundary between the two countries is provided by Article 1 of the Treaty and the area of the JRA is defined in Article 3. In the latter it is provided that "Pending the delimitation of the jurisdictional limits of each Party in the area designated below, the Parties agree to establish therein a zone of joint management, control, exploration and exploitation of the living and non-living resources . . ." The legal nature of the JRA is that of a joint exclusive economic zone in relation to third Parties. The Parties can carry out the following activities in the JRA:

(a) Exploration and exploitation of the natural resources, whether living or non-living, of the waters superjacent to the seabed and the seabed and its subsoil, and other activities for the economic exploitation and exploration;
(b) The establishment and use of artificial islands, installations and structures;
(c) Marine scientific research; and
(d) The protection and preservation of the marine environment.[224]

It is to be noted that under the Agreement only the Parties are entitled to carry out a wide range of activities in the JRA and third States are not permitted to carry out any of those activities in the JRA.[225] This form of complete exclusion of third Parties in the JRS seems incompatible with the provisions of the LOS Convention,[226] in the light of the fact that a coastal State has to share surplus fisheries resources, if any, with third States.[227] It is also to be recalled that a coastal State's right to refuse to give consent to a marine scientific research plan of a foreign vessel is quite restricted.[228]

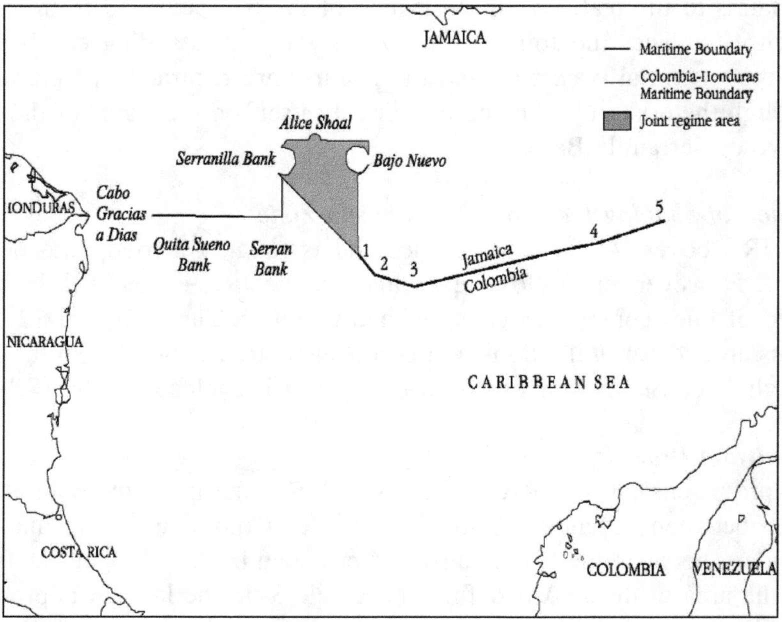

Map 14: Delimitation and Joint Exploitation between Jamaica and Colombia (Source: M. Miyoshi, "The Joint Development of Offshore Oil and Gas in Relation to Maritime Boundary Delimitation", IBRU Maritime Briefing)

Operational Mechanism and the Effectiveness of the Arrangement

Activities relating to exploration and exploitation of non-living resources, marine scientific research, and protection and preservation of the marine environment are to be carried out on a joint basis.[229] The role of defining the "joint basis" falls to a Joint Commission. It is provided that a Joint Commission is to be established to elaborate the modalities of co-operation activities.[230] The Joint Commission is to be set up with one representative for from each party.[231] It can make either non-binding recommendations or binding decisions.[232]

5.3. *Guinea-Bissau-Senegal*

Background

In 1960 Portugal and France, who were then the colonial powers of Guinea-Bissau and Senegal respectively, agreed on the delimitation of the maritime boundary between their respective colonies.[233] They set, for the demarcation of territorial sea, and for the contiguous zone and continental shelf between their colonies, a straight line running at 240° starting from the intersection of the extension of the land boundary and the low-water mark.[234] Guinea-Bissau challenged the legal force of the agreement of 1960 between Portugal and France as applicable between Guinea-Bissau and Senegal.[235]

After Guinea-Bissau and Senegal failed to settle the issue they agreed to submit the dispute to an arbitral tribunal. In 1989 the arbitral tribunal decided that the agreement of 1960 between the colonial powers has the force of law between Guinea-Bissau and Senegal.[236] However, Guinea-Bissau argued, *inter alia*, that the award was void because one of the arbitrators who voted for the award appended an opinion which was in substance contradictory to the award.[237] Guinea-Bissau also asserted that there had been no delimitation of exclusive economic zones as such in the agreement of 1960 and the tribunal had failed to address this issue.[238]

The International Court of Justice, which examined the issues raised by Guinea-Bissau, stated in 1991 that "the elements of the disputes that were not settled by the Arbitral Award of 31 July 1989 be resolved as soon as possible, as both Parties desire".[239] Following the decision by the Court, the two countries embarked on the negotiations for an EEZ delimitation. In 1993 they reached an agreement not on a boundary but on a joint development zone of EEZs, which straddles the boundary of 1960.

Method of Shaping the Joint Development Zone

The joint development zone between Guinea-Bissau and Senegal is an area surrounded by 268° azimuth and 220° azimuth starting from Cape Roxo and a circle with a radius of 200 N.M. from Cape Roxo excluding the territorial waters from the zone.[240]

Forms and Provisions

It is not difficult to figure out the reason why 268° azimuth and 220° azimuth were chosen for the limits of the joint development zone, since it can be assumed that 268° azimuth has been argued by Guinea-Bissau and 220° azimuth has been argued by Senegal and thus the two countries agreed to use the two lines as limits of their joint development zone. Note that 268° azimuth is favourable to Guinea-Bissau as it bisects the angle formed by coastal lines of the two countries giving full account of the large Bijagos Archipelago off the Guinea-Bissau's coasts, and disregarding Cape Vert which protrudes from the Senegal coasts in the form of a sharp beak. Notice also that 220° azimuth is favourable to Senegal as it bisects the angle formed by the coastlines of the two countries disregarding Guinea-Bissau's Bijagos Archipelago and taking into account Cape Vert.

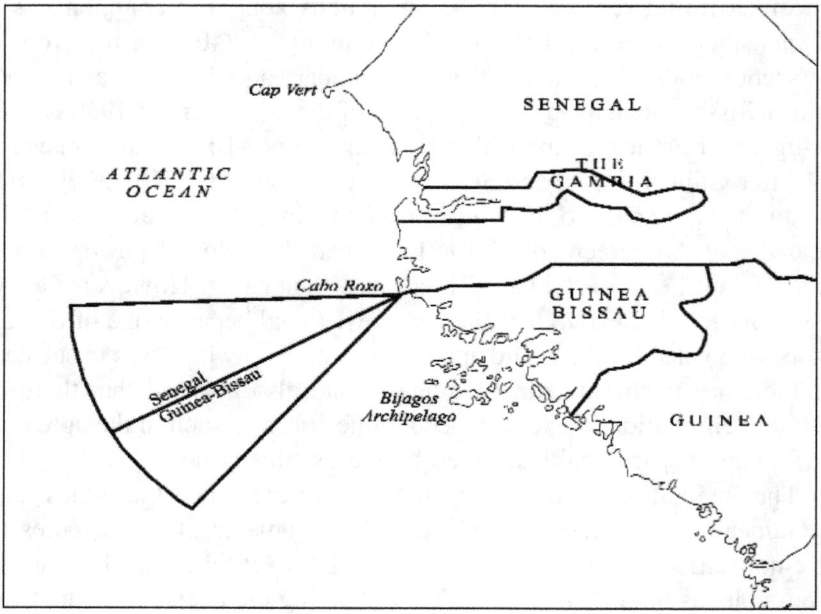

Map 15: Joint Exploitation Zone between Senegal and Guinea-Bissau

(Source: M. Miyoshi, "The Joint Development of Offshore Oil and Gas in Relation to Maritime Boundary Delimitation", IBRU Maritime Briefing)

The joint development zone is established by the "Management and Co-operation Agreement" signed on 14 October 1993. The Agreement was in turn supplemented by a "Protocol Relating to the Organisation and Operation of the Agency for Management" signed on 12 June 1995.[241] The Agreement makes it clear that the Parties' legal titles and claims to non-delimited areas are reserved.[242] The Agreement is effective for an initial period of 20 years, after which the Parties may choose to keep the Agreement, negotiating any unsettled delimitation or agree to refer to the matter to arbitration or the ICJ.[243]

It is to be noted that the applicable law with regard to gas and oil exploration and exploitation in the joint development zone is that of Senegal and the law applicable to fisheries in the zone is that of Guinea-Bissau.[244]

Operational Mechanism and the Effectiveness of the Arrangement
The joint development zone in the east of the Atlantic Ocean between Guinea-Bissau and Senegal is an EEZ joint development zone and thus relates to not only oil and gas but also to living marine resources. It is inter-

esting to note that they agreed to share the fishery resources equally between them, whereas the continental shelf resources are to be shared in the proportion of 85 per cent to Senegal and 15 per cent to Guinea-Bissau. The reason for the unequal partitioning of the non-living resources lies in the fact that prospective oil and gas in the zone have been explored and developed by Senegal in co-operation with an Irish oil company.[245] The two States agreed to establish an International Agency for exploitation of gas and oil resources. The Agency is to discharge its function through an Enterprise. And the Agency, consisting of the heads of the State Parties or persons delegated by the heads will provide general policy for the Enterprise.[246] The Agency was established in Dakar in 1995. It holds exclusive mineral or oil titles and fishing rights in the joint development zone.[247] The Agency has competence, *inter alia*, to promote mineral and petroleum resource prospecting, exploration and exploitation in the joint development zone, and to market the mineral or petroleum resources produced. The Enterprise under the direction of the Agency is to, *inter alia*, assist holders of mining or hydrocarbon resource development licenses in their administrative dealings with each State.[248]

5.4. *Saudi Arabia and Sudan*

In 1974 Saudi Arabia and Sudan concluded an agreement whereby they agreed that the lines of 1,000 meter isobath in the Red Sea between them are outer limits of each country's EEZ, and defined the maritime area left in the middle of the Sea beyond the outer limits of respective EEZs as a Common Zone.[249]

The Common Zone is a common EEZ where the two countries have exclusive "equal sovereign rights in all the natural resources".[250] They agreed to establish a Joint Commission. It is empowered, *inter alia*, to:

- undertake studies concerning exploration and exploitation of natural resources
- consider and decide on the applications for licenses and concessions concerning exploration and exploitation
- organise the supervision of the exploitation at the production stage
- determine the manner in which any accumulation or deposits of a natural resource found to extend across the boundary of the exclusive sovereign rights area of either State and the Common Zone.[251]

In the agreement the two countries noted that Sudan had already given exploration licenses to a German oil company Preussag in the area and

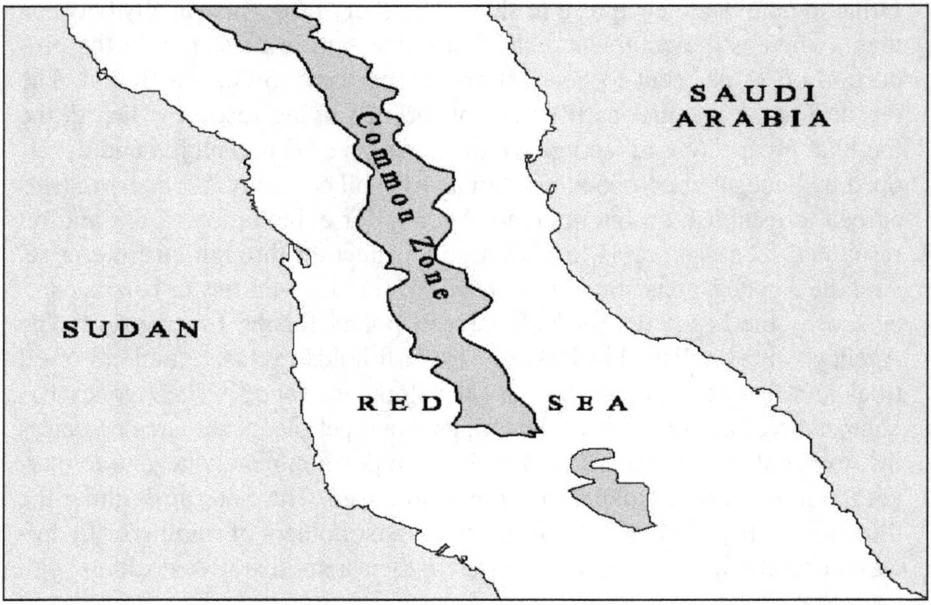

Map 16: Common Zone between Saudi Arabia and Sudan

(Source: M. Miyoshi, "The Joint Development of Offshore Oil and Gas in Relation to Maritime Boundary Delimitation", IBRU Maritime Briefing)

agreed that the Joint Commission would decide on this matter so as to preserve the rights of Sudan in the context of the regime established for the common zone.[252] The Joint Commission was established and met for the first time on 10 May 1975.[253]

6. POLICY OPTIONS 5: SINGLE PROVISIONAL FISHERIES BOUNDARY

A single provisional fisheries boundary can be utilised for the purpose of control of the exploitation of specific resources. For instance, the purpose of the provisional line between Australia and Indonesia was limited only for the enforcement jurisdiction on fishing vessels. When a State accepts a line as a provisional boundary for a specific purpose, there might be a signal to the other State that the former can be flexible about the possibility of the line being transformed into a permanent one, if the latter is willing to compromise on other issues. This was the case when Australia accepted a provisional fisheries line which was favourable to Indonesia and then agreed to utilise the boundary as a permanent one in the final delimitation of 1998.

Australia proclaimed a continental shelf in 1962, a 200 N.M. fisheries zone in 1979 and a 200-mile exclusive economic zone in 1994. Indonesia proclaimed a continental shelf in 1969 and a 200-mile exclusive economic zone in 1980.[254] The two countries negotiated on the delimitation of maritime boundaries of the continental shelf and water column together.[255] In the negotiations on the water column boundary, the main issue was the effect of the Australian islands of Ashmore in the Indian Ocean. Australia held the view that the boundary should be an equidistance line giving full effect to Ashmore, but Indonesia argued that an equidistance line should be drawn between the Indonesian archipelago and the Australian mainland.[256] In the context of the resulting stalemate, Australia and Indonesia agreed on the "Provisional Fisheries Surveillance and Enforcement Line (PFSEL)" through a Memorandum of Understanding of 1981. The PFSEL divided the disputed area in a way which delivered 70 per cent of it to Indonesia.[257] The PFSEL purported to serve as a provisional line pending final delimitation, beyond which one party could not exercise its enforcement action against the fishing vessels of the other party.[258] Article 4 of the MOU provides that:

> It is understood that in areas that overlap, and pending the permanent settlement of maritime boundaries between the two countries, neither Government will exercise jurisdiction for fisheries surveillance and enforcement purposes beyond a provisional fisheries line in respect of swimming fish species against fishing vessels licensed to fish for such species by the authorities of the other country.[259]

In 1997 the two countries reached agreement on the whole package of maritime boundaries of the continental shelf, exclusive economic zone (water column) and joint development zone. In the package, they utilised the pre-existing PFSEL as the exclusive economic zone boundary with a minor modification in the area around Ashmore and an additional boundary line further to the west of the PFSEL.[260]

7. JOINT EXPLOITATION ZONES AS ADDITIONAL ELEMENTS OF MARITIME BOUNDARIES

7.1. *Why Are Joint Exploitation Zones Needed When There Are Boundaries?*

It is to be noted that a joint development zone or a joint fishing zone can be established even after two Parties agree on delimitation, because even a successful delimitation may still require a degree of close co-operation for rational exploitation of transboundary resources.[261] Even though such a joint development zone is not a provisional arrangement under Articles 74(3)/83(3) of the LOS Convention, it also shares some common characteristics with such provisional arrangements.

In the *Gulf of Maine* Case it was pointed out that the delimitation of a maritime boundary is not necessarily a panacea for disputes over offshore resources, because petroleum does not respect national boundaries.[262] If a deposit straddles a possible boundary of neighbouring countries, exploitation of the deposit by one would damage the potential share of its neighbour. Thus the need to view the deposit of fluid petroleum or natural gas across boundaries as a single deposit is recognised. Fisheries resources are also transboundary in many cases.

This recognition of practicalities has led to a number of international agreements on joint development of gas and oil along with boundaries in the 1960s and 1970s. The Anglo-Norwegian Treaty of 1965 provides that:

> if any single geological petroleum structure or petroleum field . . . extends across the dividing line . . . the Contracting Parties, in consultation with their licensees, if any, shall seek to reach agreement as to the manner in which the structure or field shall be most effectively exploited and the manner in which the proceeds deriving therefrom shall be apportioned.[263]

In the *Tunisia/Libya Continental Shelf* Case, Judge Evensen noted in his dissenting opinion that "the underlying immediate concerns are first and foremost petroleum exploitation" but "petroleum exploitation is a mixture of know-how and luck" and thus "the drawing of a line of delimitation between States may, as far as oil potentials are concerned, be a pure gamble, an accidental fact which may leave rich structure on one side of the line and barrenness on the other".[264] Then he mentioned that: "An arrangement for joint exploration, user or even joint jurisdiction over restricted overlapping areas may be corollary to other equity consideration".[265]

7.2. *Instances of Joint Exploitation Zones as Additional Elements to Maritime Boundaries*

In 1958, Bahrain and Saudi Arabia agreed on a continental shelf boundary in the Persian Gulf which extended for a distance of 98.5 N.M. In addition to the boundary, they agreed to establish a joint zone. The whole area of the joint zone is situated on the Saudi Arabian side of the boundary.

In the joint development zone, the exploitation of oil is to be carried out in the way chosen by Saudi Arabia on the condition that it grants to the government of Bahrain half of the net revenue accruing from this exploitation.[266] The businesslike manner of both countries is notable in this case. In the delimitation agreement of 1958, they settled a long standing sovereignty dispute over the three islands: Al-Baina As-Saghiran and Al-Baina Al-Kabir came to belong to Bahrain, while Al-Baina Al-Kabir was left to Saudi Arbia.[267] Obviously, in this case the revenue-sharing scheme would have facilitated the negotiations on the maritime boundary.

The joint development zone, which France and Spain agreed to in 1974 along with a continental shelf boundary in the Bay of Biscay is also an example of joint development zones agreed as an additional element to delimitation.[268] The zone which sits astride the boundary line, appears to be a compensation for Spain because the continental shelf boundary line gives it less maritime area than the strict equidistance line would have. Notice that the continental shelf boundary is not drawn by distance criteria but from the natural boundary as the boundary line is situated between the two 4,000 metre isobaths in the middle of the Bay of Biscay. The two Parties undertake to encourage agreements between companies applying for prospecting rights in the zone in order to permit companies having the nationality of the other party to participate in such prospecting on an equal partnership basis.[269] It is also to be noted that this agreement also has provisions on straddling natural resources. If a deposit of a natural resource lies across the boundary line or if the portion of the deposit on one side of the boundary is wholly or partially exploitable by means of installations located on the other side of the boundary, then the two countries shall seek an agreement on terms for exploiting the deposit.[270]

In the *Jan Mayen* Case between Iceland and Norway, the Conciliation Commission recommended a single dividing line for both the continental shelf and the EEZ, and proposed in explicit terms that the petroleum resources across the boundary might well be jointly developed.[271] According to the recommendation, most of the joint development zone lay beyond the 200 N.M. line from Iceland.[272] In its recommendation, the Conciliation

Map 17: Joint Development Zone between Saudi Arabia and Bahrain
(Source: M. Miyoshi, "The Joint Development of Offshore Oil and Gas in Relation to Maritime Boundary Delimitation", IBRU Maritime Briefing)

Commission considered the fact that Iceland is totally dependent on imports of hydrocarbon products.[273]

The joint development zone between Iceland and Jan Mayen (Norway) was established in 1981 in accordance with the recommendation of the Conciliation Commission.[274] The agreement of 1981 also provides that any petroleum situated partly within the limits of the joint development zone is to be regarded as lying wholly within the development zone.[275]

The joint development zone in the Mediterranean Sea between Tunisia and Libya was established together with the continental shelf boundary agreement signed in 1989. It was *ad hoc* judge Evensen who proposed a joint development zone in his dissenting opinion.[276] Following the opinion given by Evensen, the Parties agreed on joint development along with the continental shelf boundary, which is divided into two parts: one part is the north-east and the other is south-east. For the joint development in the northeast part of the zone, a joint Tunisia-Libya exploration company is assumed to be established for equal benefit between the two Parties. And in the south-east part of the zone, Libya is entitled to have 90 per cent of the revenue from production giving Tunisia 10 per cent of the revenue.[277]

Italy and Yugoslavia agreed to establish a joint fishing zone in the Gulf of Trieste which is located at the north-east corner of the Adriatic Sea in

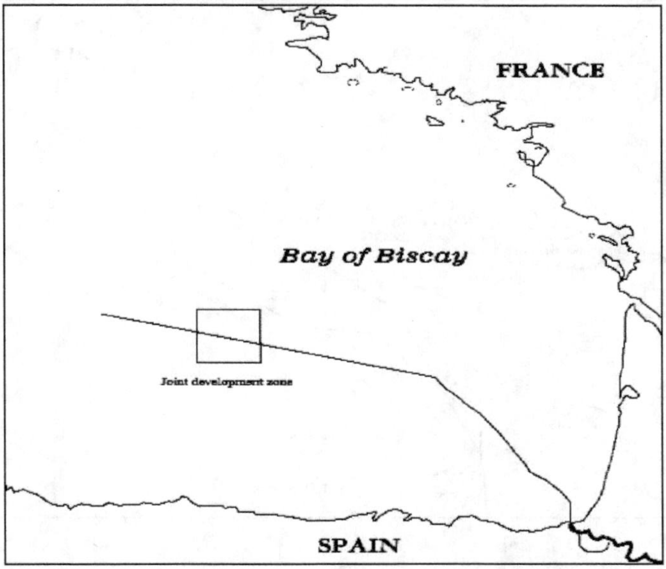

Map 18: Joint Development between France and Spain

(Source: M. Miyoshi, "The Joint Development of Offshore Oil and Gas in Relation to Maritime Boundary Delimitation", IBRU Maritime Briefing)

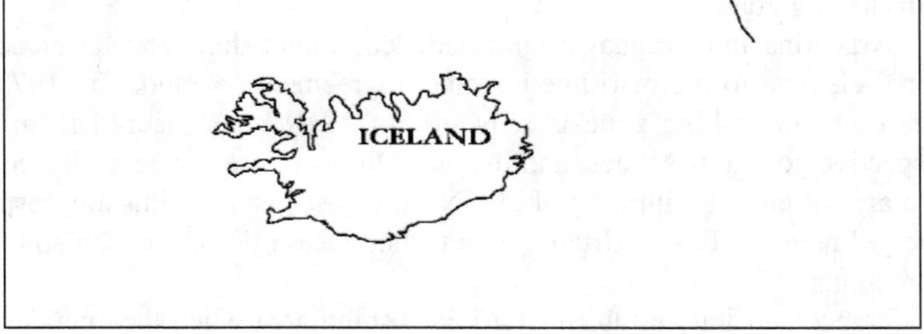

Map 19: Joint Development Zone between Norway and Iceland

(Source: M. Miyoshi, "The Joint Development of Offshore Oil and Gas in Relation to Maritime Boundary Delimitation", IBRU Maritime Briefing)

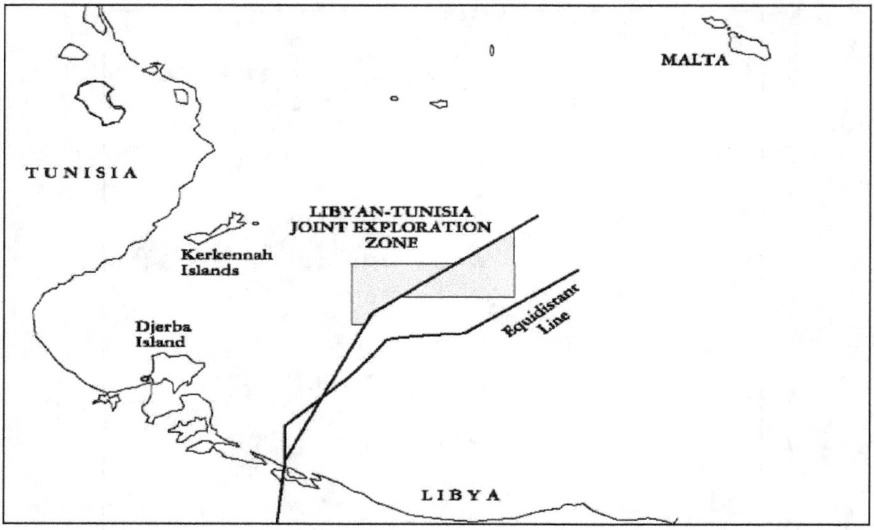

Map 20: Joint Development Zone between Libya and Tunisia

(Source: M. Miyoshi, "The Joint Development of Offshore Oil and Gas in Relation to Maritime Boundary Delimitation", IBRU Maritime Briefing)

June 1987. Notice that the joint zone straddles not the exclusive economic zone but the territorial seas of the two countries facing each other in the Gulf of Trieste where the opposing coasts lie within 24 N.M. distance. This joint fishing zone was intended to accommodate the reality of continued illegal fishing by Italian vessels on the Yugoslavian side of the territorial sea boundary which was agreed in 1977. Only limited numbers of fishing vessels registered in the local ports of each party are allowed to fish in the joint fishing zone.[278]

Argentina and Uruguay established a common fishing zone as an additional element to the maritime boundary agreement concluded in 1973.[279] The common fishing zone is seawards of 12 N.M. measured from the respective coastal baselines, and the outer limits of the zone are shaped by two arcs of circles with radii of 200 N.M. whose centre points are, respectively, Punta del Easte (Uruguay) and Punda Rasa del Cabo San Antonio (Argentina).[280]

France and Italy established a joint fishing area when they concluded a maritime boundary agreement in 1986. In the joint fishing zone, professional fishermen of both countries can continue to carry out their fishing activities.[281]

In 1954 Chile, Ecuador and Peru issued a tripartite declaration for creating a special maritime frontier zone of 10 N.M. breadth on each side of the boundaries.[282] Following the declaration, Chile and Peru concluded an agreement in 1952 as did Ecuador and Peru on the same day to establish buffer zones in which enforcement restraint for fisheries violation is mandated.[283] In 1985 Costa Rica and Ecuador concluded a similar agreement to establish a buffer zone between them.

When Colombia and Dominica agreed on the maritime boundary in 1978, they also agreed to establish a "Joint Scientific Research and Common Fishing Exploitation Zone" which straddles the maritime boundary.[284] This is not a provisional arrangement under the Articles 74(3)/ 83(3) but an additional element to the boundary agreement.[285]

8. OBSERVATIONS

We have seen that there are a number of instances where provisional arrangements are in place pending ultimate maritime delimitation, even though Articles 74(3)/83(3) do not provide illustrative examples of such arrangements. We have analysed: 5 joint development zones; 8 grey zones; 4 white zones; 2 fisheries arrangements based upon *de facto* boundaries; 5 comprehensive exploitation zones; 1 instance of provisional fisheries arrangement based upon a singe provisional line; and 11 instances of joint exploitation zones agreed as additional elements to maritime boundaries.[286] It is to be noted that there are variations between the arrangements even in the same category of the provisional arrangements.

The analysis shows that the joint exploitation zones either of the continental shelf or for fishing purposes have been the most widely used form of co-operative arrangements even before the adoption of the LOS Convention. It is interesting to note that there are not many instances of States having recourse to a comprehensive joint development, which seem to be ideal for addressing all issues in disputed areas. The reason might be that policy makers of coastal States tend to seek solutions to problems which have already arisen or are going to arise in the near future, and also because some issues which pose problems to coastal States in one region may not necessarily do so to other coastal States in other parts of the world. Furthermore, it is also notable that there are a several arrangements to deal with marine scientific research and the protection of the marine environment in disputed areas.

States practice concerning the joint exploitation zones can provide a practical guide to other States that are searching for provisional arrangements under the provisions of Articles 74(3)/83(3) of the LOS Convention in disputed areas in the other parts of the world. Some important observations can be drawn from the analysis of State practice regarding provisional arrangements.

First, it is notable that by and large formal treaties have been utilised for these arrangements. This means that most of the provisional arrangements have been sent to parliaments for approval. The reason appears to be that States usually attach a significant importance to the exploitation of gas and oil and fisheries resources. This tendency of States in giving greater importance to the provisional arrangements and using formal treaty forms rather than informal MOU can find its justification in the fact that many of the provisional arrangements usually have a very long *ratione temporis* such as renewable every forty or fifty years and many of the provisional arrangements have not been replaced by boundary agreements. As we have seen in Chapter 2, the term "arrangement" has been used sometimes to indicate co-operative mechanisms between States or the instruments which provide legal basis for such mechanisms regardless of their nomenclatures. And when the term "arrangement" is used in this context, then the titles and formalities of instruments providing such arrangements can vary. But this does not mean that an informal instrument has not been used at all for provisional arrangements. For instance, Australia and Indonesia used an MOU in 1981 to establish the Provisional Fisheries Surveillance and Enforcement Line between them.

Secondly, in relation to the first observation it is notable that it is hard to find a case where a joint development zone or a comprehensive exploitation zone has been replaced by a delimitation agreement. This might be a reason why States are usually not flexible in shaping the outer limits of the joint zones and thus accommodate conflicting positions on delimitation. We can also find that management of fisheries resources in disputed areas has some impact on the timing of the conclusion of maritime boundary agreements, although it does not mean that provisional fisheries arrangements affect Parties' positions on the ultimate boundaries themselves. This demonstrates that the choice between white zones and grey zones is important in terms of the timing of the delimitation as well as management of fisheries resources. This means that coastal States have to choose between effective management of fisheries resources in disputed areas and early delimitation of disputed areas. Note that all the white zone arrangements we have examined have been replaced by ultimate delimitation agreements.

Thirdly, it was observed that a "without prejudice clause" has normally been utilised in provisional arrangements. This fact demonstrates that the without prejudice clause is regarded as an important factor in making it clear that an arrangement is intended to be of a practical nature and not to create any legal basis for either party to reinforce its claim over the area in dispute.

Finally, joint commissions are usually set up to effectively implement arrangements, although the power of joint commissions differ from case to case. We will see later in this book that these common characteristics, with some variations, are also found in provisional fisheries arrangements in North East Asia.

NOTES

1. The *North Sea Continental Shelf* Cases, 1969 *ICJ Reports*, para. 101(c)(2). It is not clear why the ICJ mentioned in the last part of the judgement the possibility of "a regime of joint jurisdiction". Maybe the concept of natural prolongation which dominated the consideration of the Court led it to emphasise "a regime of joint jurisdiction". It is because according to the concept of natural prolongation a natural prolongation belongs to a coastal State *ipso jure* and *ab initio*, and therefore where there is one natural prolongation which belongs to two States at the same time, the common natural prolongation belong *ipso jure* and *ab initio* to two coastal States at the same time. However, in the subsequent delimitation following the judgement between the United Kingdom, Netherlands and Germany, there was no need to set up a joint development zone, because they succeeded in delimiting their boundaries in the North Sea.

2. *ICJ Reports, ibid.*, p. 83.

3. H. Fox ed., *Joint Development of Offshore Oil and Gas: A Model Agreement for States for Joint Development with Explanatory Commentary*, British Institute of International Law (1989), p. 45.

4. R. Lagoni, "Oil and Gas Deposits Across National Frontiers", 73 *A.J.I.L.* (1979), pp. 233-239.

5. R. Churchill, "Joint Development Zones: International Legal Issues", *Joint Development of Offshore Oil and Gas*", H. Fox ed., British Institute of International Law (1990), pp. 55-67.

6. I. Brownlie, *Principles of Public International Law*, Oxford Clarendon Press (1998), p. 114.

7. H. Lauterpacht, ed., *Oppenheim's International Law* (8th ed.), Longman (1955), vol. I, p. 453: see also R. Jennings and A. Watts, eds., *Oppenheim's International Law* (9th ed.), Longman (1992), pp. 566-567.

8. H.M. Al-Bahara, *The Legal Status of the Arabian Gulf States: A Study of their Treaty Relations and their International Problems*, Manchester University Press (1968), pp. 267-268.

9. M.T. El Ghoneimy, "The Legal Status of the Saudi-Kuwaiti Neutral Zone", 15 *I.C.L.Q.* (1966), pp. 690-717.

10. Article 8(2) of the Agreement of 1965. The text is reproduced in 6 *I.L.M.* (1965), No. 6, pp. 1134-1137.

11. For the maritime delimitation between Saudi Arabia and Iran, see R.F. Piertrowski, Jr., "Report Number 7-7: Iran-Saudi Arabia", *International Maritime Boundaries*, pp. 1519-1532.

12. W.T. Onorato, "Apportionment of an International Common Petroleum Deposit", 17 *I.C.L.Q.*, (1968), pp. 85-101; H. Fox *et al.* eds., *op. cit.* (1989), p. 55.

13. Masahiro Miyoshi, *The Joint Development of Offshore Oil and Gas In Relation to Maritime Boundary Delimitation*, Maritime Briefing IBRU, University of Durham, 1999, pp. 9-10.

14. H. Fox *et al.* eds., *op. cit.* (1989), p. 55.

15. Choon-ho Park, "Report Number 5-13(2): Malaysia-Thailand (Gulf of Thailand Continental Shelf)", *International Maritime Boundaries*, pp. 1111-1123.

16. Article 1 of Annex I to the Memorandum of Understanding on the Delimitation of the Continental Shelf Boundary confirm the fact. It provides that "Both Parties agree that as a result of overlapping claims made by the two countries regarding the boundary line of their continental shelves in the Gulf of Thailand, there exists an overlapping area, . . ."; for text see Choon-ho Park, *ibid.*

17. Article II, *Memorandum of Understanding between the Kingdom and Thailand and Malaysia on the Establishment of a Joint Authority for the Exploitation of the Resources of the Sea-Bed in a Defined Area of the Continental Shelf of the Two Countries in the Gulf of Thailand*; for the text, see Choon-ho Park, *op. cit.* (Report No. 5-13(2) in *International Maritime Boundaries*).

18. Article III of the Memorandum of Understanding.

19. Article VII of the Memorandum of Understanding.

20. Article III of the Memorandum of Understanding.

21. Article 8 of the *Agreement between the Government of Malaysia and the Government of the Kingdom of the Thailand on the Constitution and Other Matters Relating to the Establishment of the Malaysia-Thailand Joint Authority*; for text see Choon-ho Park, *op. cit.* (Report No. 5-13(2) in *International Maritime Boundaries*).

22. M. Valencia and Van Dyke, "Southeast Asian Seas: Joint Development of Hydrocarbons in the Overlapping Claims Areas", 16 *O.D.I.L.* (986), pp. 218-219.

23. M. Valencia and Van Dyke, *idem.*

24. Ted L. McDorman, "Number 5-19: Malaysia-Vietnam", *International Maritime Boundaries*, Vol. III on CD Rom.

25. Preamble to the *Memorandum of Understanding between Malaysia and the Socialist Republic of Vietnam for the Exploration and Exploitation of Petroleum in the Defined Area of the Continental Shelf Involving the Two Countries*; the text of the Memorandum is reproduced in Ted L. McDorman, *idem.*

26. Articles 5 and 7 of the Memorandum of Understanding.

27. Article X of the Memorandum of Understanding.

28. R.V. Prescott, "Report Number 6-2 (5); Australia-Indonesia", *International Maritime Boundaries*, p. 1246.

29. A. Bergin, "The Australian-Indonesian Timor Gap Boundary Agreement", 5 *I. J.E.C.L.* (1990), pp. 383-385.

30. A. Bergin, *ibid.*

31. R.V. Prescott, *op. cit.* (Report Number 6-2 (5) in *International Maritime Boundaries*), p. 1208.

32. *The Treaty between Australia and the Republic of Indonesia on the Zones of Co-operation in an Area between Indonesia's Province of East Timor and Northern Australia*; the text of the treaty is reproduced in R.V. Presott, *ibid.*, pp. 1256-1328.

33. Article 2(3) of the Treaty.

34. Article 33 of the Treaty.

35. Article 2 of the Treaty.

36. *The Christian Science Monitor*, 21 May 2110, at <www.csmonitor.com/durable/2001/05/21/p7s1.htm.> According to *The Christian Science Monitor*, under a new treaty the Timorese will get 85-90 percent of what could prove to be hundreds of millions of dollars in annual royalties.

37. Article 3 and 4 of the Treaty.

38. For the analysis of the 1997 Treaty, see Max Herrman and Martin Tsamenyi,

"The 1997 Australia-Indonesia Maritime Boundary Treaty: A Secure Legal Regime for Offshore Resource Development", 29 *O.D.I.L.* (1998), pp. 361-396.

39. For the sovereignty issue of the islands, see L.S. Gustafsen, *The Sovereignty Dispute over the Falklands (Malvinas) Islands*, New York Oxford University Press (1988); for a brief summary of the issue, see Patric Armstrong and Vivian Forbes, *The Falkland Islands and their Adjacent Maritime Area*, Maritime Briefing Vol. 2 No. 3, IBRU, University of Durham (1997).

40. Falkland Islands (Continental Shelf) Order 1950, S.I. 1950 No. 2100.

41. R.R. Churchill, "Falkland Islands-Maritime Jurisdiction and Co-operative Arrangements with Argentina", 46 *ICLQ* (1997), p. 463. However, during and following the 1982 conflict, there was a "naval protection zone" around the Falklands Islands by the British government which aimed at excluding Argentine ships and aircraft from the Falklands region; see P. Armstrong and V. Forbes, *op. cit.* (1997), p. 38.

42. British government, *Declaration on the Conservation of Fish stocks and on Maritime Jurisdiction around the Falkland Islands*, 29 October 1986, Reproduced in *Law of the Sea Bulletin* No. 9. (1987), pp. 18-19.

43. British Government Declaration, *ibid.*

44. UKM.L. in 63 *B.Y.I.L.* (1991), p. 645.

45. This Ordinance was replaced by the Offshore Minerals Ordinance of October 1994, which has more detailed provisions for exploration and exploitation.

46. R.R. Churchill, *op. cit.* (1997/*I.C.L.Q*), p. 468.

47. Foreign and Commonwealth Office, *Press Release No. 241* (Wednesday 28 November 1990).

48. The text of the "Joint Declaration on the Co-operation over Offshore Activities in the South West Atlantic of 27 September 1995" is reproduced in P. Armstrong and V. Forbes, *op. cit.* (1997), pp. 36-39.

49. R.R. Churchill, *op. cit.* (1997/*I.C.L.Q.*), p. 469.

50. The term "Madrid Formula" denotes a formula on sovereignty agreed by Argentina and the United Kingdom agreed on 19 October 1989. It means that the two countries pursue co-operation on the understanding that the position of either side on the question of sovereignty over the Falkland Islands is reserved. After negotiations on the basis of the formula the two governments issued a Joint Statement in February 1990, which contained a range of co-operation measures between the two States; see M. Evans, "The Restoration of Diplomatic Relations between Argentina and the United Kingdom", *40 I.C.L.Q.* (1991), pp. 476- 477.

51. R.R. Churchill, *op. cit.* (1997/*I.C.L.Q.*), p. 468.

52. See *Declaration of the British Government with regard to the Joint Declaration Signed by the British and Argentine Foreign Ministers On Co-operation Over Offshore Activities in the South West Atlantic*; the text of the declaration is reproduced in P. Armstrong and V. Forbes, *op. cit.* (1997), p. 36: Apart from the Special Zone between the UK and Argentina, the UK government has undertaken exploration activities in the continental shelf around the Falkland Islands. In October 1996 the Falkland Islands government issued seven offshore production licences to five consortia, and the five consortia have conducted seismic exploration and have drilled six wells so far. They found minor quantities of oil and evidence of gas in five of the six wells. Although large accumulations of oil have not yet been found in the first six wells, analysis of the geology in this area is on going and the data from exploration and drilling have been promising: see Falkland Islands government, *ibid.*, p. 4: see also Falkland Islands government "Falkland Islands government: Offshore Oil Exploration", p. 3, at <http://www.falklands.gov.fk/oildept.htm>.

53. See paragraph 2 of the Joint Declaration; for text, P. Armstrong and V. Forbes, *op. cit.* (1997), p. 38.

54. The Joint Declaration, paragraph 4.

55. R.R. Churchill, *op. cit.* (1997/*I.C.L.Q.*), p. 469.

56. Falkland Islands government, "Falkland Islands Offshore Oil Exploration", pp. 1-6 at <www.falklands.gov.fk/9bgt.htm>.

57. Falkland Islands government "Falkland Islands Government: Offshore Oil Exploration", p. 3, at <www.falklands.gov.fk/oildept.htm>.

58. B.H. Oxman, "The Third United Nations Conference on the Law of the Sea: The 1976 New York Convention", 71 *A.J.I.L.* (1977), p. 261.

59. Statement made by the Chairman of Negotiating Group 7 on 12 September 1978, Conf. Doc. NG7/23.

60. R.R. Churchill, "Fisheries issues in Maritime boundary delimitation", 17 *M.P.* (1993), pp. 48-49.

61. R. Lagoni, "Interim Measures Pending Maritime Delimitation Agreements", 78 *A.J.I.L.* (1984), p. 360.

62. R.R. Churchill, *op. cit.* (1993/*M.P.*), p. 45.

63. R. Lagoni, *op. cit.* (1984/*A.J.I.L.*), p. 78.

64. Thomas Wemyss Fulton, *The Sovereignty of the Sea*, William Blackwood and Sons (Edinburgh and London: 1911), pp. 606-607.

65. Thomas Wemyss Fulton, *ibid.*, p. 611.

66. *Convention between Her Majesty and the King of the French on the Limits of*

the Exclusive Rights of the Oyster and other Fishing on the Coasts of Great Britain and of France. Signed at Paris 2 August, 1839.

67. Thomas Wemyss Fulton, *op. cit.* (1911), p. 612. Only in the area in the Bay of Granville on the French coast *ad hoc* straight lines were drawn far beyond three miles.

68. Thomas Wemyss Fulton, *ibid.*, p. 614.

69. *Ibid.*

70. Thomas Wemyss Fulton, *ibid.*, p. 615. In 1867, the two countries concluded a new fisheries agreement which kept the exclusive fishery limits of the two countries the same as those in the Convention of 1839. However, the new Convention of 1867 was not ratified by the French government and its provisions, including the Article on the exclusive fishery limits, were repealed by the Sea Fisheries Act 1883: see Thomas Wemyss Fulton, *ibid.*, p. 620.

71. The *Minquiers and Ecrehos* Case, 1953 *ICJ Reports*, p. 58.

72. 1953 *ICJ Reports, idem.*

73. R. Churchill and G. Ulfstein, *Marine Management in the Disputed Areas: The* Case *of the Barents Sea*, Routledge (1992), p. 60.

74. Alex G. Oude Elferink, *The Law of Maritime Boundary Delimitation: A* Case *Study of the Russian Federation*, Martinus Nijhoff Publishers (1994), pp. 232-233,

75. Alex G. Oude Elferink, *ibid.*, pp. 146 and 229.

76. The sector line is the line of longitude running from the terminus of the existing Norwegian-Soviet frontier to the North Pole. The sector line was first introduced through a decree in 1926. The USSR proposed boundary line in the negotiation with Norway was identical with the sector line except the deviation of the line in the Svalbard area; see Alex G. Oude Elferink, *ibid.*, pp. 74-75, and p. 239.

77. R. Churchill and G. Ulfstein, *op. cit.* (1992), p. 61; the Soviet Union established a provisional 200 mile fisheries zone on 10 December 1976 through the Edict of the Presidium of the Supreme Soviet of the USSR, "On Provisional Measures for the Conservation of the Living Resources and of Regulation of Fisheries in the Maritime Areas Adjacent to the Coast of the USSR"; see W.E. Butler, "Soviet Fisheries Jurisdiction", 27 *I.C.L.Q.* (1978), pp. 442-446.

78. The sovereignty of the Svalbard and exploitation of the resources in the Svalbard area were addressed by the 1919 Peace Conference in Versailles, and the Svalbard Treaty was concluded in 1920. The treaty gave the sovereignty over the archipelago to Norway and allowing other States to have access to the Svalbard area for economic activities: see R. Churchill and G. Ulfstein, *ibid.*, pp. 64-65.

79. R. Churchill and G. Ulfstein, *ibid.*

80. Alex G. Oude Elferink, *op. cit.* (1994), p. 246; an English translation of this arrangement does not appear to have been published; the Norwegian text can be found in *Overenskomster med fremmede Stater* (Norwegian Treaty Series, 1978).

81. R. Churchill and G. Ulfstein, *op. cit.* (1992), pp. 110-130.

82. R. Churchill and G. Ulfstein, *op. cit.* (1992), pp. 66-67.

83. Note that there was some discontent in the Norway as to the Grey Zone agreement. The Atorting (the Norwegian Parliament) approved the agreement by a majority of only four votes and a Norwegian official (Mr. Treholt) who was directly involved in the negotiation with the Soviet Union was convicted as a Soviet spy and sentenced to 20 years in prison in Norway: see Oude Elferink, *op. cit.* (1994), p. 239.

84. R. Churchill and G. Ulfstein, *op. cit.* (1992), p. 113.

85. However, there are arguments that the Agreement has undermined Norway's position on the delimitation; see D. Scrivener, "Oil, Fish and Strategy: The USSR, Svalbard and the Barents Sea", *Aberdeen Studies in Defense Economics* No. 27 (1987), p. 18.

86. R. Churchill, *op. cit.* (1993/*M.P.*), p. 46.

87. R. Churchill and G. Ulfstein, *op. cit.* (1992), p. 113.

88. R. Churchill and G. Ulfstein, *op. cit.* (1992), p. 68. We will see in Part II that the issue of the management of the living resources and jurisdiction in the joint fishing zone in the East Sea was one of the difficult one in the negotiations between Korea and Japan on a fisheries agreement as Korea tried to avoid a tight management scheme in the area fearing its possible impact on the territorial sovereignty over the Dok-do.

89. R. Churchill and G. Ulfstein, *op. cit.* (1992), p. 114.

90. *Ibid.*

91. Alex G. Oude Elferink, *op. cit.* (1994), p. 242.

92. Alex G. Oude Elferink, *op. cit.* (1994), p. 238.

93. According to the Act on the Amendment of the Act concerning Fisheries Rights of 1950 of 9 June 1977, the Swedish fishing zone comprised the sea areas beyond the outer limit of the territorial sea as defined by the government.

94. Denmark became a party to the 1958 Convention on the Continental Shelf on 10 June 1964 and extended its fisheries zone beyond 12 N.M. in 1977.

95. Alex G. Oude Elferink, *op. cit.* (1994), p. 201.

96. Alex G. Oude Elferink, *idem.*

97. Leaso is 101 km^2 with 2,700 inhabitants, Anholt is 22 km^2 with 150 inhabitants, Hesselo is 0,7 km^2.

98. Utklippan is 0,12 km^2 and uninhabited.

99. Exchange of Notes, produced in *U.N.T.S.* Vol. 1088, No. 16669, pp. 205-209.

100. See R. Churchill, *op. cit.* (1993/*M.P.*), p. 46.

101. Alex G. Oude Elferink, *op. cit.* (1994), p. 206.

102. Alex G. Oude Elferink, *op. cit.* (1994), pp. 201-202

103. However, the Danish baseline connecting Hesselo to Sjaelland received only half effect; see Erik Franckx, *ibid.*, p. 1935.

104. Erik Franckx, "Report Number 10-2: Denmark-Sweden", *International Maritime Boundary*, p. 1935.

105. Erik Franckx, *ibid.*, p. 1934.

106. Erik Franckx, *idem.*

107. Act of 1977 December 1977 on the Polish Fishing Zone; see Alex G. Oude Elferink, *op. cit.* (1994), p. 195.

108. Act of 17 December 1977 concerning the Continental Shelf.

109. Alex G. Oude Elferink, *op. cit.* (1994), p. 196.

110. Chile, Ecuador and Peru met in 1952 and signed the Santiago Declaration on the Maritime Zone which they subsequently ratified and which was acceded to by Costa Rica in 1955. They proclaimed, *inter alia*, as a principle of their international policy that each of them possesses sole sovereignty and jurisdiction over the area of sea adjacent to the coast of their respective countries to a minimum distance of 200 N.M. from the said coasts; see D.J. Attard, *The Exclusive Economic Zone*, Oxford Clarendon Press (1987), pp. 7-9.

111. Kaldone G. Nweihed, "Report Number 2-13(2): Trinidad and Tobago-Venezuela", *International Maritime Boundaries*, pp. 656-658.

112. *Ibid.*

113. Kaldone G. Nweihed, "EZ (uneasy) Delimitation in the Semi-enclosed Caribbean Sea: Recent Agreement between Venezuela and Her Neighbours", 8 *O.D.I.L.* (1980), p. 17. The text of the Agreement reproduced in 2 *I.J.C.E.L.*, 1987, pp. 101-102.

114. Article 4 of the Fishing Agreement of 1985.

115. Article 2 of the Fishing Agreement of 1985.

116. Article 4 (2) of the Fishing Agreement of 1985.

117. Article 5 of the Fishing Agreement of 1985.

118. Article 3 of the Fishing Agreement of 1985.

119. The Joint Declaration is reproduced in 2 *I.J.C.E.L.* (1987), p. 102.

120. The Fishing Agreement was to be effective only for 2 years with a possibility of an extension of one year. This very limited *ratione temporis* appears to have reflected parties' very strong intention to reach an early delimitation. R.R. Churchill, *op. cit.* (1993/*M.P.*), p. 49.

121. Kaldone G. Nweihed, *op. cit.* (1980/*O.D.I.L.*), pp. 14-19.

122. Article 11 of the Agreement.

123. Jonathan I. Charney "Report Number 1-2 Addendum: Canada-France (St. Pierre and Miqueon)", *International Maritime Boundaries*, p. 399.

124. The Proces-Verbal of 1989 is reproduced in 29 *I.L.M.* (1990), pp. 7-11.

125. Article 4 of the Proces-Verbal; the Note-Verbal is reproduced in 29 *I.L.M.* (1990), pp. 12-13.

126. Canada's Note-Verbal dated 30 March 1990; reproduced in 29 *I.L.M.* (1990), p. 12.

127. France's Note-Verbal dated 30 March 1989; reproduced in 29 *I.L.M.* (1990), pp. 13-14.

128. For the award, see Jonathan I. Charney, *op. cit.* (Report No. 1-2, in *International Maritime Boundaries*); see also Charney "Progress in International Maritime Boundary Delimitation Law", 88 *A.J.I.L.* (1994), pp. 227-233.

129. Section 1 of the Fisheries Limits Act of 1976 provides that "British fisheries limits extend 200 miles from the baseline from which the breadth of the territorial sea adjacent to the United Kingdom, the Channel Islands, and the Isle of Man is measured". Nautical charts attached to the Notice to Mariners No. 2611 (1976) showed 200 mile fisheries zone around Rockall; regarding the continental shelf, the British government's position was either that Rockall was an extension of the continent upon which the United Kingdom rests or, alternatively, that Rockall had its own continental shelf; see C. Symmons, "The Rockall Dispute Deepens: An Analysis of Recent Danish and Icelandic Actions", 35 *I.C.L.Q.* (1986), pp. 164-165, and p. 261.

130. The UK government declared Rockall as British territory in 1955 and placed Rockall under the administration of Scotland by the Island of Rockall Act of 1972; see Jon M. Van Dyke *et al.*, "The Exclusive Economic Zone of the Northwestern Hawaiian Islands: When Do Uninhabited Islands Generate an EEZ", 25 *S.D.L.R.* (1988), pp. 453-454.

131. Clive R. Symmons, *op. cit.* (1986/*I.C.L.Q.*), pp. 345-346.

132. Clive R. Symmons, *ibid.*, pp. 352-355.

133. D.H. Anderson, "British Accession to the UN Convention on the Law of the Sea", 46 *I.C.L.Q.* (1997), 778-779: Article 121(3) of the LOS Convention provides that:

> Rocks which cannot sustain human habitation or economic life of their own shall have no exclusive economic zone or continental shelf.

134. UK Government, *Command Paper* No. Cm 4373.

135. UK Government, *Explanatory Memorandum on the Agreement between the United Kingdom and Denmark on the Delimitation of the Maritime Boundary between the United Kingdom and the Faroe Islands*, 14 June 1999.

136. Article 5(a) of the Agreement; *Command Paper* No. Cm 4373.

137. Article 5 of the Agreement; *Command Paper* No. Cm 4373.

138. Article 8 of the Agreement; *Command Paper* No. Cm 4373.

139. Article 6 (b) of the Agreement; *Command Paper* No. Cm 4373.

140. Article 7 of the Agreement; *Command Paper* No. Cm 4373.

141. Choon-ho Park, "Fishing under Troubled Water: The North East Asia Fishery Controversy", 2 *O.D.I.L.* (1974), pp. 93-135.

142. Article (1) of the Agreement. The Agreement was produced in English in *New Directions in the Law of the Sea*, Oceana Publication (1973), Vol. I, pp. 492-495.

143. R.R. Churchill, *op. cit.* (1993/*M.P.*), p. 47.

144. M.M. Whiteman, *Digest of International Law*, US Department of State (1965), Vol. 4, p. 1018; see also the internet home page of International Pacific Halibut Commission, <www.iphc.washington.edu/halcom/>.

145. M.M. Whiteman, *ibid.*, Vol. 4, p. 1017.

146. Article 1 of the Convention and Article 3 of the Protocol of 1979.

147. Article 2 (1) of the Convention provides that: "Every national or inhabitant, vessel or boat of the United States of America or of Canada engaged in fishing on the high sea in violation of this Convention or of any regulation adopted pursuant thereto may be seized by duly authorised officers of either Contracting Party and detained by the officers making such seizure and delivered as soon as practicable to an authorised official of the country to which such person, vessel or boat belongs, at the nearest point to the place of seizure or elsewhere as may agreed upon. The authorities of the country to which such persons, vessel or boat belongs alone shall have jurisdiction to conduct prosecutions for the violation of the provisions of this Convention . . ."

148. The Protocol is titled "The Protocol Amending the Convention Between the United States of America and Canada for the Preservation of the Halibut Fishery of the Northern Pacific Ocean and Bering Sea" and reproduced with its annex in M. Nordquist and K.R. Simmonds, eds., *New Directions in the Law of the Sea*, Oceana Pubications (1980), Vol. IX, pp. 233-249.

149. See E. Hey, *The Regime for the Exploitation of Transboundary Marine Fisheries Resources*, Martinus Nijhoff (Dordrecht, 1989), pp. 156-158.

150. Article 2 of the Annex to the Protocol.

151. Article 16 of the Annex provides that: "Pending delimitation of maritime boundaries between Canada and the United States in the Convention area, the following principles shall be applied as interim measures in the boundary region:
(a) as between the Parties, enforcement of the Convention shall be carried out by the flag State;
(b) neither Party shall authorise fishing for halibut by vessels of third parties;
(c) either Party may enforce the Convention with respect to fishing for halibut, or related activities, by vessels of third parties.

152. D. Nowell, "The Canada-United States dispute in Dixon Entrance: history, current issues and prospects for conflict resolution" in C. Grundy-Warr ed., *International Boundaries and Boundary Conflict Resolution*, University of Durham (1990), pp. 267-277.

153. *Overenskomster med fremmede Stater* (Norwegian Treaty Series, 1989) No. 904.

154. Churchill, *op. cit.* (1993/*M.P.*), p. 48.

155. It is not clear what is the regime on the enforcement jurisdiction in the area as there is no provision on the enforcement jurisdiction in the agreement among them.

156. The U.N., *National Maritime Claims* (1989).

157. Erik Franckx, "Report Number 10-9: Sweden-Soviet", *International Maritime Boundaries*, p. 2061.

158. Alex G. Oude Elferink, *op. cit.* (1994), pp. 207-208.

159. Alex G. Oude Elferink, *ibid.*, p. 207; Gotland is 3,200 km^2 in area, and is 125 km long and 52 km wide. Gotland has a population of 5,5000.

160. Alex G. Oude Elferink, *ibid.*, p. 208.

161. Alex G. Oude Elferink, *ibid.*, p. 149, and p. 208.

162. R. Churchill, *op. cit.* (1993/*M.P.*), p. 49; It is not clear from the literature whether this obligation not to take any measure which prejudices the result of

future negotiation on maritime delimitation was also intended to be applied to the delimitation of continental shelf or not.

163. Alex G. Oude Elferink, *op. cit.* (1994), p. 149.

164. R. Churchill, *op. cit.* (1993/*M.P.*), p. 49.

165. M. Fitzmaurice-Lachs, "The Legal Regime of the Baltic Sea Fisheries", 24 *N.I.L.R.* (1982), p. 205; see also the internet homepage of the IBSFC (<www.ibsfc.org/>).

166. Article 1 of the Gdansk Convention. The text of the Convention is reproduced in the internet homepage of the IBSFC.

167. Article 9 of the Gdansk Convention; M. Fitzmaurice-Lachs, *op. cit.* (1982/*N.I.L.R*), pp. 208-209.

168. See R. Churchill *op. cit.* (*1993/M.P.*), p. 49.

169. Alex G. Oude Elferink, *op. cit.* (1994), p. 213.

170. Erik Franckx, *op. cit.* (Report No. 10-9 *in international Maritime Boundaries*), pp. 2058-2061.

171. Paragraph 2 of the Agreement concerning the Delimitation of the Continental Shelf and of the Swedish Fishing Zone and the Soviet of 18 April 1988; reproduced by Erik Franckx, *ibid.*, p. 2067.

172. Erick Franckx, *Maritime Boundaries in the Baltic Sea: Past, Present and Future*, Maritime Briefing Vol. 2, No. 2, University of Durham (1996), pp. 12-16.

173. Alex G. Oude Elferink, *op. cit.* (1994), pp. 219-243.

174. Alex G. Oude Elferink, *op. cit.* (1994), pp. 192-193.

175. R. Churchill, *op. cit.* (1993/*M.P.*), p. 49.

176. Erik Franckx "Report Number 10-10: Poland-Sweden", *International Maritime Boundaries*, p. 2080.

177. U.N., *The Law of the Sea: National Claims to Maritime Jurisdiction*, 1992, pp. 45-46.

178. See Alex G. Oude Elferink, *op. cit.* (1994), pp. 181-182.

179. Alex G. Oude Elferink, *op. cit.* (1994), p. 182.

180. Bogskar consists essentially of two rocks, measuring together about 4-5 km^2 and is not inhabited.

181. Erik Franckx, "Report Number 10-3: Finland-Sweden", *International Maritime Boundaries*, pp. 1945-1952.

182. For the history of maritime delimitation between Finland and the Soviet Union, see Erik Franckx, *op. cit.* (1996/*Maritime Briefing*), pp. 6-10.

183. Alex G. Oude Elferink, *op. cit.* (1994), p. 189.

184. Erik Franckx, "The 1998 Estonia-Sweden Maritime Boundary Agreement: Lessons to be learned in the area of continuing and/or succession of States", 31 *O.D.I.L.* (2000), p. 270.

185. Erik Franckx, "Baltic Sea: Finland-Sweden Delimitation Agreement", 11 *I.J.M.C.L* (1996), p. 394.

186. Erik Franckx, "Report Number 10-3: Finland-Sweden", *International Maritime Boundaries*, pp. 1945-1957.

187. Alex G. Oude Elferink, *op. cit.* (1994), p. 189.

188. *Decision No. 163 of the Council of Ministers of the USSR on the Introduction of Provisional Measures to Protect the Living Resources and Regulate Fishing in the Areas of the Pacific and Arctic Oceans adjacent to the Coast of the USSR*. This Decision defines the boundary of exclusive fisheries zone of the USSR as follows:

> ***in the Pacific Ocean and the southern group of the Kuril Islands- the line equidistant from these islands and the territory of Japan***; in the Sovestskii and Kunashirskii Straits- the State boundary of the USSR; in the Seas of Okhotsk and Japan-the median line or the line equidistance from the coastline of the USSR and the coasts of the adjacent States (emphasis added).

189. Alex G. Oude Elferink, *op. cit.* (1994), p. 310.

190. Alex G. Oude Elferink, *idem.*

191. Alex G. Oude Elferink, *idem.*

192. Article 7 of the Agreement; Unofficial translation of the Agreement was produced in the *Japanese Annual of International Law* No. 28 (1985), pp. 297-299.

193. Statement of the Japanese government of 25 February 1977, reproduced in 28 *Japanese Annual of International Law*, p. 86. The map by the Japanese fisheries organisation for fishermen used the equidistance line as a fisheries boundary between Japan and Soviet Union.

194. Unofficial translation of the Agreement was produced in the *Japanese Annual of International Law* No. 28 (1985), pp. 297-299.

195. Newspaper.

196. Falklands Government Office, *Falklands Islands' Interim Conservation and Management Zone: Fisheries Report 87/88*, p. 5.

197. Falklands Government Office, *ibid.*, p. 7.

198. The FICMZ was established on 29 October 1986 by a Proclamation of the Falkland Islands Governor. According to the Proclamation, "the zone has (as) an inner boundary the outer limits of the territorial sea of the Falkland Islands and has (as) its seawards boundary the line formed by the circumference of a circle which has a radius of 150 nautical miles and its centre at Latitude 51° 40'S,

Longitude 59° 30' W, except that between those points in the circumference situated at 52° 30' S, Longitude 63° 19.25 'W and 54° 08.68' S, Longitude 60° 00' W seawards boundary shall be a rhumb line". (Proclamation No. 4 of 1986).

199. C.R. Symmons, "The Maritime Zones around the Falkland Islands", 37 *I.C.L.Q.* (1988), pp. 289-291.

200. According to the Falkland Islands Government, each year 250,000-300,000 tonnes of fish are taken in the FCIMZ and the fishery generates *around £20m per annum in licence fees for the government.* See Falkland Islands Government, "Fishing" at <www.Falklands.gov.fk/2e.htm>.

201. R.R. Churchill, "Falkland Island-Maritime Jurisdiction and Co-operation Arrangement with Argentina", 46 *I.C.L.Q.* (1997), p. 464.

202. R.R. Churchill, *ibid.,* p. 464.

203. M. Evans, "The Restoration of Diplomatic Relations between Argentina and the United Kingdom", 40 *I.C.L.Q.* (1991), pp. 476-481.

204. Foreign and Commonwealth Office, *Press Release No. 241* (Wednesday 28 November 1990).

205. *Ibid.*, see also M. Evans, *op. cit.* (1991/*I.C.L.Q.*), p. 481; R.R. Churchill, *op. cit.* (1997/*I.C.L.Q.*), p. 464.

206. According to Churchill, this proclamation has a good cause because "at that time of the Joint Statement, the area to which the prohibition applied was high seas and it obviously would have been very difficult, if not impossible, to justify enforcing the prohibition against foreign vessels"; see R.R. Churchill, *op. cit.* (1997/*I.C.L.Q.*), p. 464. The OFCZ was established by Falkland Islands Governor's Proclamation on 28 November 1990. The OFCZ is shaped as a half donut and surrounds the half of the FICMZ to the eastside. The OFCZ is defined by a complex set of co-ordinates. The line connecting point 8, and point 9 of the OFCZ is an arc line drawn along the line of 200 N.M. distance from the nearest points on the baseline of the territorial sea of the Falkland Islands.

207. The British government did not include this area, so-called "Gap" into the FICMZ in 1990, because it was not clear whether the area would fall within the 200-mile line from Argentina. However, it became clear in 1991 that the Gap lay outside Argentina's 200-mile exclusive zone from its baselines. The Gap is important for the conservation of fisheries resources because it serves as part of migration route of the illex squid: see R.R. Churchill, *ibid.*, pp. 465-466.

208. R.R. Churchill, *ibid.,* p. 466.

209. Falkland Islands Government, "Fishing" <www.falklands.gov.fk/e2.htm>.

210. When the United Arab Emirates was being formed in 1971, Sharjah became a part of the U.A.E along with Abu Dabi, Dubai, Umm al Qaiwan. Note that even

the 12 N.M. zone around Abu Musa was also contested by Umm al Qaywayn before the establishment of the U.A.E. The question of the legal status of Abu Musa's continental shelf arose because Umm al Qaywayn had granted concessions from its mainland up to 3 N.M. of Abu Musa, whilst Sharjah had granted concessions in the 12 N.M. zone of Abu Musa. However, immediately after the formation of the U.A.E. in 1971, Umm al Qaywayn acknowledged Sharjah's right to a 12 N.M. zone around Abu Musa; see Robert F. Pierytrwski, Jr., "Report Number 7-10: Sharjah-Umm al Qaywayn", *International Maritime Boundaries*, p. 155.

211. M.H. Hlais Neyadi. *The Maritime Zones of the United Arab Emirates with Particular Reference to Delimitation*, Ph.D. Thesis, Edinburgh University (1997), p. 261.

212. M.H. Hlais Neyadi, *ibid.*, pp. 261-262.

213. M.H. Hlais, Neyadi, *idem.*

214. A.A. El-Hakim, *The Middle Eastern States and the Law of the Sea*, Manchester University Press (1979), pp. 127-128.

215. A.A. El- Hakim, *op. cit.* (1979), pp. 122-128. C.G. MacDonald, *Iran, Saudi Arabia, and the Law of the Sea: Political Integration and Legal Development in the Persian Gulf*, Greenwood Press (Connecticut USA, 1980) pp. 131-134.

216. Robert F. Pierytrwski, Jr., *op. cit.* (Report Number 7-10 in International Maritime Boundaries).

217. Article 1 of the Memorandum of Understanding; for the text of the Memorandum, see *Middle East Economic Survey* Vol. 15, No. 28, Supplement (5 May 1972).

218. Article 2 of the Memorandum of Understanding. There was a failed attempt by the Iranian side to exercise its jurisdiction in the waters under the U.A.E. jurisdiction in 1992: see Matar Hamed Hlais Al Neyadi, *op. cit.* (1997/Ph.D. Thesis), p. 263.

219. Article 4 of the Memorandum of Understanding provides that "Exploitation of petroleum resources of Abu Musa and of the seabed and subsoil beneath its territorial sea will be conducted by Buttes Gas and Oil Company under the existing agreement, which must be acceptable to Iran. Half the oil revenues hereafter attributable to the said exploitation shall be paid direct by the Company to Iran and half to Shajah. The Memorandum of Understanding is reproduced in Ali A. El-Hakim, *op. cit.* (1979), pp. 208-211.

220. Honduras and Colombia concluded a maritime boundary agreement in 1986, whereby part of Serranilla Bank came to belong to the maritime area of Honduras. However, the agreement stirred up a controversy over the status of Serranilla Bank and Bajo Neuvo in Honduras. As the time of writing, the Agreement has yet to enter into force because the Honduras Government has not got approval from the Legislature.

221. K.G. Nweihed, "Report Number 2-18: Colombia-Jamaica", *International Maritime Boundaries*, pp. 2179-2198.

222. K.G. Nweihed, *ibid.*

223. The treaty does not mention in explicit terms that the boundary is an EEZ boundary. However, from the provisions of the text, we can understand that the boundary is about an EEZ. The text of the Agreement is reproduced in K.G. Nweihed, *ibid.*, pp. 2179-2198.

224. Article 3(2) of the Colombia-Jamaica Treaty.

225. Article 3(4) of the Colombia-Jamaica Treaty.

226. It is ironic that Jamaica which had opposed to the institution of the 200 N.M. maritime zone advancing the concept of "matrimonial sea" as opposed to the concept of "patrimonial sea" and had voted against the Santo Domingo Declaration in 1972, now established a JRA which is closed to third States.

227. See Article 62 of the LOS Convention.

228. See Article 249 of the LOS Convention.

229. Article 3(3) of the Colombia-Jamaica Treaty.

230. Article 4(1) of the Colombia-Jamaica Treaty.

231. Article 4(2) of the Colombia-Jamaica Treaty.

232. Article 4(3) of the Colombia-Jamaica Treaty.

233. The agreement between Portugal and France was done in a form of Exchange of Notes dated 23 August 1960. The Exchange of Notes is reproduced by A.O. Adede, "Report Number 4-4: Guinea-Bissau-Senegal", *International Maritime Boundaries*, pp. 872-874.

234. A.O. Adede, *ibid.*, pp. 865-866.

235. A.O. Adede, *ibid.*, p. 869.

236. 20 *R.I.A.A.* (1989) pp. 153-160.

237. A.O. Adede, *op. cit.* (Report No. 4-4 in *International Maritime Boundaries*), p. 868.

238. A.O. Adede, *idem.* Notice that the Agreement of 1960 mention the territorial waters, contiguous zone and the continental shelf but not the EEZ.

239. 1991 *ICJ Reports*, paras. 19, 30. On 23 August 1989, Guinea-Bissau filed an application in the Registry of the International Court of Justice regarding the validity of the arbitral award of 1989. Pending the decision of the ICJ, Guinea-Bissau filed second application, on 12 March 1991, arguing that "all the maritime territories appertaining respectively be delimited". The Court rejected the application in November 1991.

240. J.R.V. Prescott, "Report No. 4-4(4) and (5): Guinea-Bissau-Senegal", *International Maritime Boundaries* (vol. III), pp. 2251-2255.

241. The text of the Agreement and Protocol are reproduced in R.V. Prescott, *idem.*

242. Article 4 of the Agreement.

243. Article 8 and 9 of the Agreement.

244. Article 24 of the Protocol.

245. R.V. Prescott, *op. cit.* (Report No. 4-4(4) and (5) in *International Maritime Boundaries*), p. 2253.

245. Masahiro Miyoshi, *op. cit.* (1999/*Maritime Briefing*), p. 40.

247. Article 5 of the Protocol of Agreement Relating to the Organisation and Operation of the Agency for Management and Co-operation of 12 June 1995; partly reproduced in M. Miyishi, *op. cit.* (1999/*Maritime Briefing*), p 40.

248. Article 6 of the Protocol.

249. Ali A. El-Hakim, *op. cit.* (1979), p. 185.

250. Ali A. El-Hakim, *idem.*

251. Article 7 and 14 of the Saudi Arabia-Sudan Agreement of 16 May 1974; partly reproduced in M. Myoshi, *op. cit.* (1999/*Maritime Briefing*), pp. 32-33.

252. Ali A. El-Hakim, *op. cit.* (1979), p. 186.

253. Ali A. El-Hakim, *op. cit.* (1979), p. 187.

254. Victor Prescott, "Report Number 6-2(2): Australia-Indonesia (Timor and Arafura Seas)", *International Maritime Boundaries*, p. 1213.

255. Max Herriman and Martin Tsameny, "The 1997 Australia-Indonesia Maritime Boundary Treaty: A Secure Legal Regime for Offshore Resource Development?", 29 *O.D.I.L* (1998), pp. 361-396.

256. R.V. Prescott, "Report Number 6-2(4): Australia-Indonesia (Fisheries)", *International Maritime Boundaries*, p. 1232.

257. R.V. Prescott, *idem.*

258. The MOU is reproduced in Victor Prescott, *ibid.*, pp. 1238-1243.

259. The term "swimming species" was used *vis-à-vis* the term "sedentary species". For the sedentary species, they agreed to apply the existing seabed boundaries between them, and in the areas where there were no sea-bed boundaries, they agreed to consult as necessary with a view to avoiding difficulties arising between them with regard to the exercise of their respective jurisdiction; see Article 5 of the MOU.

260. In the PFSEL, Ashmore has a 12 N.M. arc and later in the exclusive economic zone boundary it has 24 N.M. arc.

261. The *Gulf of Maine* Case, 1984 *ICJ Reports*, para. 240.

262. 1984 *ICJ Reports*, para. 240.

263. Agreement between the UK and Norway Relating to the Delimitation of Continental Shelf, 551 *U.N.T.S.* (1965), p. 214.

264. 1982 *ICJ Reports*, p. 320.

265. The *Tunisia/Libya Continental Shelf* Case, 1982 *ICJ Reports*, p. 320.

266. Second Article of Bahrain-Saudi Arabia Boundary; the text is reproduced in R. Pietrowski Jr., "Report Number 7-3: Bahrain-Saudi Arabia", *International Maritime Boundaries*, pp. 1495-1497.

267. Paragraphs 7 and 9 of the Bahrain-Saudi Arabia Agreement; for the text, see R. Pietrowski Jr., *idem.*

268. The text of the Convention on the Delimitation of the Continental Shelf in the Bay of Biscay and Convention on the Delimitation of Territorial Sea and Contiguous Zone in the Bay of Biscay are reproduced in D.H. Anderson, "Report Number 9-2: France-Spain", *International Maritime Boundaries*, pp. 1728-12734.

269. Annex II to the Convention on the Continental Shelf Boundary; for text, see D.H. Anderson, *ibid.*

270. Article 4 of the Convention on the Continental Shelf Boundary.

271. E.L. Richardson *et al.*, "Report and Recommendations to the Governments of Iceland and Norway of the Conciliation Commission on the Continental Shelf Area between Iceland and Norway", 20 *I.L.M.* (1981), p. 791.

272. Report and Recommendations to the Governments of Iceland and Norway of the Conciliation Commission on the Continental Shelf Area between Iceland and Jan Mayen, 20 *I.L.M.* (1980), pp. 825-826.

273. Report and Recommendation, *idem.*

274. D.H. Anderson, "Report No. 9-4: Iceland-Norway (Jan Mayen)", *International Maritime Boundaries*, pp. 1756-1760. In 1980, Norway (Jan Mayen) and Iceland agreed on the EEZ boundary and also agreed to refer the continental shelf boundary question to the Conciliation Commission.

275. Article 8 of the Agreement on the Continental Shelf between Iceland and Jan Mayen. The text of the Agreement is reproduced in D.H. Anderson, "Report Number 9-4: Iceland-Norway (Jan Mayen)", *International Maritime Boundaries*, pp. 1762-1765.

276. He stated that "My starting point for such a proposal would be an adjusted equidistance line starting at the point where the 26° line from Ras Ajdir intersects the 12-mile territorial sea limit. I venture to propose an adjusted line of delimitation veering from this point in a direction some 46°-47° (north-east). In addition, the following system of joint exploitation of petroleum resources may be indicated. On both sides of the straight line a line veering some 10-15° from the delimitation should be drawn. The area thus indicated should be of approximately the same size. The two areas thus indicated should constitute a joint exploration. For this joint exploitation zone, the Parties should establish a joint policy of exploration and exploitation". 1982 *ICJ Reports*, p. 321.

277. H. Fox *et al.*, eds., *op. cit.* (1989), p. 64.

278. Tulio Scovazzi, "Report Number 8-7(2): Italy-Yugoslavia", *International Maritime Boundaries*, pp. 1640-1641.

279. Jimenez de Arechaga, "Argentina-Uruguay (Report Number(3-2)", *International Maritime Boundaries*, pp. 757-762.

280. Article 73 of the Maritime Boundary Agreement; for the text of the agreement, Jimenez de Arechaga, *ibid.*, pp. 764-776.

281. Tullio Scovazzi and G. Francalanci, "France-Italy (Report Number 8-2)", *International Maritime Boundaries*, pp. 1571-1576.

282. Jimenez de Arechaga, "Report Number 3-5: Chile-Peru", *International Maritime Boundaries*, p. 795.

283. David Colson, "The Legal Regime of Maritime Boundaries", *International Maritime Boundaries*, p. 60.

284. K.G. Nweihed, "Report Number 2-2: Colombia-Dominican Republic", *International Maritime Boundaries*, pp. 479-480.

285. The text of the Agreement is reproduced by K.G. Nweihed, *ibid.*, pp. 488-490.

286. In these numbers, the common oyster fisheries zone established in 1839 between the UK and France and agreements in North East Asia are not counted.

CHAPTER FOUR

Troubled Waters in North East Asia

1. INTRODUCTION

In Chapter 1 we discussed what makes delimitation difficult and which law governs in the absence of maritime boundaries. Then we examined some of the possible provisional arrangements in the disputed areas.

Now, I will examine the provisional maritime regime in North East Asia where the delimitation of maritime boundaries between the littoral States is elusive and the provisional maritime regime is unstable. In the discussion the focus will be on the legal relations between South Korea, China and Japan, although North Korea and Taiwan will be also mentioned when such a need arises.

The complicated legal disputes in North East Asia became more complicated after the three countries ratified the LOS Convention in 1996 and then subsequently proclaimed their respective exclusive economic zones. On top of that, the adoption of straight baselines for measuring the breadth of the territorial seas by Japan and China in 1996 further complicated the legal disputes on the law of the sea in the region. Territorial disputes over islands in the region are also factors that contribute to the complexity of the legal disputes in the region. In this regard, as the three coastal States are traditionally reluctant to submit their disputes to a third party institution to resolve them, the prospect of resolving the disputes between the coastal States on the law of the sea through a third party institution appears to be very unlikely.

For a discussion on the provisional maritime orders regime, I will examine the unilateral maritime claims in the region, disputes regarding the unilateral claims, delimitation disputes between the coastal States, the factors that make the delimitation of maritime boundaries difficult, and characteristics of the provisional maritime regime in the region that is unstable and requires further development. I will also discuss possible solutions to the disputes in the region and suggest ways for further development of the maritime regime in the region.

2. NATIONAL CLAIMS IN THE REGION

2.1. *The Republic of Korea (South Korea)*

2.1.1. *The Territorial Sea*

In 1910, Korea was placed under the Japanese colonial rule and was then liberated in 1945 at the end of World War II.[1] The southern part of the Korean peninsula was placed under the control of the United States Army Military Governor who headed the "South Korean Interim Government" until 1948 when the independent government of the Republic of Korea was established in the southern part of the Korean peninsula.[2] The legislation of South Korea that first mentioned the "territorial waters" is the "Ordinance on the Functions of the Korean Coast Guard" promulgated on 10 May 1948 by the South Korean Interim Government.[3] The Ordinance defined the extent of the territorial sea of South Korea. Paragraph (b) of section III of the Ordinance provided that:

> Territorial waters include the ports, harbours, bays and other enclosed areas of the sea along the coast of Korea and a marginal belt of the sea extending from the coastline outward one marine league, or three geographical miles.

It is understood that the three miles was adopted for the extent of the Korean territorial waters by the United States Military Governor in Korea because the Governor's own country, the United States, had adopted three mile limits for her own territorial seas.[4]

Although the South Korean Interim Government led by the US military Governor was replaced by the independent government of the Republic of Korea in 1948, most of the laws enacted by the interim government remained valid in the independent Korean government until 20 January 1962.[5] The "Ordinance on the Functions of the Korean Coast Guard" by the US Governor was one of the Ordinances which survived until 20 January 1962 and thus the extent of the territorial seas of South Korea was three N.M. from 1948 to 1962. However, the extent of the territorial waters of the Republic of Korea became blurred on 20 January 1962 when the

Ordinance on the Function of the Korean Coast Guard was abolished, without being replaced with new legislation, by the "Act on Special Measures regarding Old Laws and Regulations" along with other many Ordinances. This vagueness to the extent of the territorial waters of South Korea lasted 16 years until 31 December 1977 when the "Act on the Territorial Seas", which specifies the extent of the Korean territorial waters, was promulgated.[6]

The Territorial Sea Act of 1977 and its Enforcement Decree of 1978 provide twelve miles for the extent of the Korean territorial waters generally, but provides only three N.M. for the extent of the territorial waters in the Western Channel of the Korea Strait, where the coasts of Korea and Japan are facing each other with only 22.75 N.M. distance at the narrowest points. Article 1 of the Act provided that: "The territorial sea of the Republic of Korea shall be the area of the sea up to a limit not exceeding twelve N.M. from baselines: Provided that the breadth of the territorial sea in specified area shall be otherwise determined within the limit of twelve N.M. in accordance with the Presidential Decree". Following Article 1 of the Act, the Enforcement Decree of the act contains special provisions for the extent of the territorial waters in the Western Channel of the Korea Strait. The main reason for the adoption of the three-mile extent in the Korea Strait is that the Korea Strait is used for international navigation.[7] Article 3 of the Enforcement Decree is clear on this point. It provides that: "In accordance with the provisions of the provision of Article 1 of the Act, the territorial sea in the seas forming ***the Korea Strait used for international navigation*** shall be the area of the sea on the landwards side of the line joining the lines. . .". (emphasis added)

In this regard, it appears that the Korean government also took into account the United States' concern on leaving a high seas corridor in the Korea Strait.[8] However, as the LOS Convention is now in force and Korea is a Party to it, questions can arise as to whether Korea is willing to extend the extent of its territorial waters up to 12 miles as it is entitled to a maximum of 12 mile territorial waters. Note that even if Korea extends its territorial sea to 12 miles in the Korea Strait, it would not cause real inconvenience for other States in terms of navigation. It is because other States now under the LOS Convention, unlike under the Geneva Convention regime, can enjoy "transit passage" in straits used for international navigation where there are no high seas left. But, the possibility seems very high that Korea will prefer to keep the high sea /EEZ corridor in the Korea Strait because if there is no high sea/EEZ corridor left in the Korea Strait then the other States are entitled to transit passage in the whole area of the Strait except

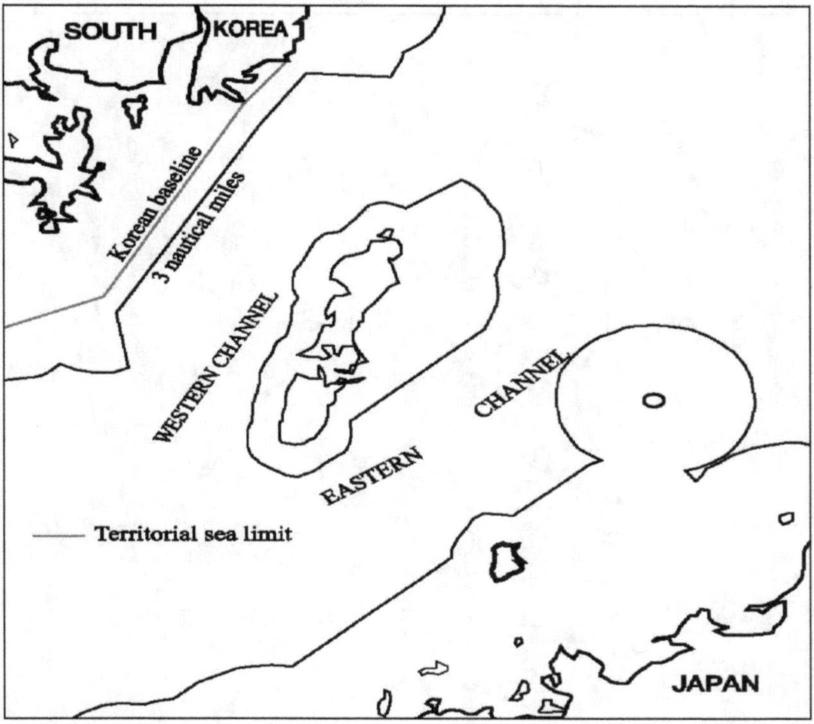

Map 21: The Korea Strait

the internal waters in the Korea Strait. Note that the regime of transit passage is not applicable to the straits through which there is a high seas route, or a route through an exclusive economic zones according to Article 36 of the LOS Convention.[9] Note also that under the current regime in the Korea Strait, Korea can keep the territorial sea belt where only innocent passage is applicable and encourage the ships of other countries to take routes through the high seas/EEZ corridor rather than the territorial sea.[10]

2.1.2. *Baselines*

The Territorial Sea Act of 1977 adopted straight baselines as well as normal baselines for measuring the breadth of the territorial sea of Korea. Article 2 of the Act provides that:

> (1) The normal baselines for measuring the breadth of the territorial sea shall be the low-water line along the coast as marked on large-scale charts officially recognised by the Republic of Korea.

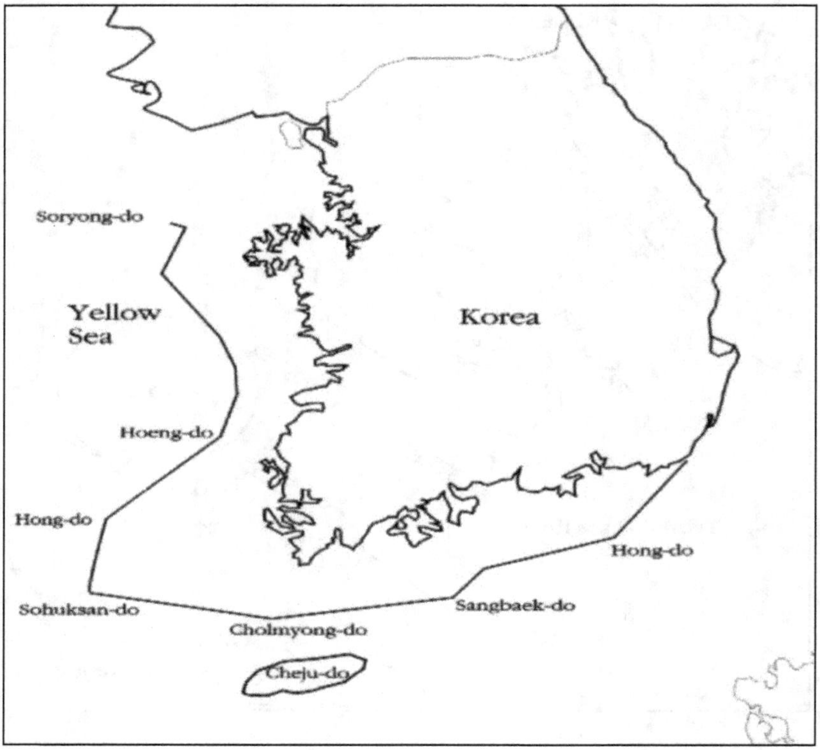

Map 22: Korea's Straight Baselines

> (2) In the area of the sea where special geographical circumstances exist, straight lines joining points as provided for in the Presidential Decree may be employed.

Following Article 2 of the above-mentioned act, the Enforcement Decree specified 19 points selected from the islands and rocks off the Korean peninsula in the south and the west. Along with the straight baselines in the south and the west of the Korean peninsula, the Korean government drew closing lines for two small bays in the East Sea. It is to be noted that many of the straight baseline points under the Territorial Sea Act of 1977 are identical with the points chosen for the measurement of the twelve-mile fisheries zone of Korea under the fisheries agreement between Korea and Japan signed in 1965.[11] We will discuss in detail the fisheries agreement between Korea and Japan of 1965 in the next chapter.

With regard to the validity of the straight baselines of Korea, none of Korea's neighbouring countries, i.e. Japan and China have challenged them so far. When the Korean government was preparing the legislation on the territorial seas, Japan, UK, and the US showed their wishes through diplomatic notes that straight baselines were not to be drawn to connect the Korean peninsula and the island of Cheju. In particular, the Japanese government asked for consultation on this matter with the Korean government and some consultation was carried out between the two governments. The Korean government did not draw straight baselines to connect the island of Cheju and the Korean peninsula.[12]

It is generally understood that the geographical situation of the Korean coastline in the south and west are deeply indented and fringed with islands, and thus meet the basic conditions for drawing straight baselines provided for in Article 7 of the LOS Convention.[13] However, the fact that the coast is deeply indented and cut into and fringed with islands does not mean that any points chosen from the coast for drawing straight baselines would meet the other conditions provided for in Article 7 of the LOS Convention. For example, Article 7 of the LOS Convention provides that straight base lines should join the "appropriate point" and that "the drawing of straight baselines must not depart to any appreciable extent from the general direction of the coast". Reisman and Westerman noted that the southern part of the Korean peninsula is generally deeply indented and cut into and fringed with islands, but they expressed a critical view on the way some of the Korea straight baselines are drawn.[14] The US State Department expressed similar views on the Korean straight baselines.[15]

2.1.3. *The Peace Line*

On 18 January 1952 the President of the Republic of Korea declared Korea's sovereignty over the waters and the continental shelf adjacent to the Korean peninsula through the "Proclamation of Sovereignty over the Adjacent Seas".[16] Although the Proclamation used the term "sovereignty", the Proclamation is generally somewhat similar to that of the Truman Proclamations of 1946. The Proclamation purported to forestall the foreseen influx of Japanese fishing vessels into the waters around the Korean peninsula as the MacArthur Line which restricted the fishing activities of Japanese vessels to the designed area around Japan was then expected to be lifted soon.[17] Although the Proclamation used the term "sovereignty" rather than "sovereign right", it is clear from the context of the Proclamation that its purpose was not to declare extended territorial waters but to declare limited rights for preserving and utilising the natural resources in, on or under the seas adjacent to Korea.[18] The Proclamation specified the zone where it was to be applied. The outer limits of the zone were to reach as far as 150 N.M. from the coast of Korea.

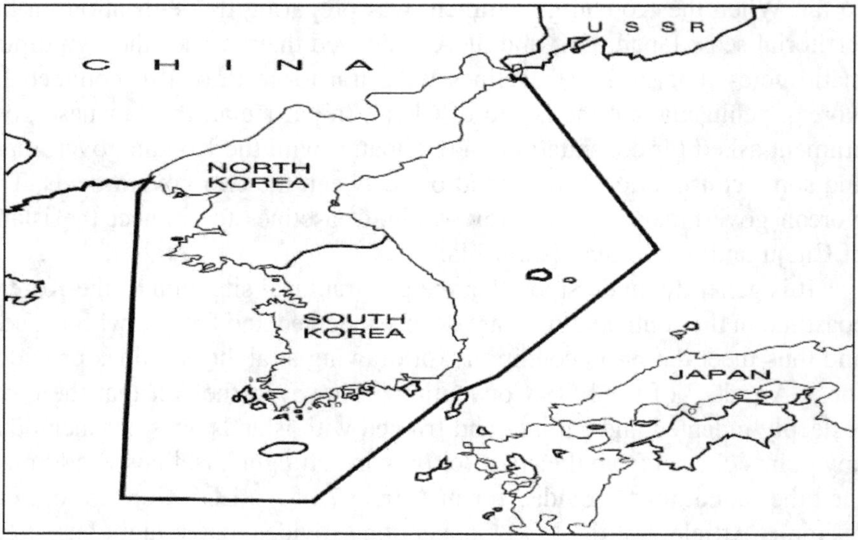

Map 23: The Peace Line

The Japanese government has strongly challenged the legality of the Peace Line.[19] The zone established by the Proclamation came to be called "the Peace Line" because Dr. Rhee, then the Korean President, argued in response to Japan's protests that the zone was intended to maintain peace between Korea and Japan.[20] In 1954 Korea enacted the Law on Conservation of Fisheries Resources with a view to implementing the Proclamation.[21] As the Korean coast guard seized a number of Japanese fishing vessels found fishing within the Peace Line, the disputes between Korea and Japan regarding the Peace Line escalated.[22] However, the dispute between Korea and Japan regarding the so-called Peace Line was shelved in 1965 when Korea and Japan concluded a fisheries agreement, whereby both countries recognised the rights of either Party to establish an exclusive fisheries zone up to 12 N.M. from the coasts and also guaranteed the freedom of fishing beyond the outer limit of the exclusive fisheries zones, and thus the Peace Line became inapplicable to Japanese fishermen.[23]

2.1.4. *The Continental Shelf*

The above-mentioned President's Proclamation of 18 January 1952 also claimed the Republic of Korea's rights over the continental shelf although the primary purpose of the Proclamation was to prevent Japanese fishermen from fishing in the waters around the Korean peninsula. Paragraph 1 of the Proclamation stated that:

> The Government of the Republic of Korea holds and exercises the national sovereignty over the shelf adjacent to the peninsula and insular coasts of the national territory, no matter how deep it may be, protecting, preserving and utilising, therefore, to the best advantage of national interests, all the natural resources, mineral and marine, that exist over the said shelf, on it and beneath it, known, or which may be discovered in the future.

However, the Korean government has never enforced the provisions of the Proclamation on the continental shelf. Later in 1970 the Korean government legislated the "Submarine Mineral Resources Development Act" with a view to exploiting petroleum and natural gas existing under the waters adjacent to the coasts of the Korean peninsula.[24] Following the Act its Presidential Decree designated the so-called "submarine mineral exploitation area" in 1971. Korea's continental shelf claim that was embodied in the "submarine mineral exploitation area" overlapped with the continental shelf claims made by Japan and Taiwan.[25]

In 1974, Korea and Japan concluded two treaties on the continental shelves to solve the disputes: one is on delimitation of the boundary of the continental shelf in the East Sea and the other is on joint development of the continental shelf in the East China Sea. We will discuss the two treaties in detail at a later stage.

2.1.5. *The EEZ*

The Korean government promulgated the "Exclusive Economic Zone Act" on 8 August 1996. Generally, the provisions of the act follow relevant provisions on EEZ in the LOS Convention on the Law of the Sea. Article 1 of the said act makes it clear that the EEZ, which is to be established by the act, is the one that is referred to in the LOS Convention on the Law of the Sea. Article 1 of the act provided that "The Republic of Korea shall establish an exclusive economic zone referred to in the United Nations Convention on the Law of the Sea under this Act".

With regard to the boundary of EEZ, the Act on the Exclusive Economic Zone provides that "the boundary shall be established by agreement with the relevant States on the basis of international law".[26] However, paragraph 2 of Article 5 of the act implies that the Korean government would use a median line as a geographical limit in the exercising of rights in the absence of agreed delimitation. It provides that:

> The rights of the Republic of Korea in the exclusive economic zone referred to in the provisions of Article 3, unless otherwise agreed upon between the Republic of Korea and the States concerned, shall not be

> exercised in the area of the sea beyond the median line between the Republic of Korea and the States concerned . . .[27]

Even if paragraph 2 of Article 5 of the act employs prohibitive language, it in fact hints that the Republic of Korea might exercise its EEZ rights in the waters within the median line pending an ultimate EEZ delimitation with its neighbouring States. There is doubt as to whether this provision of paragraph 2 of Article 5 can be justifiable in the light of the provision of paragraph 3 of Article 74 of the LOS Convention. It is to be recalled that we have noted in an earlier discussion in Chapter 2 that the LOS Convention, unlike the Geneva Convention on the Continental Shelf, does not contain provisions for justification of a unilateral median line pending an ultimate delimitation of the EEZ/continental shelf boundary.

It is also notable that Korea has designated, through the "Law on the Exercise of Sovereign Rights with respect to Foreigners' Fishing in the Republic of Korea's EEZ, Special Prohibition Zones in its EEZ" (the so-called "EEZ Fisheries Law"), where fishing of foreign fishing vessels are not permitted at all.[28] Article 4 of the EEZ Fisheries Law provides that: "Foreigners shall not conduct fishing activities in the zones which are to be designated by the President for the conservation of fisheries and restructuring of fisheries industries". And Article 5 of the EEZ Fisheries Law provides that: "Those foreigners who wish to conduct fishing activities in the Exclusive Economic Zone, except in the Special Prohibition Zones, must obtain permits per vessel issued by the Minister of the Ministry of Maritime Affairs and Fisheries".[29] The Special Prohibition Zones were established in the East Sea, the Yellow Sea and in the western Channel of the Korea Straits.

The Special Prohibition Zone in the western Channel of the Korea Strait extends from the Korean coast up to 12 miles or to the median line with the Japanese coasts. The Special Prohibition Zone in the western Channel of the Korea Strait is justifiable in light of the fact that the extent of the territorial sea in the western Channel of the Korea Strait is only 3 miles and Korea has a right, under its title to the territorial sea, to exclude foreigner's fishing in the zone within 12 miles from the Korean coasts. However, it is notable that the Special Prohibition Zone in the East Sea extends as far as 120 miles from the Korean coast and the zone in the Yellow Sea extends as far as 60 miles from the Korean coast.

It is doubtful whether the total prohibition of foreigners fishing in the Special Prohibition Zones in Korea's EEZ in the Yellow Sea and the East Sea is justifiable under Korea's sovereign right based upon the LOS Convention.

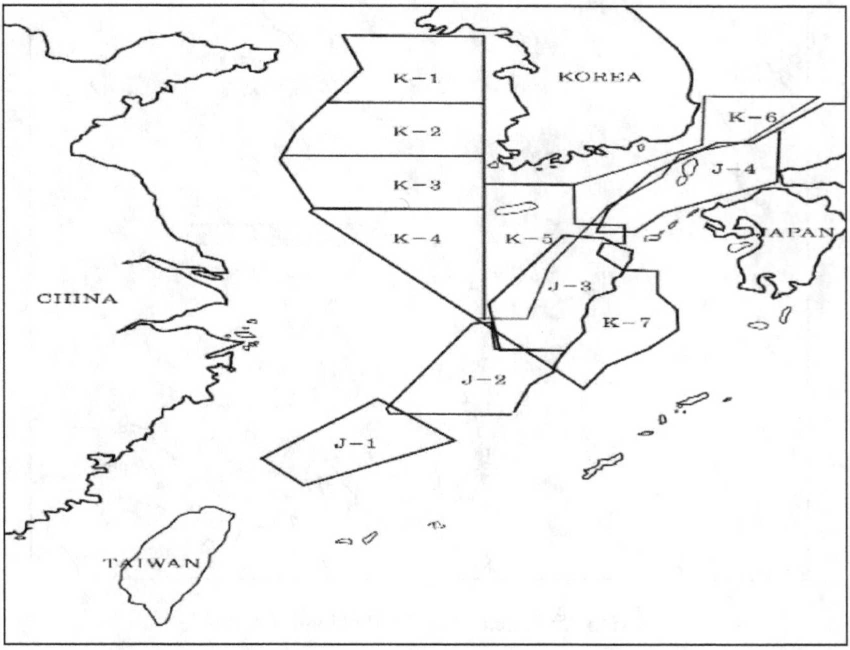

Map 24: Overlapping Continental Shelf Claims of Korea and Japan in 1970s

2.2. *Japan*

2.2.1. *The Territorial Sea*

Japan traditionally has kept a 3-mile territorial sea since the 19th century.[30] The Japanese official position on the extent of its territorial waters was first shown in 1870 when the Japanese government made a proclamation of neutrality during the Franco-Prussian war indicating that it kept a 3-mile distance for its territorial sea.[31] Since then, Japan kept the 3-mile territorial waters until 1977. In 1977 Japan enacted the Law on the Territorial Sea thereby extending its territorial sea to 12 miles.[32] However, Japan kept the breadth of its territorial seas 3 miles at its major five straits used for international navigation.[33] The Law on the Territorial Sea provides for the median line for the delimitation of the territorial waters. Article 1 of the law provides that:

> The territorial sea of Japan comprises the areas of the sea extending from the baseline to the line twelve miles seawards thereof. Provided that, where any part of that line as measured from the baselines beyond the median line, the median line (or the line which may be agreed upon

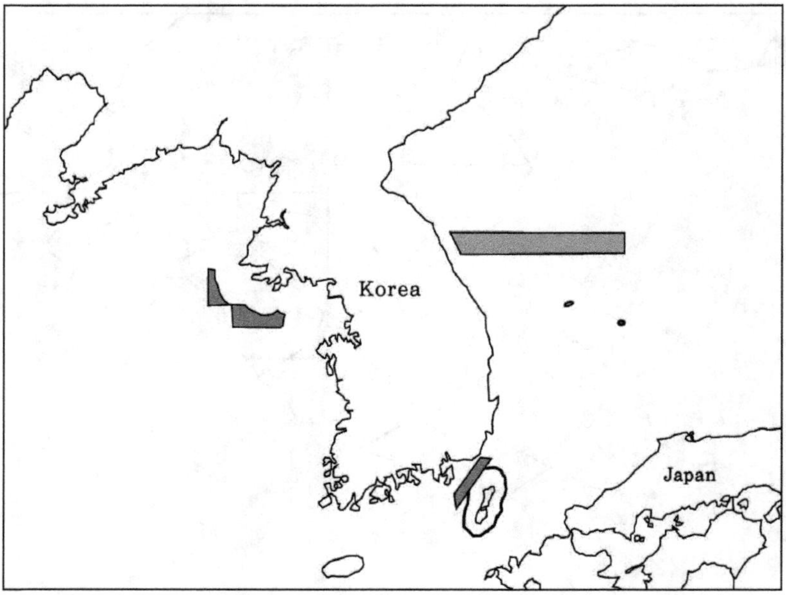

Map 25: Korea's Special Prohibition Zones

between Japan and a foreign country as a substitute for the median line) shall be substituted for that part of the line.

From the provision of Article 1 of the Japanese law on the territorial sea we can see that the boundary of Japan's territorial waters is to be automatically the median line for the area where there are overlapping claims unless there is an agreement with its neighbouring States on delimitation. As the LOS Convention allows room for unilateral application of the median line for the demarcation of territorial seas, the above-mentioned provision seems somewhat justifiable under the provisions of the LOS Convention. Article 15 of the LOS Convention provides that: "Where the coasts of two States are opposite or adjacent to each other, neither of the two States is entitled, failing agreement between them to the contrary, to extend its territorial sea beyond the median line every point of which is equidistant from the nearest points on the baselines from which the breadth of the territorial seas of each of the two States is measured".

2.2.2. *Baselines*

The Japanese Law on the Territorial Sea of 1977 adopted the method of straight baselines along with normal baselines for measuring the breadth of the territorial sea. However, the Enforcement Order of the law of 1977 relied in principle on the method of normal baselines and exceptionally adopted the straight baselines only in Seto Naikai and in five straits which are used for international navigation.[34] In 1996, however, Japan adopted straight baselines for measuring the breadth of the territorial sea for most parts of its coasts. The Enforcement Order of 1996 of the Law on the Territorial Sea and the Contiguous Zone provided 162 straight baselines connecting a total of 194 base points.[35]

The United States Department showed very critical views on the legality of the Japanese straight baselines. It commented that: "Japan's coastline in many locations does not meet the LOS Convention's geographical conditions required for applying straight baselines. And, for the most part, the waters enclosed by the new straight baseline system do not have the close relationship with the land, but rather reflect the characteristics of the territorial sea or high seas".[36]

The Japanese straight baselines adopted in 1996 sparked off a diplomatic row with Korea as the Korean government challenged the legality of some of the Japanese straight baselines in light of the relevant provisions of the LOS Convention. Korea also argued that prior consultation for drawing the straight baselines had not been met in accordance with Article 1 of the Fisheries Agreement of 1965 between Korea and Japan and thus the newly adopted straight baselines could not be applied to fisheries relations between Korea and Japan. As the Japanese authorities arrested a number of Korean fishing vessels applying the newly adopted straight baselines, the dispute between Korea and Japan escalated.[37]

The main point here is whether the new straight baselines adopted by the Japanese government can be applicable to Korean fishermen. Articles 1 and 4 of the Fisheries Agreement of 1965 contain provisions relevant to our analysis here. Korea and Japan agreed in Article 1 that they recognise the rights of each other to establish a 12 miles fisheries zone where each Party has exclusive jurisdiction with respect to fisheries *vis-à-vis* the other Party and then went on to agree that "in case either High Contracting Party uses the straight baseline for the establishment of its fishery zone, the straight baseline shall be determined through consultation with the other High Contracting Party".[38] And the two countries agreed in Article 4 that, "the rights of control (including the right to halt and inspect vessels and jurisdiction) in waters outside the exclusive fisheries zone shall be exercised only by the High Contracting Party to which the ships belong".

In this regard, Japan argued that the provisions in Article 1 of the agreement is only about the straight baseline for the fisheries zone and not for

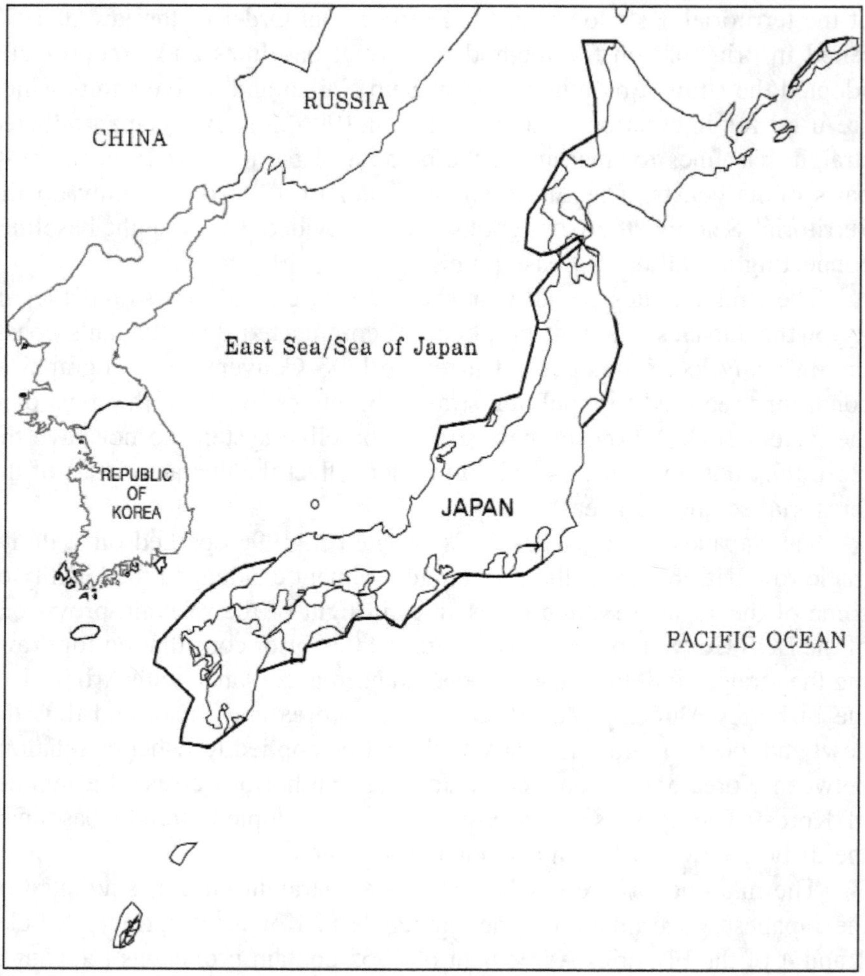

Map 26: Japan's Straight Baselines

the territorial sea, and thus the need for prior consultation envisaged in the provisions did not arise when Japan adopted straight baselines for measuring the breadth of the territorial waters.[39]

The question as to whether the new straight baselines adopted by the Japanese government can be applicable to Korean fishermen actually arose in Japanese domestic courts. The Hamada branch of the Matsue District in Shimane Prefecture of Japan, when faced with the question of whether the Japanese authorities can exercise enforcement jurisdiction over a Korean fishing vessel, Daodong-Ho No. 999, ruled on 15 August 1997 that the

Japanese authorities has no jurisdiction over the Korean vessels' fishing activities in the Japan's newly expanded waters.[40]

As the disputes persisted and the arrest of Korean fishing vessels continued, several rounds of talks were held between the two governments in 1997 and 1998. The talks, however, were not successful in resolving the disputes. Although the Fisheries Agreement of 1965 contains a compulsory arbitration clause, neither Korea nor Japan has asked for arbitration.[41] Japan notified its intention to terminate the Fisheries Agreement on 23 January 1998 and thus the agreement was terminated a year after the notification in accordance with Article 10 of the agreement. Now it remains to be seen whether the Korean government wishes to refer to arbitration under the arbitration clause, with regard to the dispute that arose under the Fisheries Agreement of 1965, which was terminated.[42]

2.2.3. *The Continental Shelf and the EEZ*

Japan denied even in the mid-1980s the continental shelf regime embodied in the Geneva Convention on the Continental Shelf because it feared the regime might negatively affect its sedentary species fisheries overseas.[43] Japan therefore never ratified the Geneva Convention on the Continental shelf. However, with regard to mineral resources, the Japanese government began to recognise the coastal State right over the continental shelf in the late 1960s. Since then the Japanese government indirectly, without specific legislation on the continental shelf, has laid claims to the continental shelf by way of recognising and approving applications for mining by Japanese companies in the continental shelf around Japan.[44] In 1969 four Japanese oil companies applied mining rights in four blocks, which were delineated by Japanese companies in the East China Sea and the Japanese government approved the mining rights by Japanese oil companies in the four blocks.[45] In this way, the mining areas of Japan designated by companies have increased in number and in total size.[46]

In 1977, Japan proclaimed a 200-mile exclusive fisheries zone.[47] It is generally understood that the Japanese proclamation of the 200-mile exclusive fisheries zone was prompted mainly to match its position with the Soviet Union, which proclaimed such a zone in the previous year.[48] However, Japan did not apply its exclusive fisheries zone regime to the nationals of Korea and China because of its fisheries agreements with them, which provided the freedom of fishing in the area beyond certain distant from the shore.[49]

In 1996 Japan proclaimed an exclusive economic zone and a continental shelf through the Law on the Exclusive Economic Zone and the Continental Shelf.[50] The law came into effect on 20 July 1996, on the same day when the LOS Convention came into effect for Japan.[51]

Map 27: Japan's Continental Shelf Claims
(Source: Japanese Ministry of Industry and Trade)

The law makes it clear that the sovereign rights and other rights of Japan in its exclusive economic zone and continental shelf are based upon the relevant provisions of the LOS Convention.[52] However, as Japan was not in a position to apply the fisheries regime in the law of 1996 to the nationals of Korea and China, because of the old fisheries agreements with them, Japan began to negotiate new fisheries agreements with Korean.

It is notable that Japan established five Special Prohibition Zones in its EEZ, through the Law on the Exercise of Sovereign Rights with respect to Foreigners' Fishing in Japan's EEZ. The Special Prohibition Zones are the zones where fishing of foreign fishing vessels is not permitted at all. The Special Prohibition Zones are designated in the five straits – the Soya Strait, the Tsugaru Strait, the eastern channel of the Korea/Tsushima Strait, the western channel of the Korea/Tsushima Strait and the Osumi Strait – which are used for international navigation and where the extents of the territorial waters are only 3 miles. The Special Prohibition Zones extend from the coasts to 12 miles or to the median line with its neighbouring countries. It seems that the Special Prohibition Zones are justifiable under Japan's entitlement to a 12-mile territorial sea in the five straits where it claims only 3-mile territorial waters.

Although the Law on the Exclusive Economic Zone and the Continental Shelf of Japan follows the concept of the zone and the shelf prescribed by the LOS Convention, the provisions of the law on the delimitation do not seem to fully follow the letter and spirit of the relevant provisions of the LOS Convention. With regard to the delimitation of EEZ, the Law on the Exclusive Economic Zone and Continental Shelf provides for the median line where there are overlapping claims. It seems appropriate to scrutinise Article 2 of the law. It provides that:

> . . . where any part of that line [i.e., 200 N.M. line form the bàseline] as measured from the baseline of Japan lies beyond the median line . . ., the median line (or the line which may be agreed upon between Japan and a foreign country as a substitute for the median line) shall be substituted for that part of the line.[53]

An identical provision is found in Article 2 of the law on the delimitation of the continental shelf. As we have discussed earlier, the median line provision of Article 2 of the law on the continental shelf might be justified under Article 6 of the Geneva Convention on the Continental Shelf which provides the median line as a boundary in the absence of agreement unless another boundary line is justified by special circumstances. It is, however, notable here that Japan has never ratified the Geneva Convention and has ratified the LOS Convention, which dropped any mention of median or equidistance line for the delimitation of EEZ or the continental shelf.[54]

Now a question arises as to whether these provisions can be justified under customary international law. It appears that 27 or so countries have domestic legislation mentioning an equidistance line and many of the delimitation of boundaries have been settled applying the equidistance line.[55] Therefore, the main problem in the Japanese legislation is not the fact that it mentions the equidistance line. The problem of the Japanese legislation is the fact that it upholds an equidistance line which is to be drawn unilaterally when there is no agreement. This means that the legislation actually empowers the Japanese law enforcement authorities to exercise Japanese jurisdiction at sea against foreign vessels within unilaterally drawn equidistance lines.[56] It is to be recalled that States are under the obligation of mutual restraint pending the delimitation of EEZ/continental shelf boundaries as we have discussed. If the Japanese authorities exercise its jurisdiction in the disputed areas against foreign vessels, then it might constitute an act which can "jeopardise or hamper the reaching of the final delimitation". Therefore, the Japanese legislation which presupposes that a unilateral and mechanical median line is justified in the absence of an agreed boundary is not in accordance with the provisions of Paragraph 3 of Articles 74 and 83 of the LOS Convention which places primary emphasis on the co-operation and mutual restraints between relevant States in the absence of a boundary.

2.3. *The People's Republic of China (China)*

2.3.1. *The Territorial Sea*

China's position on the breadth of its territorial sea was unclear until 1958.[57] On 25 October 1958 the Chinese government announced through its Declaration on China's Territorial Sea that the breadth of its territorial sea was 12 miles.[58] Later in 1992 China enacted the Law on the Territorial Sea and the Contiguous Zone confirming its position on a 12-mile territorial sea and newly establishing a contiguous zone.[59] The law of 1992 employed only the method of straight baselines for measuring the breadth of the territorial sea and did not mention normal baselines.[60] However, it was not until 1996 that the Chinese government specified its straight baselines by issuing the Declaration on the Baselines of the Territorial Sea of the People's Republic of China on 15 May 1996.[61]

2.3.2. *Baselines*

The Declaration on the Baselines of the Territorial Sea of the People's Republic of China on 15 May 1996 specified its straight baseline points.

The straight baselines are divided into two groups: one is for the territorial sea of the mainland and the other is for the separate territorial seas of the Xisha Islands (Paracel Islands) that are an object of territorial dispute with Vietnam.[62] The Chinese declaration sparked off protest from neighbouring countries. Vietnam has protested against the Chinese straight baselines around the Paracel islands, arguing for sovereignty over the islands.[63] Korea challenged, on the basis of the provisions in the LOS Convention, the legality of some of the Chinese baselines drawn in the Yellow Sea and the East China Sea where Korea is also a coastal State.[64] It is notable that the United States showed critical views of the Chinese baselines. It mentioned that:

> Much of China's coastline does not meet either of the two LOS Convention geographic conditions required for applying straight baselines. And for the most part, the waters enclosed by the new straight baselines system do not have the close relationship with the land, but rather reflect the characteristics of high seas or territorial sea.[65]

2.3.3. *The Continental Shelf and the EEZ*

In the early 1970s, China showed interest in its rights over its continental shelf without specific positions on the geographical limits of the continen-

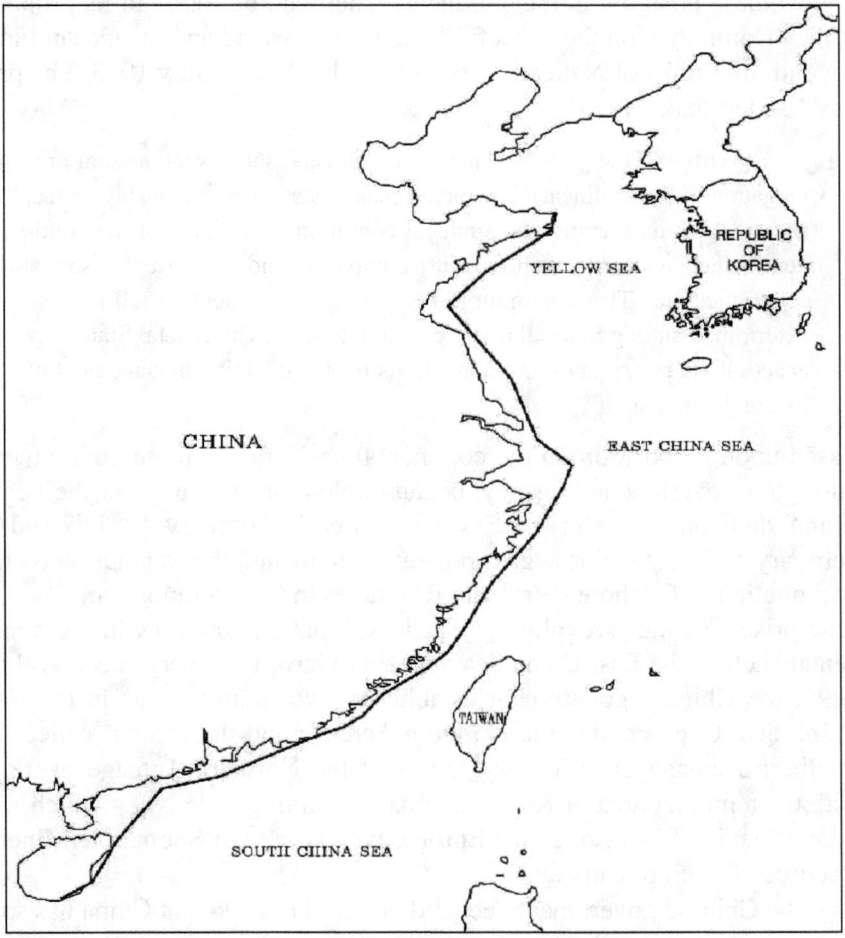

Map 28: China's Straight Baselines

tal shelf when Korea, Japan and Taiwan showed attempts to exploit the resources in the continental shelf in the East China Sea.[66] The *Ren-min Ribao*, a daily newspaper published by the Chinese government asserted on 29 December 1970 that:

> Taiwan Province and the islands appertaining thereto . . . are China's sacred territories. The resources of the sea-bed and subsoil of the seas around these islands and of the shallow seas adjacent to other parts of China all belong to China, their owner, and we will never permit others to lay their hands on them. The People's Republic of China has the right to explore and exploit the resources of the sea-bed and subsoil of these areas . . .[67]

The Chinese position on the continental shelf can be found in its proposal at the "Committee on the Peaceful Uses of the Seabed and the Ocean Floor beyond the Limits of National Jurisdiction" held on 16 July 1973. The proposal stated that:

> . . . By virtue of the principle that the continental shelf is the natural prolongation of the continental territory, a coastal State may reasonably define, according to its specific geographical conditions, the limits of the continental shelf under its exclusive jurisdiction beyond its territorial sea or economic zone. The maximum limits of such a continental shelf may be determined among States through consultation. . . . The coastal State may enact all necessary laws and regulations for the effective management of its continental shelf.[68]

Based upon its position on the continental shelf, the Chinese government, through a governmental agency, began exploratory drillings on the continental shelf off China in the East China Sea in February 1981.[69] And in February 1982 the Chinese government promulgated the Regulation on the Exploitation of Offshore Petroleum Resources in Co-operation with Foreign Enterprises.[70] China's exploratory and exploitation activities in the continental shelf in the East China Sea appears to have been very successful. In 1992, the Chinese government established two mining areas in the East China Sea. One is called the Northern Acreage and the other is called the Southern Acreage. And it was found that the Northern Acreage overlaps with the a mining area of Korea, Submarine Mining Area No. 4 which was established in 1971 through the Enforcement Decree of Submarine Mineral resources Development Act.[71]

The Chinese government declared on 15 May 1996 that China has sovereign rights and jurisdiction in its exclusive economic zone and continental shelf.[72] On 26 June 1998, the Chinese government promulgated the Act on the Exclusive Economic Zone and the Continental Shelf.[73] Note that it took several years for the Chinese government to legislate a law on its EEZ and continental shelf even after its declaration of 15 May 1996. It was because there were worries from the local Chinese governments in the coastal areas that the legislation might cause a negative impact on the Chinese fisheries industries. The act provides for China's sovereign rights and jurisdiction in its EEZ and continental shelf generally in accordance with provisions of the LOS Convention.[74] With regard to the breadth of the EEZ and the continental shelf, the act provides that China's EEZ extends to a distance of 200 miles and China's continental shelf extends to the outer edge of the continental margin.[75] With regard to the delimitation, the act provides equitable principles. Article 2 of the act provides that:

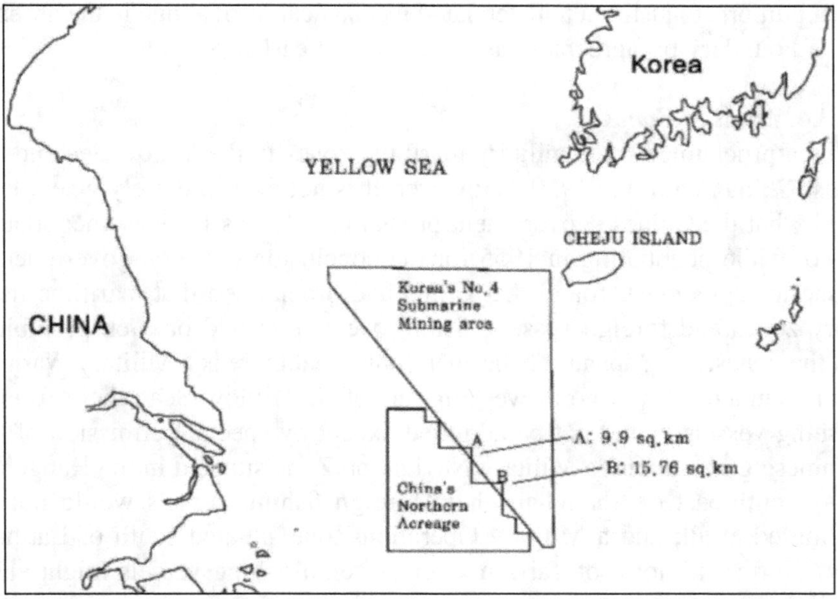

Map 29: Overlap of the Continental Shelf Claims between Korea and China

> . . . Conflicting claims regarding the exclusive economic zone and the continental shelf by the People's Republic of China and States with opposite and adjacent coasts shall be settled, on the basis of international law and in accordance with the principle of equity, by an agreement delimiting the areas so claimed.

As we have discussed earlier, Articles 74(1) and 83(1) the LOS Convention are the outcome of a compromise which purported to please the conflicting groups, i.e., one group which upheld the equidistance principle and the other group which vindicated equitable principles and thus it can be argued that Article 2 of Chinese Act on the EEZ and the continental shelf which specifically mentions "the principle of equity" for delimitation is slightly different from the provisions of the LOS Convention. Nevertheless, however, it is to be noted that the provision upheld some of the essential elements of Articles 74(1) and 83(1) of the LOS Convention such as "on the basis of international law", and "by an agreement". It might be the Chinese provisions are in fact corresponding to the customary international law on delimitation.[76] In this sense, it can be said that the Chinese provision on the delimitation of EEZ and continental shelf is much more in accordance with the relevant provisions of the LOS Convention than the Japanese one that purports to justify a unilateral and mechanical median line in the absence of a boundary by agreement as we discussed earlier.

2.3.4. *Military Zones*

China proclaimed three military maritime zones in the Yellow Sea and the East China Sea in the 1950s. However, it is not clear precisely when, how and what the Chinese government proclaimed. It was Japanese negotiators who, while negotiating in 1955 for the conclusion of a non-governmental fisheries agreement, found that China had proclaimed that maritime military zones and foreign vessels' fishing were restricted or even prohibited in the zones. The Japanese fishermen found that there is a Military Warning Zone situated in the north-western part of the Yellow Sea where foreign fishing vessels would not be admitted except by special permission of the Chinese government; a Military Navigation Zone situated in the Hangchou Bay south-east of Shanghai where foreign fishing vessels would not be admitted at all; and a Military Operation Zone situated south of Latitude 29° North and north of Taiwan where foreign fishing vessels might enter only at their own risk.[77]

As the Geneva Conventions of 1958 and the LOS Convention have no specific provisions on military zones, there can be differing views on the legality of maritime zones or security zones. In times of armed conflict, however, there is room for justification of the military zones such as self-defence or necessity.[78] O'Connell once observed that "operational zones, if permitted on the high seas at all in cases of limited wars, are permitted only for the purpose of belligerent operations among the protagonists and not for the purpose of molesting neutrals".[79] The 200-mile Total Exclusion Zone established by the British government at the time of the armed conflict between the United Kingdom and Argentina over the Falkland Islands/Islas Malvinas is an example of this.[80] However, in light of the freedom of the high seas provided in the LOS Convention and also in light of the fact that the security rationale was finally rejected as a basis for the contiguous zones by the ILC in 1956, a justification for the military zones in times of peace is highly difficult to find.[81]

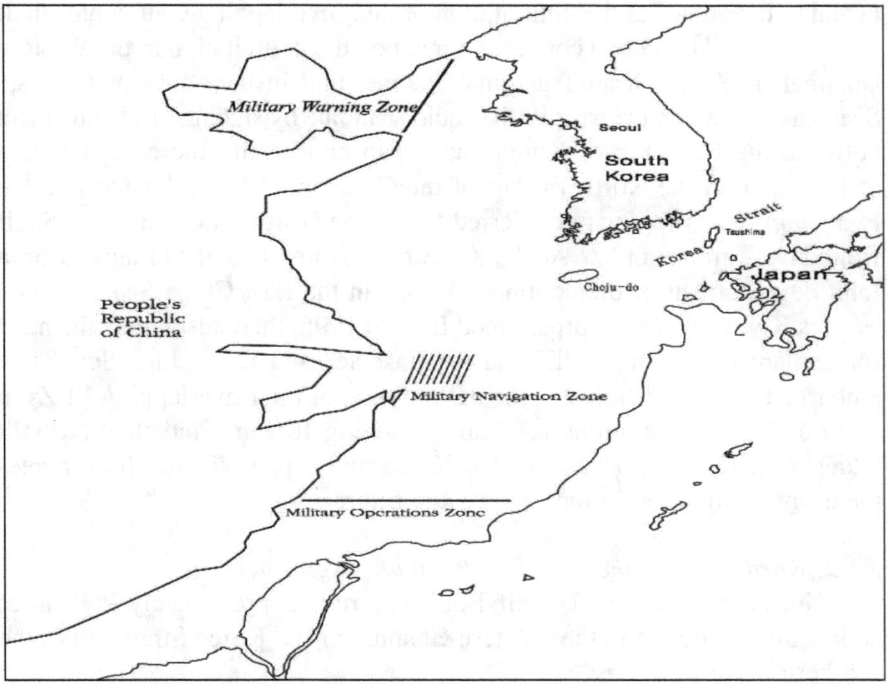

Map 30: China's Military Zones

(Source: Based on the US State Department's Map 503732-789543842)

3. DELIMITATION OF BOUNDARIES OF THE EEZ AND THE CONTINENTAL SHELF IN THE REGION

3.1. *Delimitation between Korea and Japan in the East Sea*

3.1.1. *Background*

Korea and Japan face each other over the East Sea.[82] Korea laid claims to the continental shelf in the East Sea and the East China Sea through the enactment of "The Submarine Mineral Resources Development Act" of 1970 and its Enforcement Decree of 1971. In 1968, when the Korean government was drafting the act, the Japanese government showed interest in the act and asked the Korean government to notify the Japanese government of the contents of the bill.[83] As the so-called "submarine mineral exploitation area" of Korea was *de facto* the continental shelf extended to the area where the Japanese government thought it part of Japan's continental shelf, the need for negotiation arose and thus the two governments began to negotiate for the delimitation of their overlapping continental shelf in 1970. In 1974, the two countries agreed on the delimitation in the Western Channel of Korea Strait/Tsushima Channel and in the south of the East Sea/Sea of Japan on the basis of the equidistant line by signing the "Agreement between the Republic of Korea and Japan concerning the establishment of boundary in the Northern Part of the Continental Shelf Adjacent to the two Countries" (hereinafter referred to as "the Northern Continental Shelf Boundary Agreement").[84] At the same time Korea and Japan agreed on a joint development of the continental shelf in the East China Sea.

As Korea and Japan proclaimed EEZ in 1996, there also arose the need for demarcation of their EEZs in the East Sea and East China Sea. They embarked on negotiations for the delimitation of their overlapping EEZs in 1996 and the negotiations are hardly moving forward and thus are still under way. It does not appear that Korea and Japan will reach an agreement on delimitation in the foreseeable future.

3.1.2. *Northern Continental Shelf Boundary Agreement*

The Northern Continental Shelf Boundary runs approximately 260 miles connecting 35 points in the Western Channel of the Korea Strait and in the southern part of the East Sea.

There was no difficulty for Korea and Japan in agreeing on applying the equidistance principle in drawing the boundary. Numerous islands and islets in the area were taken into consideration as basepoints. An analysis shows that there are slight deviations from the strict equidistance in the boundary.[85] It is interesting to see that the boundary line stops at Point 35 leaving vast areas of the East Sea undelimited. Here arises a question of why the two countries did not delimit the East Sea further to the north. One can surmise that the territorial issue between Korea and Japan over Dok-do might be a reason for the incomplete delimitation. In sharing with the view, Professor Park wrote that:

> Since 1952, Japan and South Korea have been engaged in a serious territorial dispute over the ownership of Dokto (in Korean), Takeshima (in Japanese), or Liancourt Rocks (in the West), a cluster of rocks and two uninhabited islands (under South Korean control) situated North East of the terminus (Point No. 35). Consequently, the boundary line provisionally ends at 71.3 N.M. southwest of Dokto, pending settlement of this dispute.[86]

In this regard, a close analysis of the charts shows that Point No. 35 is situated closer to Dok-do than to Changki-gap on the Korean coast and to

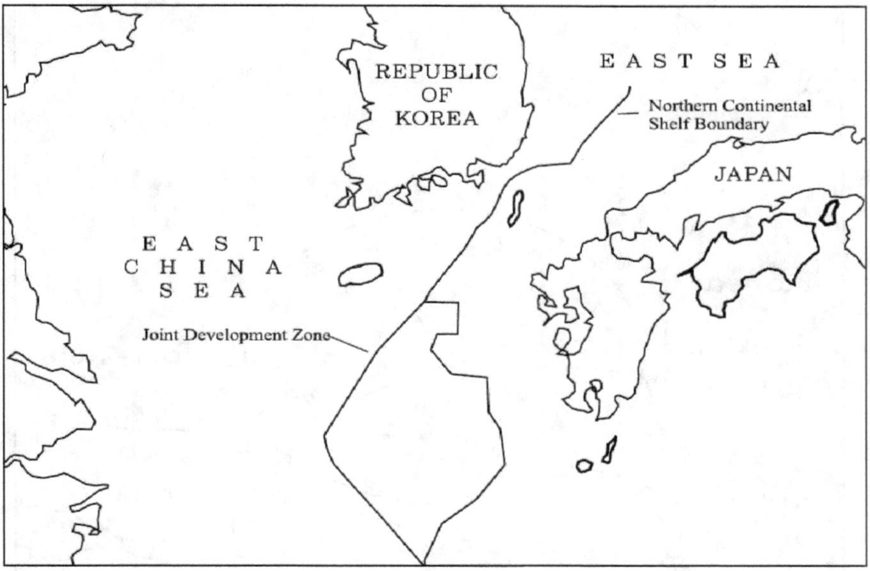

Map 31: Northern Continental Shelf Boundary between Korea and Japan

(Source: J.R.V. Prescott, Maritime Jurisdiction in East Asian Seas)

Hinomi-saki on the Japanese coast both of which were used as basepoints. Note that Dok-do is 71.3 miles from point No. 35 and Changki-gap and Hinomisaki are 80 miles from Point No. 35.

It means that Dok-do is located in a position which might have affected the northern continental shelf boundary but it was disregarded as a basepoint. In other words, the northern continental shelf boundary agreement is a precedent where Dok-do is ignored as a basepoint in a maritime delimitation.

Now a question arises of why Dok-do was ignored as a basepoint in the northern continental shelf boundary agreement. It might be the case that the island was ignored because there was no continental shelf in terms of the provisions of the 1958 Geneva Convention on the Continental Shelf.[87] It is to be recalled that the Geneva Convention on the Continental Shelf defined the continental shelf in terms of the depth of the waters superjacent to the submarine areas and exploitability of the submarine areas.[88] Note that the area around Dok-do is more than 2,000 metres deep and thus did not allow exploitation by the technology of the 1970s. In this regard, it is to be noted that Korea's domestic legislation, the Submarine Mineral Resources Development Act of 1970 and its Enforcement Decree of 1971, did not establish any submarine mineral exploitation area around Dok-do

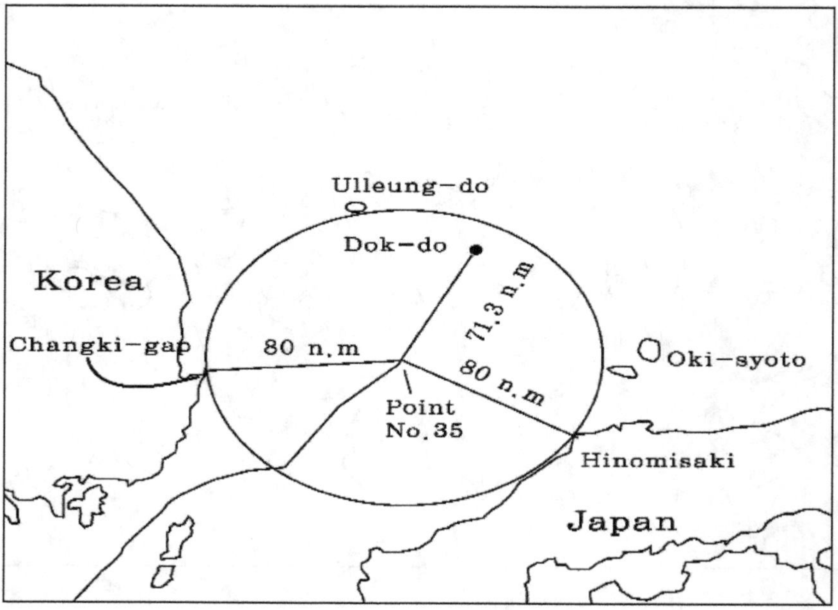

Map 32: Point 35 in the Northern Continental Shelf Boundary and Dok-do

and that Point No. 35 of the northern continental shelf boundary is identical with the Northernmost point (Point No. 1 in the seabed mining block6-1) in Korea's submarine mineral exploitation area in the East Sea.

Now let us turn to the southern terminus of the Northern Continental Shelf boundary. Point No. 1 of the boundary was chosen as an equidistant point from Korea's basepoint Cheju-do and Japan's basepoint Tori-Shima.[89] It is to be noted that Torishima is comprised of three tiny rocks that might be classified as rocks under the provisions of paragraph 3 of Article 121 of the LOS Convention. However, it appears that Korea did not challenge the legality of Tori-Shima as Japan's basepoint in the negotiation on the delimitation.[90]

3.1.3. *Applicable Principles of Delimitation of EEZ in the East Sea*

It would not be difficult for Korea and Japan to agree on applying the equidistance principle in the East Sea. In the negotiation at the UNCLOS III, Japan joined the group which supported the principle of "equidistance-special circumstance" for the delimitation of EEZ or the continental shelf between adjacent or opposite States.[91] Japan's adherence to the equidistance principle is well-illustrated by the provisions in its Law on Exclusive Economic Zone and the Continental Shelf referring to equidistance lines as we have discussed earlier. According to Japan's domestic legislation, its boundary of EEZ and continental shelf is the equidistance line with its neighbouring States when there is no agreement with the neighbouring States. It is also to be recalled that Japan argued for the equidistance line in the negotiations with Korea on the delimitation of the continental shelf in the 1970s. In the same manner, Japan has been arguing for application of the strict equidistance line for the boundary of EEZ between Korea and Japan in the East Sea and in the East China Sea.

Korea did not voice its position on the delimitation of EEZ and continental shelf in the negotiation at the UNCLOS III. The reason why Korea kept silent on the issue of delimitation might be that any uniform approach would not serve the interests of Korea. That is to say that the equidistance line best serves Korea's interests in the East Sea with Japan and in the West Sea with China whereas equitable principles can best serve the interests of Korea in the East China Sea. Thus Korea agreed with Japan to delimit the boundary of the continental shelf in the East Sea on the basis of an equidistance line in 1974, but argued for the principle of natural prolongation for the delimitation of the continental shelf in the East China Sea in the 1970s. Therefore, it would not be difficult for Korea to agree in principle with Japan to apply the equidistance principle for the delimitation of EEZ in the East Sea, while arguing for application of equitable principles for the delimitation of EEZ in the East China Sea. However, even if Korea and Japan can agree to draw a boundary in the East Sea following an equidistance line, there are some obstacles which make the delimitation difficult. We will discuss these obstacles in detail now.

3.1.4. *Disputes regarding the Legality of Japan's Straight Baselines*

As mentioned earlier, Japan's new straight baselines promulgated in 1996 sparked off serious disputes with Korea. At first glance, Japan's straight baselines might appear to have been drawn in a modest way, because 116 segments of the baselines are less than 24 miles long and 31 segments are between 24 to 48 miles long.[92] However, the problems of Japan's straight baselines appear to be in the fact that in many cases they were employed in the locations where neither the coastline is deeply indented and cut into nor is there a fringe of islands along the coast in its vicinity. The States Department of the United States commented that:

> Generally, the coastal geography of the Japanese islands along which the straight baselines have been drawn do not conform to the requirement called for in Article 7, paragraph 1 of the LOS Convention. For the most

> part, the coastline of these Japanese islands are neither "deeply indented and cut into", nor is there a "fringe of islands" in the immediate vicinity.

It can be said that the conditions provided in paragraph 1 of Article 7 are primary ones which serve as criteria to see whether the use of straight baselines in certain localities is permissible, and conditions such as the length of straight baselines and the extent of deviation of straight baselines from the general direction of the coast are secondary.[93] And thus if a straight baseline satisfies the secondary conditions while failing to meet the primary one, it cannot be valid under the relevant provisions of the law of the sea. In many cases, the Japanese straight baselines seem to meet the secondary conditions while failing to meet the first one. The Korean government challenged the legality of some of the straight baselines employed by the Japanese government.[94] Three rounds of talks were held on the legality of Japan's straight baselines between the Korean government and the Japanese government without any resolution of their disputes. As the provisions of the LOS Convention do not provide any numerical conditions for the straight baselines,[95] the relevant provisions of the LOS Convention would be of little help in solving a dispute on the legality of straight baselines between two littoral States unless they agree to submit the dispute to a third party for resolving their dispute.

There is also another important aspect of the implication of the dispute on the delimitation with regard to the relationship between the EEZ boundary and the continental shelf boundary in the area where the northern continental shelf boundary was drawn in 1974 between Korea and Japan. The northern continental shelf boundary was drawn following the equidistance line between the coasts of the two countries in the area. Then Korea employed straight baselines in the area in 1978, and Japan employed straight baselines in 1996. Thus, if an equidistance line is to be drawn in the area taking into account the straight baselines of both countries, the equidistance line would not be the same as the northern continental shelf boundary. Interestingly, some amount of natural gas was found in the part of the area which is now on the Korean side but would fall onto the Japanese side if straight baselines were taken into account in drawing the boundary.[96] Of course, the issue of re-drawing of continental shelf boundary would not arise unless both Parties agree to do so because of the northern continental shelf boundary agreement. No doubt that the agreement was concluded to provide a permanent continental shelf boundary since the agreement does not contain a termination clause. Furthermore Japan did not challenge the legality of Korea's straight baseline in the area whereas Korea challenged the legality of Japan's straight baselines in the area. However, the problem here is whether Korea and Japan prefer to draw an EEZ boundary that is identical to the northern continental shelf boundary. If they prefer to draw a different line for the purpose of a boundary of the water column in the area where the northern continental shelf boundary is drawn, then they have to resolve the dispute over the Japanese straight baselines.

3.1.5. *Question of Sovereignty over Dok-do and Delimitation Issues*

The question of ownership over Dok-do (known as Liancourt Rocks in the western world and Takeshima in Japan) was raised by Japan in January 1952 on the occasion when the Korean President, Mr. Lee proclaimed the so-called "Peace Line" adjacent to the Korean peninsula, thereby placing Dok-do on the Korean side of the line. In a diplomatic note of 1952 sent by the Japanese government to the Korean government, it noted the fact that Dok-do was placed on the Korean side of the line and argued that the island is without question Japanese territory and it does not recognise any claim by Korea over the island.[97] Thereafter a series of diplomatic notes containing lengthy and detailed arguments on the sovereignty of Dok-do were exchanged between the two countries. It is notable that in 1954 Japan proposed and Korea rejected that the question of sovereignty over Dok-do be brought to the International Court of Justice.[98] The basic position of the Korean Government regarding Dok-do is that there is no dispute as such over Dok-do. In fact, there is a sentiment among the Korean people that Japan's territorial claim over Dok-do was caused by greed and old imperialism and that Japan tries to create a dispute of Dok-do because Japan has nothing to lose by doing so. Sharing this line of thinking, Dr. Hee Kwon Park, a former director of international legal affairs division at the Foreign Ministry of the Republic of Korea wrote that:

> It is interesting, however, to observe that Japan tries to avoid or even objects to submitting other territorial disputes involving Japan to the ICJ. Neither in the dispute with China over Senkaku (Diaoyuta) Islands nor in the case against Russia over the so-called "Northern Territories" has Japan shown any willingness to submit these issues to the ICJ. This apparently contradictory position on the part of Japan seems to stem from its belief that it has nothing to lose in the case of Tokdo, whatever the judgement of the ICJ might be.[99]

Even if the Korean government denies that there is a dispute over the ownership of Dok-do, a third party institution would rule that there is a legal dispute over the ownership of Dok-do on the basis of jurisprudence of the PCIJ in the *Mavrommatis* Case where the PCIJ defined a dispute as "a disagreement on a point of law or fact, a conflict of legal views or of interests between two persons",[100] if the question of sovereignty over Dok-do is ever to be brought before an international judicial institution. However, it is to be recalled that the ICJ can establish its jurisdiction on the basis of an agreement between the parties and Korea has neither accepted the compulsory jurisdiction of the ICJ provided in Article 36 of the Statute of the ICJ nor is it willing to agree to submit the issue to the ICJ. Furthermore, in 1965 when the two governments normalised their diplomatic relations, they agreed, through an Exchange of Notes concerning the Settlement of Disputes between the government of the Republic of Korea and the government of Japan, to exclude any compulsory dispute settlement. The Exchange of Notes provides that:

> Unless agreed otherwise, both Governments shall try to resolve disputes between the two countries first through diplomatic channels and then through conciliation in accordance with procedures agreed upon by the two Governments when they cannot resolve disputes through diplomatic channels.[101]

Therefore, it is hard to imagine that both Korea and Japan will stand in the courtroom of the ICJ to resolve the question of territorial sovereignty over Dok-do. Also, Korea would not negotiate with Japan over the question of sovereignty over Dok-do as long as Korea has effective control over Dok-do. As such, the presence of Dok-do in the middle of the East Sea stands in the way of negotiations on the delimitation of the EEZ boundary between Korea and Japan.

Now a difficult question arises in relation to the dispute settlement procedures under the LOS Convention for resolving the disputes of delimitation of EEZ in the East Sea. The question is whether a third party institution can deal with the dispute on delimitation of EEZ in the East Sea between Korea and Japan on the basis of compulsory procedure of Section 2 of Part XIV of the LOS Convention. It is to be recalled that the disputes on maritime delimitation are subject to compulsory procedures of the LOS Convention unless a State Party declares in writing that it does not accept the compulsory procedures regarding disputes concerning the interpretation or application of Articles 15, 74 and 83 of the Convention relating to sea boundary delimitation.[102] As neither Korea nor Japan has opted out of the compulsory procedures of the LOS Convention for the disputes on maritime delimitation, it seems *prima facie* that the delimitation dispute in the East Sea between Korea and Japan is subject to the compulsory procedures of the LOS Convention. However, it remains to be seen whether the abovementioned Exchange of Notes concerning the Settlement of Disputes between the two countries can be used by one of the two countries in order to exclude the compulsory procedures of the LOS Convention relying on Article 281 of the Convention. Paragraph 1 of Article 281 provides that:

> If the States Parties which are parties to a dispute concerning the interpretation or application of this Convention have agreed to seek settlement of the dispute by a peaceful means of their own choice, the procedures provided for in this Part apply only where no settlement has been reached by recourse to such means and the agreement between the parties does not exclude any further procedure.

It seems important to discuss the implication of Article 281 of the LOS Convention in connection with the Exchange of Notes in the delimitation dispute in the East Sea. Interestingly, Article 281 of the LOS Convention was relied upon along with Articles 280 (Settlement of disputes by any peaceful means), 282 (Obligations under general, regional or bilateral agreement) and 283 (Obligation to exchange views) by Japan in the *Southern Bluefin Tuna Arbitration* to challenge the jurisdiction of the arbitration court constituted under Section 2 of Part XV of the LOS Convention on the request of Australia and New Zealand, which challenged the legality of Japan's experimental fishing activities in the light of the provisions of the LOS Convention and the Convention for the Conservation of Southern Bluefin Tuna (hereinafter referred to as "the CCSBT").[103] Japan noted to the fact that Article 16 of the CCSBT provides for its own dispute settlement procedures and argued that Article 16 of the CCSBT excludes the compulsory procedures of the LOS Convention without the consent of all parties to the dispute.[104] It appeared to Australia and New Zealand that Article 16 of the CCSBT does not provide any compulsory dispute settlement procedures for disputes concerning the interpretation and application of the CCSBT, but merely provides a list of various means of dispute settlement procedures and that any such disputes shall be referred to the International Court of Justice or arbitration with the consent of all parties to the disputes.[105] However, the arbitral tribunal noted to the importance of the term "consent" used in Article 16 of the CCSBT in referring any disputes to dispute settlement procedures saying that: "the ordinary meaning of these terms of Article 16 makes it clear that the dispute is not referable to adjudication by the International Court of Justice (or, for that matter, ITLOS), or to arbitration, at the request of any party to the dispute".[106] The arbitral tribunal went on to conclude that "the intent of Article 16 is to remove proceedings under that Article from the reach of the compulsory procedures of section 2 of Part XV of UNCLOS, that is, to exclude the application to a specific dispute of any procedure of dispute resolution that is not accepted by all parties to the dispute".[107] We can see that the tribunal in the *Southern Bluefin Tuna Arbitration* recognised the dispute settlement procedure in the CCSBT as "peaceful means of their own choice" under Article 281 of the LOS Convention, which excludes the compulsory dispute settlement procedure under the LOS Convention.[108]

Now let us turn to the question as to whether the Exchange of Notes in 1965 between Korea and Japan can exclude the compulsory procedures of the LOS Convention for the settlement of the dispute over EEZ delimitation in the East Sea. As neither Korea nor Japan have exercised their rights provided by Article 298 of the LOS Convention to opt out of the compulsory procedures of the LOS Convention and neither of them have chosen any of the procedures in Article 287 of the LOS Convention, the maritime delimitation dispute might be brought before an arbitration tribunal constituted in accordance with Annex VII of the LOS Convention upon the request of either country after exhaustion of "exchange of views". However, there is a possibility that the other country might successfully challenge the jurisdiction of the arbitral tribunal relying on the Exchange of Notes as "peaceful means of their own choice" excluding the compulsory dispute settlement procedure under the LOS Convention, if the Party does not wish to stand before the arbitral tribunal constituted under the LOS Convention.

What if either Korea or Japan declares to opt out of the compulsory procedures of the LOS Convention for the delimitation issue in accordance with Article 298 of the Convention? In that situation, can either Japan or Korea alone bring the delimitation dispute between them to a compulsory conciliation? This question is of practical importance, because the opt-out declaration can be made in writing at any time.[109] Article 298 provides that even if a State opted out of the compulsory procedures entailing binding decisions for the disputes on delimitation, the disputes, nevertheless, shall be referred to conciliation on the request of the other State, unless the dispute necessarily involves the concurrent consideration of any unsettled dispute concerning sovereignty or other rights over continental or insular land territory. Now a difficult question arises as to whether the delimitation disputes in the East Sea between Korea and Japan necessarily involve the concurrent consideration of unsettled sovereignty dispute over Dok-do. It would be the conciliation committee which will decide as to whether the delimitation disputes in the East Sea between Korea and Japan necessarily involve the concurrent consideration of the unsettled sovereignty dispute over Dok-do. If the conciliation committee is inclined to draw the boundary in the

East Sea without any consideration of the presence of Dok-do and thus without touching upon the question of territorial sovereignty over Dok-do, then the committee would decide to carry out conciliation. However, if the conciliation committee is inclined to give any effect to Dok-do in the delimitation, then the committee would face the question of sovereignty over Dok-do and it should declare that it has no competence to carry out conciliation because the delimitation dispute in the East Sea between Korea and Japan necessarily involves the concurrent consideration of unsettled sovereignty dispute over Dok-do.

3.1.6. *Article 121(3) of the LOS Convention*

As Dok-do is a small island situated in the middle of the East Sea, an important question arises as to whether Dok-do is a rock in terms of Article 121(3) of the LOS Convention, which provides that:

> Rocks which cannot sustain human habitation or economic life of their own shall have no exclusive economic zone or continental shelf.

If Korea and Japan can agree that Do-do is a rock in terms of Article 121(3) of the LOS Convention and thus disregard it in the delimitation, then delimitation of EEZ boundaries in the East Sea would be attained without raising the sovereignty issue. Before examining whether Dok-do is a rock, it will be appropriate to briefly examine the drafting history and issues of interpretation of the provision of Article 121(3).

In the two Geneva Conventions on the Territorial Sea and Contiguous Zone, and on the Continental Shelf, there is no distinction between islands and rocks.[110] However, from the early 1930s, there have been arguments that an island should meet certain standards to be entitled to maritime zones. For example, Gidel from France put forward the following definition of an island in the Hague Codification Conference of 1930: "An island is a natural elevation of the sea-bed, surrounded by water, which is above water at high tide and the natural conditions which permit the ***stable residence of organised groups of human beings*** (emphasis added)".[111] In a similar context the United Kingdom argued in the same conference that an island should be "capable of occupation and use".[112]

In the UN Sea-Bed Committee which was the forerunner to the third UN Conference on the Law of the Sea, the Maltase Ambassador, Pardo, raised the need for restricting maritime zones from small islands, saying that: "If a 200 mile limit jurisdiction could be founded on the possession of uninhabited, remote or very small islands, the effectiveness of international administration of ocean space beyond national jurisdiction would be gravely impaired".[113] With this line of reasoning, various proposals were put forward to set the standards for qualification of islands which are entitled to the exclusive economic zone and the continental shelf.[114] However, it is to be noted that there were also States such as Greece, the United Kingdom, Japan and island States in the Pacific which advocated for the same legal status for all islands and thus for deleting the rock-provision.[115] The problem in Paragraph 3 of Article 121 of the LOS Convention is that it was adopted without further elaboration on the terms used in the paragraph once it was introduced into ISNT/Part II in the Second Session of the UNCLOS III in 1974.[116] Thus, there is now vagueness in the terms of "human habitation" and "economic life of their own". For instance, it is not clear at all as to how many people are required to meet the "human habitation" and what level of economic life can be seen as "economic life of their own".[117]

Noting the problem, Hodgson and Smith argued in 1976 that: "this paragraph should be eliminated for geographical reasons as being impossible to administer, If for other reasons, this paragraph must remain it is essential that a rock be defined objectively so as to remove all doubts as to which rocks would be affected by this provision".[118] E.D. Brown said that the provision is "intolerably imprecise" on the meaning of "cannot sustain human habitation or economic life of their own", and thus the provision is "difficult to apply".[119] Kwiatkowska and Soons pointed out that: "the meaning of the apparently simple provision of Article 121, paragraph 3 is not easily identifiable".[120] For J.I. Charney the paragraph is "a laconic provision".[121]

It is not clear whether prior customary international law denies the entitlement of rocks and whether the rock-provision in the LOS Convention is declaratory of the customary law.[122] State practices are diverse in this regard. The diversity in the State practice might be explained not only by the fact that the rock-provision was not known in the Geneva Convention regime but also by the fact that States tend to take into account various factors such as the delimitation methods, distance from the mainland and nearby islands, general geography in the relevant coastlines, besides size of the feature, when they consider whether or not, or how much effect is given to a feature.[123] For instance, Rockall played no role in the UK-Ireland delimitation agreement of 1988.[124] Similarly, Paulau Perak, a rock some 80 N.M. from the mainland of Malaysia played no role in the Indonesia-Malaysia-Thailand delimitation agreement of 1971.[125] However, the small Venezuelan island, Aves Island was given full effect in the US-Venezuela agreement of 1978[126] and in the Netherlands(Antilles) – Venezuela agreement of 1978.[127] As the State practice does not appear to be consistent in giving effect or no-effect to small features in the maritime delimitation, it is not appropriate to draw a conclusion from the practice that rocks have always been disregarded or that rocks which have been disregarded in the delimitation were disregarded always because they are rocks.[128]

However, no matter how vague the provision might be and how diverse the State practices might be, it cannot now be denied that the provision has a binding force on the Parties to the LOS Convention, and there is a convincing assertion that the provision is now general international law binding even on States that are not parties to the LOS Convention.[129] It is also to be recalled that the LOS Convention does not allow any reservation to be made.[130] It seems that there are a number of islands which might fall into the category of "rocks" in terms of Article 121(3) of the LOS Convention and some States might wish to rely on the rock-provision. For example, the Dominican Republic, St. Vincent and Grenadines have been arguing that the Venezuelan Aves is a rock in terms of Article 121(3) of the LOS Convention and thus it would be very difficult for Venezuela to use Aves as its basepoint in delimitation with these countries.[131] It is notable that the UK government renounced its 200 N.M. fisheries zone around Rockall in 1997 when she acceded to the Convention, indicating that Rockall is a rock under Article 121(3) of the LOS Convention and thus it is incompatible with her obligation under the Convention to keep Rockall as a basepoint for her fisheries zone.[132]

3.1.7. *Issue of Entitlement of Dok-do*

It would not be difficult for Korea and Japan to agree to apply the equidistance principle for delimiting the EEZ boundary in the East Sea, because they have a precedence of drawing an equidistance line for a maritime boundary in the Northern Continental Shelf Boundary Agreement in 1974. However, even if they agree to draw the boundary following an equidistance line in the East Sea, they have to face a difficult question as to whether Dok-do is a rock in terms of Paragraph 3 of Article 121 of the LOS Convention. If Dok-do is not a rock and thus has some effect in the delimitation, then the question of which country benefits from Dok-do would arise. However, if Korea and Japan agree that Dok-do is a rock and thus Dok-do would not have any effect, then the question of ownership over Dok-do would not hamper the process of negotiations.[133] Therefore, the way of interpretation and application of Paragraph 3 of Article 121 of the LOS Convention seems very important in the delimitation in the East Sea.

It appears that Korea prefers to disregard Dok-do in the delimitation relying on Article 121 (3) and thus to avoid the question of sovereignty

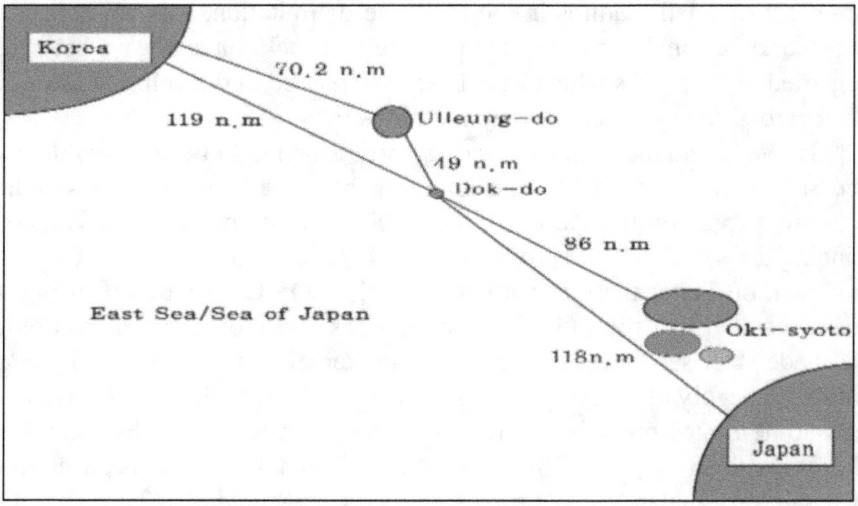

Map 33: Islands and Delimitation in the East Sea

over Dok-do in the negotiations on the delimitation. It is not difficult to see why Korea prefers to do this. Firstly, by doing so the Korean government can avoid the sovereignty question being raised. Note that the Korean government denies even the presence of territorial disputes, whereas the Japanese government wishes to raise the question. If an equidistance line is to be drawn between Korea's Ullung-do and Japan's Oki-syto[134] disregarding Dok-do, then Dok-do would be placed 18 N.M. from the Korean side of the boundary.

Also, by relying on Article 121(3) in the delimitation of the East Sea, Korea can reinforce its argument for disregarding a number of Japanese islets in the East China Sea in the delimitation. If the Korean government insists that Dok-do is not a rock in term of Article 121(3), then it would be difficult for the Korean government to argue to disregard Japanese islets in the East China Sea, where the prospect for sizeable oil reserves is very high.

Sharing this argument, the Korean Prime Minister said in Parliament on 21 November 1998 that:

> The Government does not exclude the possibility of arguing that Dok-do can have its own EEZ. However, the Government is of the view that it is good for Korea in terms of cause and real interests to regard Dok-do as a rock. By doing so, Korea can have a good justification in terms of international law and the possibility of including Dok-do into Korea's

> EEZ is enhanced. . . . We need to note the fact that China and Japan have a number of rocks in the East China Sea and they wish to use those rocks as their basepoints. We need, therefore, to note the possibility that we might end up with shrunken EEZ in the East China Sea where prospects for oil and other economic values are high if we provide justification for China and Japan in their attempts to use their rocks in the delimitation.[135]

However, it would not be easy at all for Korea to persuade Japan to agree on the delimitation of boundaries where Dok-do is given no effect in accordance with Article 121(3) of the LOS Convention and thus Dok-do is placed within Korea's EEZ. It is to be recalled that Japan is one of the countries which argued for the deletion of Paragraph 3 of Article 121 of the LOS Convention in the UNCLOS III negotiations.[136] And, furthermore, Japan might wish to place Dok-do in its EEZ or at least to straddle the boundary line.[137] Therefore, Japan is very much tempted to make use of the vagueness of the terms "economic life of their own" and "human habitation" in Paragraph 3 of Article 121 of the LOS Convention in order to argue that Dok-do is not a rock but an island and it should be used as a Japanese basepoint.

Here it will be useful to examine briefly the natural situation of Dok-do to see whether Article 121(3) is applicable. Dok-do is not a single rock, but it is a group of two main islets and a number of tiny rocks. The combined area of these islets and rocks is only 0,187 km^2. There is one family that has an address in Dok-do, but the family does not live there all-year round. Only the chief of the family spends three to four months for seasonal fishing in Dok-do. Fresh waters were found in Dok-do; however, the quality of the water is not good enough for drinking in its natural State. Dok-do has a manned lighthouse and there is a platoon of 30 or so Korean para-military policemen stationed in Dok-do.[138] Dok-do itself does not have any natural resources and the seabed around it is as deep as 2,000 metres, thus the prospect of oil reserves there is almost non-existent.[139] However, the waters off Dok-do are rich in squid and sardine.

Since there are diverse opinions on the interpretation and application of Paragraph 3 of Article 121 of the LOS Convention, and there is no case law on the question of interpretation and application of Article 121(3) of the LOS Convention, answers to the question as to whether Dok-do is a rock would be different from person to person.[140] For instance, J.I. Charney would say that Dok-do is not rock but an island. He stresses that Article 121(3) uses the word "or" between "human habitation" and "economic life of their own", and then argued that a feature does not need both human habitation and an economic life of its own for getting away from the restrictions of Article 121(3).[141] He went on to argue that human habitation does not require that people reside permanently on the feature or the economic life be capable of sustaining a human being throughout the year.[142] He also supports the view that a feature may serve as a base of seasonal fishery operations and its status depends upon its actual economic worth rather than a classical agrarian concept of viability and thus a feature need not to have tillable soil and sufficient portable water to sustain human habitation to qualify as an island and not a rock.[143] However, Van Dyke, for instance, would say that Dok-do is a rock. Van Dyke emphasises that Paragraph 3 of Article 121 requires a stable human community resided on a feature to qualify as an island, not as a rock. We can imagine here that the opposing views on the interpretations of Article 121(3) would be exchanged between Korea and Japan in the negotiations on the delimitation of EEZ boundary in the East Sea. If Dok-do is to be regarded as an island and not as a rock and thus to have any effect in the delimitation then the question as to who will benefit from Dok-do should inevitably arise in the negotiations.[144]

Apart from the question of interpretation and application of Article 121(3) of the LOS Convention which concerns the issue of entitlement, the question whether Dok-do can be used as a basepoint can also be examined in terms of delimitation principles. Although there is no case law on the interpretation and application of Article 121(3) of the LOS Convention, there has developed a rule in case law which restricts or ignores minor features lying far from the coast in the delimitation. Note that Dok-do is located in the middle of the East Sea; it is 119 N.M. from the Korean mainland, 49 N.M. from Korean Ullung-do, 117 N.M. from the Japanese mainland and 86 N.M. from Japanese Oki-syto. Giving Dok-do full effect would produce an inequitable solution.

In fact, international courts have been trying to achieve an equitable solution in delimitation by giving limited or zero effect to small features situated far from the coastal line. In the *North Sea Continental Shelf* Cases, the ICJ, in mentioning the role of equidistance line, said that:

> The continental shelf area off, and dividing, opposite States, can be claimed by each of them to be a natural prolongation of its territory. These prolongations meet and overlap, and can therefore only be delimited by means of a median line; and, *ignoring the presence of islets, rocks and minor coastal projections*, the disproportionally distorting effect of which can be eliminated by other means, such a line must effect an equal division of the particular area involved (emphasis added).[145]

The above-mentioned consideration in the *North Sea Continental Shelf* Case has been respected for equitable solutions in the subsequent cases. In *Anglo-French Arbitration*, the tribunal completely ignored the small rocks and islets in the Channel Islands when it gave 12 N.M. enclave of the continental shelf to the Channel Islands.[146] The tribunal gave only half effect to Britain's Scilly Isles lying some 21 miles from the mainland for "an appropriate abatement of the inequitable effects of the distorting geographical feature", even if the Isles are inhabited.[147] The ICJ in the *Tunisia/Libya Continental Shelf* Case gave only half effect to the Kerkennah islands 11 miles off the Tunisia coast, although the islands are inhabited and 180 km^2 in area.[148] In the *Gulf of Maine* Case, the Chamber, when considering a provisional starting line for delimitation, referred to the above-mentioned passage of the 1969 judgment of the ICJ saying that:

> In pursuance of this remark, the Chamber likewise would point out the potential disadvantage inherent in any method which takes tiny islands, uninhabited rocks or low-tide elevations, sometimes lying at a considering distance from terra firma, as basepoint for the drawing of a line intended to effect an equal division of a given area.[149]

Taking this line of thinking, the Chamber in *the Gulf of Maine* Case gave only half effect to Canada's Seal island off the southwest coast of Nova Scotia although the Seal island is inhabited.[150] In the *Libya/Malta Continental Shelf* Case, the ICJ ignored Filfla which is uninhabited and situated 3 N.M. south of Malta, saying that "the Court finds it equitable not to take into account of Filfla in the calculation of the provisional median line between Malta and Libya", and thus ignored Filfla even though the rock was used for a basepoint drawing straight baselines by Malta.[151] In the *Eritrea-Yemen Arbitration*, the tribunal decided not to give any effect to the Yemeni mid-sea islands, namely the small sole island of Jabal al-Tayr and the group of islands and islets called Zubayr, noting that these features are "small and uninhabitable" and far away from the Yemeni main coastline.[152]

In the *Qatar/Bahrain* Case, the ICJ decided that a tiny feature, Qit'at Jaradah, belongs to Bahrain and an island which is above the waters at high tides, but the Court noted that the island is "very small, uninhabited and without vegetation" and thus was denied the right to use it as a basepoint in drawing a provisional median line to eliminate disproportionate effects in the delimitation.[153] All these cases confirm the following observation made by Van Dyke: "In summary, recent arbitrations, judicial decisions and negotiations have been relatively consistent in refusing to give full effect to small insular formation in delimiting maritime boundaries".[154] It appears from the various sources that Dok-do is not larger than Scilly, Kerkennahm and Seal island.[155] It also seems that Dok-do is more or less similar to Filfla in size. Note that Dok-do is further away from its mainland than Scilly, Kerkennah, Seal island and Filfla are from their respective mainlands.[156] Thus, it is highly probable that an international court would give no-effect or very limited effect to Dok-do for an equitable solution, if the dispute on delimitation in the East Sea is ever brought to an international court for settlement.[157]

However, even if Dok-do would seem, *prima facie*, to be a rock, and the Korean government wishes to draw an EEZ boundary disregarding Dok-do, the disputes on delimitation of an EEZ boundary would remain unsolved for a long time unless one of the Parties asks for a compulsory conciliation provided for in Article 298 of the LOS Convention.

3.2. *Delimitation between Korea and China in the Yellow Sea*[158]

3.2.1. *Different Positions on the Delimitation Principles*

As both Korea and China proclaimed EEZs and the continental shelf in the Yellow Sea and in the East China Sea where the width between the nearest coasts of the two countries is less than 400 N.M., the need for delimitation of boundaries has arisen. It was in 1996 when the two countries embarked upon bilateral negotiations for delimitation of boundaries of their EEZs and the continental shelves. And the negotiation is still under way. Korea has argued for the application of the equidistance-special circumstances principles for the boundaries in the Yellow Sea, whereas China argued for the application of equitable principles.[159] In the negotiation the Korean government argued that an equdistance line should be drawn first for the *prima facie* boundary in the Yellow Sea and then the check can be made as to whether there are any special circumstance which require modification of the *prima facie* equidistance line. China, however, refused to draw an equidistance line even for the sake of negotiation at an earlier stage, which is subject to modification according to the special circumstances. China argued that the first step in delimiting boundaries is to list and balance all the relevant factors and then the next step of drawing the boundaries is to be followed.[160]

China's basic preference for the equitable principles is well illustrated by Article 2 of the Act on the Exclusive Economic Zone and the Continental Shelf on 26 June 1998, which provides for "the principle of equity" for delimitation of boundaries of EEZ and the continental shelf as was mentioned earlier. It is also to be noted that China has clearly supported the equitable principles in the negotiations in the UNCLOS III. A Chinese delegate at the Plenary Meeting of the 9th Session on 25 August 1980 of the UNCLOS III said that:

> . . . In our view, the delimitation of the exclusive economic zone and the continental shelf between States with opposite or adjacent coasts should be determined through negotiations on an equal footing by the parties concerned, in accordance with the principle of equity and taking into account all relevant factors and circumstances. This is the only way to attain a result which is fair and just to all parties concerned. The median or equidistance line is only one method of delimitation, which maybe adopted only when it is in conformity with the principle of equity.[161]

Even if China's adherence to the equitable principles in the negotiations with Korea on the delimitation of maritime boundaries in the Yellow Sea is understandable in the light of its basic position on the equitable principles, the difference of positions between Korea and China on the applicable principles on delimitation of maritime boundaries in the Yellow Sea is hard to be narrowed down. It is not clear as to what factors China thinks relevant in the delimitation of the Yellow Sea. Some writers mentioned that China's position in the delimitation in the Yellow Sea is to draw a maritime boundary following the silt line, which is the line where the sediments from the Chinese rivers meet.[162] If the silt line is adopted as a boundary then two thirds of the Yellow sea will fall into Chinese zones. However, it is difficult to see why a maritime boundary or shelf boundary should be drawn along the silt line because the delimitation in question is about continental shelf and water column and has nothing to do with the question of where the additional subsoil is from. Note that the two countries share a common geological continental shelf. With regard to the silt-line claims of China, Charney pointed out that: "Such arguments have attracted little contemporary support, notwithstanding the language of the *North Sea Continental Shelf* Cases".[163] The Chinese government might also argue the principle of proportionality as a relevant factor. However, it appears that there is no fundamental discrepancy between the lengths of the relevant coasts of Korea and China. And also there appears to be no substantial gap between the ratio of the lengths of the relevant coasts and the ratio of the zones that are to be divided along the equidistance line.[164]

From Korea's standpoint, the Chinese argument that the listing and balancing of all the relevant factors should come first before drawing any *prima facie* line would seem a tactic for delaying the delimitation of boundaries in the Yellow Sea and thus for keeping the traditional fishing patterns as long as possible.[165] In this regard it is interesting to note that China does not always stick to the equitable principles. Note that China vindicated equidistance lines for demarcations in the Bay of Korea with North Korea and in the Gulf of Tongkin with Vietnam.[166]

It remains to be seen how long it will take for the Korean government and the Chinese government to overcome their differences on the applicable principles on the delimitation. However, the equidistance line seems appropriate for a boundary in the Yellow Sea where Korea and China are opposite States and there is no particular factor to be taken into account. Furthermore, the Yellow Sea lies within a common geographical structure. Jeanette Greenfield expressed a similar view:

> Having regard to the geographical common shelf area between China and Korea in the Yellow Sea, it appears that the median line method would be the most suitable.[167]

Ironically, however, Korea would still have some difficulty in accepting the equidistance line for the boundary of the continental shelf, even if China agreed with Korea to draw an equidistance line for the delimitation of a boundary in the Yellow Sea, because the western limits of the Korean submarine mineral exploration areas are not drawn in accordance with the equidistant line and the western limit areas are drawn further out to the Chinese side beyond the equidistance line.

3.2.2. *Disputes over the Basepoints*

There is no territorial dispute over islands between Korea and China. However, it would not be at all easy for Korea and China to agree on each other's valid basepoints as there are several problematic islets and submerged features situated far from the shores. As mentioned earlier, Korea has challenged some of the straight baselines adopted by China in 1996 in light of the provisions of Article 7 of the LOS Convention. Among others, the disputes on the effect of Haijiao (Barren Island or Tungtao), Dongnanjiao, Machaiheng and Waikejiao which China used in drawing straight baselines would be the most problematic in the negotiations. Haijiao is a tiny rock situated about 69 miles from Shanghai and Dongnanjiao is also a tiny rock and situated further out to sea than Haijiao. Note that Machaiheng and Waikejiao appear to be submerged sand features situated about 40 miles from the Chinese coasts.

It appears that the Korean government challenged the legality of using the Haijiao, Hainanjiao, Machaiheng and Weikejiao as basepoints for drawing straight baselines. Dr. Hee-Kwon Park, a former director at the international legal affairs division of the Korean Foreign Ministry, pointed out

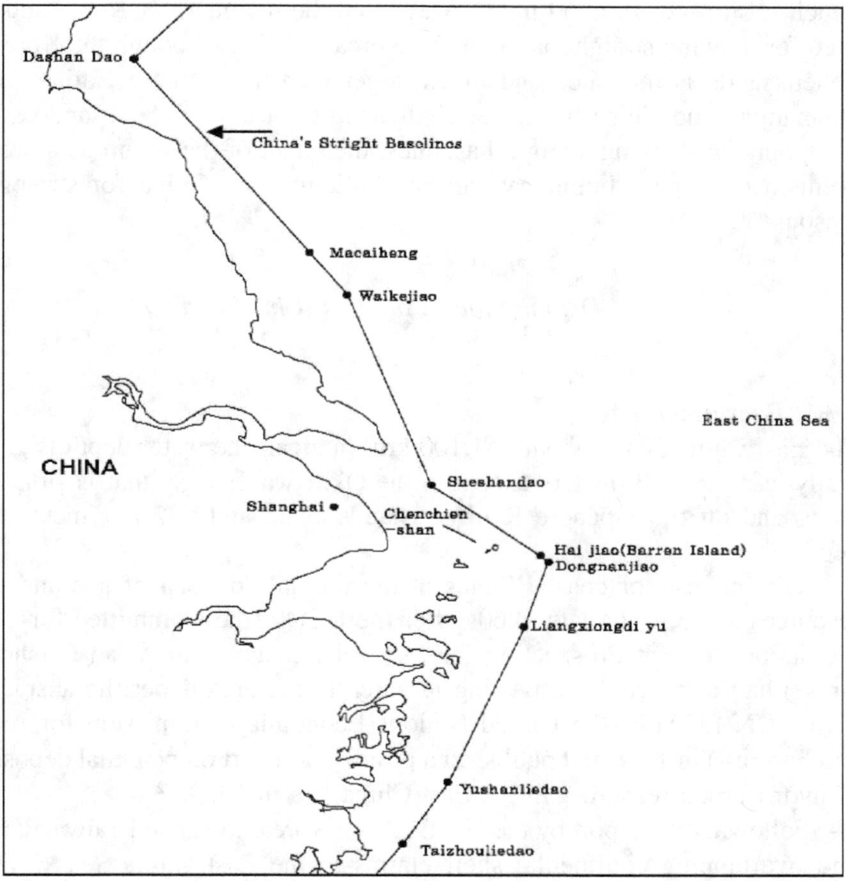

Map 34: Some Chinese Islands

that the "straight baselines running through the Haijiao and Hainanjiao depart, to an appreciable extent, from the general direction of the coast and the area lying within the baselines running through the two rocks is too distant from the mainland to be adequately subject to the regime of internal waters".[168] And he also asserted that Machaiheng and Weikejiao may be submerged sand banks or low-tide elevations and there is no fringe of islands around.[169] The US State Department similarly noted that Machaiheng is located in waters of less then 3 metres depth and Weikejiao is a low-tide elevation.[170]

However, China would try to defend the legality of the use of the problematic Haijiao, Dongnanjiao, Machaiheng and Weikejiao as basepoints for straight baselines and thus to use these features as basepoints in the delimitation. Also China might wish to challenge Korea's island Soheuksando which is situated some 60 miles away from the mainland of Korea and is used for drawing straight baselines by Korea in 1978. Although the Korean Soheuksando is inhabited and much larger than the Chinese Haijiao and Donnanjiao and China has never challenged the use of Soheuksando as a basepoint for drawing straight baselines, the effect of the island as a basepoint in drawing a boundary can be challenged by China for strategic reasons.

3.3. *Delimitation in the East China Sea*

3.3.1. *Background*

The East China Sea is about 751,100 km^2 in area. The water-depth is generally less than 349 meters except in the Okinawa Trough, that is situated along and off the Japanese Ryukyu islands chain and is 2,717 meters in depth.[171]

The interests of coastal States in the possible deposit of gas and oil resources arose in the late 1960s after the CCOP (the Committee for Coordination of Joint Prospecting for Mineral Resources in Asian Offshore Areas) had conducted prospecting research in the area under the auspices of the UN ECAFE (the United Nations Economic Commission for Asia and Far East) in 1968 and published a promising report on potential deposits of hydrocarbon resources in the East China Sea in 1969.[172]

Following the report by the UN ECAFE, Korea, Japan and Taiwan have laid overlapping continental shelf claims to the East China Sea.[173] The Beijing government did not lay down any specific continental shelf claims at that time. But when a liaison office of non-governmental nature was established by Korea, Japan and Taiwan in Seoul in November 1970 for consultation of joint development in the region, the Beijing government protested against its establishment, and then the liaison office became inactive.[174] It was later in 1992 when the Chinese government established two mining areas in the region: one is the Northern Acreage and the other is the Southern Acreage.

As mentioned earlier, when the Korean government was preparing for the enactment of the "Submarine Mineral Resources Development Act" in 1968, the Japanese government showed interest in the act and thus the two governments began to negotiate for the delimitation of their overlapping continental shelf in 1970. The negotiations led to a joint development agreement in the East China Sea.

The need for the delimitation of EEZ boundaries in the East China Sea arose in the late 1990s when Korea, China and Japan ratified the LOS Convention and enacted laws on EEZ and the continental shelf. And thus bilateral negotiations between Korea, Japan and China for delimitation of EEZ boundaries began in late 1996.

The delimitation of EEZ and the continental shelf boundaries in the East China Sea is some what more complicated than that in the East Sea or in the Yellow Sea: the coastal States are not two, but three; there is a territorial dispute regarding the Senkaku/Diaoyutai between China and Japan; there is a dispute on the baselines and basepoints; the three coastal States do not have common positions on the applicable principles on the delimitation; and the existence of the Okinawa Trough raises a difficult question with regard to the effect of the Trough on the delimitation and the relationship between the boundaries of EEZ and the continental shelf.

It can be reasonably presumed here that the negotiations between Korea, China, and Japan for delimitation of EEZ/the continental shelf boundaries in the East China Sea would not result in a solution in the near future.

3.3.2. *Different Positions of the Coastal States on the Delimitation*

Although Korea has been arguing for the equidistance principle in the Yellow Sea and in the East Sea, Korea has been upholding the equitable principles in the East China Sea relying on the natural prolongation theory.[175] Korea's reliance on the equitable principles in the East China Sea is derived from its position that the underwater natural prolongation of the Korean peninsula extends to the Okinawa Trough which is situated about 280 miles from Korea's southernmost island of Mara-do, and that Japan's natural prolongation underneath the sea ends at the Okinawa Trough which is only about 100 miles from the Japanese Ryukyu islands.[176]

China also upholds the natural prolongation theory for the delimitation of boundaries in the East China Sea relying on the presence of the Okinawa Trough.[177] In this regard, the Chinese government argued that:

> The East China Sea continental shelf is the natural extension of the Chinese continental territory. The People's Republic of China has inviolable sovereignty over the East China Sea continental shelf.[178]

The Chinese Exclusive Economic Zone and Continental Shelf Act of 1998 specifically mentions "natural prolongation" in Article 2, which provides that: "The continental shelf of the People's Republic of China comprises the seabed and subsoil of the submarine areas that extend beyond its territorial sea throughout the natural prolongation of its land territory to the outer edge of the continental margin". Japan, however, has been arguing to adopt a strict equidistance line for maritime boundaries in the East China

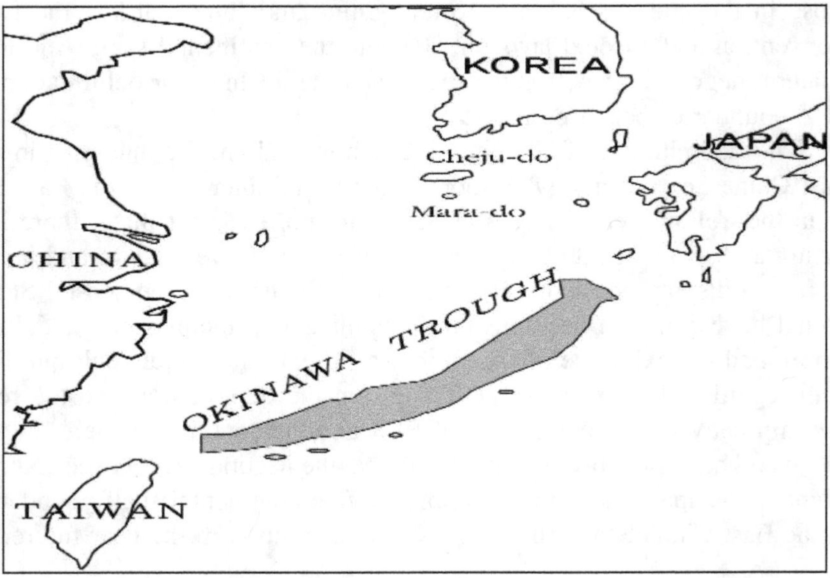

Map 35: East China Sea and the Okinawa Trough

Sea whether it be the continental shelf or the exclusive economic zone, as it best benefits Japan because Japan has a number of small islands in the East China Sea and wishes to refute the natural prolongation theory with the argument for an equidistance line.[179]

Korea and China might not stick to the natural prolongation with regard to the delimitation of the exclusive economic zone (water column) boundaries whereas they would do so with regard to the delimitation of the continental shelf boundaries in the East China Sea, because there is no doubt that Japan is also entitled to 200 N.M. economic zone under the institution of the exclusive economic zone of the LOS Convention. With regard to the Chinese position, Mr. Keyuan wrote that: "Recently, China has realised the difficulty of applying the natural prolongation principle to maritime boundary delimitation, and begun to consider the principle of proportionality which is favourable for China. . . . If China insists on natural prolongation for the delimitation of the continental shelf while agreeing to the median line as the line of delimitation of the EEZ, then there would be two different delimitation lines in the East China Sea with Japan . . ."[180] With regard to Korea's possible position, it is to be noted as will be seen in the next chapter that Korea agreed to establish a joint fishing zone in the East China Sea on the basis of two contending hypothetical equidistance lines argued by Korea and Japan respectively and most of the joint fishing zone is situated in the upper part of the joint development zone of 1974. This arguably implies that Korea might agree to apply distance criteria as long as the boundaries of water column are concerned in the East China Sea.[181]

Now therefore, a question would arise as to whether two different boundaries, that is to say, one for the water column and the other for the continental shelf, can be drawn. This question would be entered in the negotiations on the delimitation of maritime boundaries in the East China Sea. It is to be recalled that, as we examined in Chapter One, a single boundary is normally a matter of choice which is to be decided by coastal States through negotiations and thus the question is who has the stronger argument.[182] The points of argument in this regard would be whether the Okinawa Trough really disrupts the unity of the continental shelf in the area. The question of whether a geological factor such as the Okinawa Trough really disrupts the unity of the continental shelf appears to be a decisive factor in the maritime area where the distance between the nearest coasts of neighbouring States is less than 400 miles. We will discuss these questions soon, but before that we need further discussion on the delimitation of the EEZ in the East China Sea.

3.3.3. *Proportionality and Cut-off Effect in the East China Sea*

As mentioned earlier, China might argue for the application of proportionality in the delimitation of maritime boundaries in the East China Sea because China appears to have the longest coastal line among the three littoral States in the area. In particular, China appears to have a good case for arguing the application of proportionality in relation to Japan because the substantial part of the Japanese coast in the area is formed not by the coasts of the Japanese main islands but by the chain of Ryukyu islands which has a great many openings therein.

The concept of proportionality can be seen in the judgment by the ICJ in the *North Sea Continental Shelf* Cases when it mentioned "a reasonable degree of proportionality, which a delimitation carried out in accordance with equitable principles ought to bring about between the extent of the continental shelf areas appertaining to the coastal State and the length of its coast measured in the general direction of its coastline".[183] However, as the term "reasonable degree" suggests, the proportionality test does not require that the ratio of sizes of delimited areas should be the same as the ratio of coastal lengths in the relevant areas. The ICJ did not accept the German argument in the *North Sea Continental Shelf* Cases that the delimitation should be made as to ensure "a just and equitable share of the available continental shelf in proportion to the length of its coastal line or sea-frontage".[184] From this line of reasoning, the ICJ in the *Tunisia/Libya Continental Shelf* Case said that the delimitation it was going to deliver satisfied the proportionality test as the ratio of the coastal lengths of Libya and Tunisia was calculated at between 31:69 and 34:66 and the ratio of areas divided by the line that the Court was going to render was put at 40 (Libya): 60 (Tunisia).[185] Although proportionality plays a role mainly as an *ex post* factor to check the equitableness of the delimitation of boundaries in international courts,[186] it sometimes plays a decisive role in the process of drawing boundaries by States in negotiations.[187] For example, France and Spain measured their coastal lengths in the process of delimitation in the Bay of Biscay and the ratio was put at 1.549 (France): 1 (Spain) and they divided the continental shelf between them at a ratio of 1.63 (France): 1 (Spain).[188]

Regardless of when proportionality plays a role in the process of delimitation, the main problem in applying proportionality appears to be in the vagueness of the desired relationship between the relative lengths of the coastlines and the divided areas.[189] In other words, the question is how far the "reasonable degree" allows a gap between the two ratios of the coastal lengths and the divided areas. The reasonable degree appears to be different from case to case. For example, in the *St. Pierre and Miquelon Arbitration*, the tribunal put the ratio of relevant coastal lengths at 15.3 (Canada): 1 (France)[190] and then drew a boundary in such a way as to divide the maritime area at the ratio of 16.4 (Canada): 1 (France).[191] This is an instance where the ratio of the divided areas followed the ratio of coastal lengths very closely. However, in the *Jan Mayen* Case between Norway and Denmark, the ICJ left a big gap between the two ratios: it drew a line which divides the areas at the ratio of 3 (Denmark): 1 (Norway)[192] although the Court calculated that the ratio of the relevant coastal lengths is between 9:1 and 9:2 (Denmark: Norway).[193]

In the East China Sea, the ratio of relevant coastal lengths among China, Japan, and Korea seems to be at 2.1:1.2: 1 and the ratios of areas divided along the equidistance line appears to be at 3.3:3.6: 1 when the Chinese and Japanese rocks are used as their basepoints, or at 2.2:2.3: 1 when the rocks are excluded from the basepoints.[194] Note that Korea does not have any rock in the area whereas Japan has a number of rocks and China has a few rocks in the East China Sea and some of their rocks are situated far away from their main coasts. Therefore, China and Japan have a common interest in arguing for the utilisation of their rocks as their basepoints. From the rough assessment, we can also see that China and Korea have a good position for arguing for application of the proportionality factor in delimiting maritime boundaries in the East China Sea whereas Japan would argue for the application of the strict equidistance line.

Korea might be tempted to argue for remedying the cut-off effect, noting that its southernmost point in the hypothetical boundaries drawn by applying the equidistance method fall far short of the middle of the East China Sea, whether or not its neighbours' rocks are used as their basepoints. It can be recalled that Germany successfully argued for the problem of cut-off effect which was to be brought about when the equidistance method was applied.[195]

However, the hypothetical Korean argument for the need of addressing the cut-off effect begs the following question: can the German reliance on the problem of a cut-off effect also be utilised by Korea where, unlike Germany, its coastline is not concave? In other words, the question is whether the concave shape of the coastline is a necessary condition for advancing the argument of a cut-off effect. It may not be. Professor Charney observed that the international courts have shown a tendency "to delimit maritime boundaries so that all disputants are allotted some access to areas approaching the maximum distance from the coast permitted for each zone"[196] He observed that Germany was given a seaward window that approaches the middle of the North Sea in the *North Sea Continental Shelf* Cases; Honduras was given a share in the ocean seaward of the gulf closing line in the *Gulf of Fonseca* Case; France was given a corridor of 200 miles long in the *St. Pierre and Miquelon Arbitration*; and Norway (Jan Mayen) was allowed to have the middle of the disputed areas in the *Jan Mayen* Case.[197] It is not clear whether the so-called "maximum reach" is now a rule of customary international law. However, it is clear that there is a general support for the application of maximum reach from the State practice because "in virtually all maritime boundary delimitations already established (approximately 140), coastal States are not cut-off from either access to the open sea or to areas near the centre of the water body on which their coastlines are located".[198]

3.3.4. *The Question of Tri-junction in the East China Sea*

As there are three coastal States in the East China Sea, there should be a common understanding among the three States in order for the maritime boundaries in the area to be complete. Of course, the negotiations for drawing maritime boundaries can be conducted in three bilateral negotiations; namely, negotiations between Korea and Japan, between Japan and China, and between China and Korea. However, if the three boundaries between

Korea and Japan, between Japan and China and between China and Korea do not meet with each other altogether, then the boundaries cannot be complete. Because if the three boundaries do not join together somewhere in the East China Sea, it means that there is an undelimited area or that there are different views among the three States on where the boundaries should go. Therefore, the task of reaching agreement on tri-junctions among the three coastal States in the East China Sea is of vital importance for the completion of the delimitation of the maritime boundaries.

It would be ideal for the three relevant States to "get together around a table" and agree on tri-junctions. The following agreements are examples of tripartite agreements on tri-junctions: the agreement among Indonesia, Malaysia and Thailand of 1979 for the delimitation in the Malacca Strait;[199] the agreement among India, Sri Lanka and The Maldives of 1976 for the delimitation in the Indian Ocean;[200] and finally the agreement among India, Indonesia and Thailand of 1978 for the delimitation in the Andaman Sea.[201]

However, three bilateral agreements can also establish tri-junctions with prior co-ordination among the relevant coastal States. The three tri-junctions in the North Sea were all agreed by way of three bilateral agreements. These are tri-junctions among Norway, Denmark and the United Kingdom, Denmark, Germany and the United Kingdom, and Germany, the Netherlands and the United Kingdom.[202] It seems that co-ordination behind the scenes play an important role in agreeing on a tri-junction by way of three bilateral agreements. If the co-ordination is not successfully conducted then a tri-junction cannot be successfully established. For instance the United Kingdom concluded two boundary agreements in 1966: one was with Denmark and the other was with the Netherlands, thereby establishing a tri-junction among Denmark, the Netherlands and the United Kingdom in the North Sea. In the agreements, however, the United Kingdom appears to have disregarded the possibility that her continental shelf would border that of Germany. However, after Germany and Denmark and Germany and the Netherlands concluded boundaries agreements in 1971 giving Germany a corridor leading to the middle of the North Sea following the ICJ judgment on the *North Sea Continental Shelf* Cases of 1969, the United Kingdom had to conclude new agreements with Denmark and The Netherlands and also with Germany for new tri-junctions in the North Sea.[203]

Unfortunately, however, exclusive bilateralism has made its presence in the East China Sea as the joint development agreement of 1974 between Korea and Japan and the ongoing discussion between China and Japan on the prior notification scheme on the marine scientific research between China and Japan illustrates. And the fisheries agreement between China and

Japan, which will be discussed in the next chapter is also an instance where the exclusive bilateralism played a role. If therefore tripartite negotiations are difficult to take place to pinpoint a tri-junction in the East China Sea and thus the bilateralism is to be utilised, then there should be caution in bilateral agreements not to delimit the area to which a third party, which is also a coastal State, can lay claims to in good faith.[204]

Apart from the procedural aspects of agreeing on tri-junctions such as whether it is through bilateral negotiations or trilateral negotiations and also apart from the different positions on the principles applicable to the delimitation in the region, there are some substantive aspects to the valid basepoints in the East China Sea as there are a number of small islands in the East China Sea. Korea has only one island, called Mara-do which is in a position to affect the delimitation in the East China Sea. As Mara-do is big enough to accommodate about 20 families (about 80 people) and only 3 miles from the main island, Cheju island, there would not be any objection from China or Japan in using Mara-do as Korea's basepoint. However, since the Chinese tiny rocks, Haijiao and Dongnanjiao which we discussed earlier, are in a position to affect the delimitation in the East China Sea and Korea has objected to the use of these rocks as basepoints for drawing straight baselines, the question of how much effect the Chinese rocks can have in the delimitation is a point of dispute between Korea and China. It appears from the Chinese literature and charts that the island, Chen-chien-shan, which is located about 18 miles from Haijiao to the Chinese coast can be a valid basepoint as it is a big island with 900 inhabitants.[205]

It is Japan that has the largest number of islands, islets and rocks in the East China Sea. From Japanese literature and charts it appears that there are 14 or so small islands which might affect the delimitation of maritime boundaries with Korea and China in the East China Sea and that about half of these islands are not capable of sustaining human habitation or economic life of their own.[206] However, as Japan tries to play down the provisions of Article 121(3) of the LOS Convention and tries to use all the features, China and Korea have discussed with Japan the selection of Japanese basepoints and the degree of the effect of the Japanese tiny islands. Here the territorial dispute over Senkaku and Diayoutai between China, Taiwan and Japan are also factors which stand in the way of effective negotiations.[207]

Once the three coastal States agree on the valid basepoints and the effect of the basepoints, then provisional tri-equidistance lines and provisional tri-junctions can be marked on the map and from there further negotiations can be carried out for adjusting the initial equidistant tri-junction.[208] In the negotiations for adjusting the initial positioning of the tri-equidistance line, Korea and China would have a strong argument to take account into the geography as an important relevant factor in delimitation in the area.

3.3.5. *Effect of the Okinawa Trough in the Delimitation*

Even if Korea, China and Japan could agree on the valid basepoints and their effect on the boundaries, they might be in a situation where they should consider the effect of the Okinawa Trough in the delimitation. Both Korea and China have argued for the application of the natural prolongation theory relying on the presence of the Okinawa Trough in the delimitation of maritime boundaries in the East China Sea where the distance between the nearest costs of the coastal States is less than 400 miles.[209] Thus the question of whether China and Korea can benefit from the presence of the Okinawa Trough necessarily arises. If the Okinawa Trough can be accepted as an important factor in the delimitation then there is a high possibility that the EEZ boundaries would be drawn differently from the continental shelf boundaries.

As I have examined in Chapter 1, the ICJ or arbitration courts have not made different boundaries of the continental shelf and the exclusive economic zone when they have dealt with the issue of the delimitation of boundaries of the zone and the shelf at the same time. It may be recalled that the courts have made single boundaries in the *Gulf Maine* Case, *Jan Mayen* Case, *Guinea-Guinea Bissau Arbitration*, *St. Pierre and Miquelon Arbitration*, *Qatar/Bahrain* Case, and *Eritrea/Yemen Arbitration*. As was also noted in Chapter 1, the number of instances where there are two different boundaries are few. However, different boundaries can be drawn if Korea and China have a good argument *vis-à-vis* Japan relying on a possible effect of the Okinawa Trough. Because, as the ICJ puts it, "the concept of a single maritime boundary does not stem from multilateral treaty law but from State practice, and that it finds ***its explanation in the wish of States*** (emphasis added)".[210] However, Japan would not prefer different boundaries as Japan sticks to the equidistance principle both for continental shelf and the EEZ boundaries. Thus the answer to the question of whether different boundaries can be drawn depends on which States have the stronger argument. Therefore, the question of whether the Okinawa Trough can have some effect in the delimitation takes vital importance.

In the *Libya/Malta Continental Shelf* Case Libya argued for the application of the natural prolongation theory saying that there is a fundamental discontinuity between the continental shelves of the two Parties along with the rift zone situated between Libya and Malta.[211] However, the Court rejected the Libyan argument. It declared that:

> . . . since the development of the law enables a State to claim that the continental shelf appertaining to it extends up to as far as 200 miles from its coasts, whatever the geological characteristics of the corresponding sea-bed and subsoil, there is no reason to ascribe any role to geological or geophysical factors within that distance either in verifying the legal title of the States concerned or in proceeding to a delimitation as between their claims. . . . It follows that, since the distance between the coasts of the Parties is less than 400 miles, so that no geophysical feature can lie more than 200 miles from each coast, . . .[212]

Although we can understand the Court's emphasis on the distance criterion, it appears that the Court went too far when it refused to give "any role" to the geological factors, which might be one of the relevant factors in achieving an equitable solution.[213] In this regard, Evans has a point when he said that:

> The Court correctly saw "distance" as an element within the concept of natural prolongation but erred in seeing it as the primary component of it.[214]

Ironically the ICJ gave thought to the possible existence of the trough which Libya argued and then concluded that:

> The Court is not satisfied that it would not be able to draw any sufficiently cogent conclusions from it as to the existence or not of the fundamental discontinuity on which the Libyan argument relies.[215]

What does this mean? It might imply that if there had been a fundamental discontinuity between Malta and Libya, then the Court would have taken into account the fundamental discontinuity along the Rift Zone in the delimitation. Note that in the paragraph of the judgment where the Court denied any role to geological or geophysical factors within 200 miles, the Court mentioned a fundamental discontinuity, implying that the Court would have taken the rift zone into account if it constituted a fundamental discontinuity. It thus said:

> the feature referred to as the "rift zone" cannot constitute a fundamental discontinuity terminating the southward extension of the Maltese shelf and the northward extension of the Libyan as it were some natural boundary.[216]

In fact, that the ICJ in the *Tunisia/Libya Continental Shelf* Case admitted that it can into take account certain geomorphological configurations of sea-bed in the delimitation, saying that: "the physical factor constituting the natural prolongation is not taken as a legal tile, but as physical circumstances considered to be elements of an equitable solution".[217] The Court, however, decided not to give a role to the Tripolitanian Furrow as "it is not such a significant feature that it interrupts the continuity of the Pelagian Block as the common natural prolongation of the territory of both Parties".[218] Then the Court went on to make it clear that there has to be a fundamental discontinuity of the seabed in order for a submarine feature to be taken into account in delimitation. It thus said:

> unless it were such as to disrupt the essential unity of the continental shelf so as to justify a delimitation on the basis of its identification as the division between areas of natural prolongation, it would be an element inappropriate for inclusion among the factors to be balanced up with a view to equitable delimitation.[219]

The position of the ICJ in the *Tunisia/Libya Continental Shelf* Case, which looks into whether there is disruption in the essential unity of the continental shelf, appears to be in line with the consideration of the court in the *Anglo/French Arbitration.* In the *Anglo/French Arbitration* the UK proposed a boundary along the Hurd Deep and Hurd Deep Zone, as an alternative to the equidistance line, which is situated alongside 1,000 metre isobath and somewhat south to the equidistance line.[220] In consideration of the UK proposal, the court compared the Hurd Deep and Hurd Deep Fault Zone with the Norwegian Trough in the North Sea and noted that the former has "minor faults in the geological structure of the shelf" in comparison with the "deep Norwegian Trough".[221] And the court decided not to accept the UK proposal noting the "essential continuity of the continental shelf in the Channel and the Atlantic region". It thus said:

> The geological faults which constitute the Hurd Deep and the so-called Hurd Deep Fault Zone, even if they be considered as distinct features in the geomorphology of the shelf, are still discontinuity in the seabed and subsoil which ***do not disrupt the essential unity of the continental shelf*** either in the Channel or the Atlantic region.[222]

From the observation made above, the question whether the Okinawa Trough constitutes a fundamental discontinuity appears to be important in the delimitation. Unlike the Rift Zone, the Tripolitanian Furrow and the Hurd Deep and Hurd Deep Fault Zone, the Okinawa Trough appears to disrupt the unity of the continental shelf in the region, as it is generally as deep as 2,700 metres and thus could be argued as an important relevant circumstance in the delimitation in the region.[223] However, even if the three littoral States have a common understanding that a fundamental discontinuity of the shelf is a relevant circumstance, there might be different views on whether the Okinawa Trough really disrupts the unity of the seabed in the East China Sea.[224] In this regard, a question arises as to whether the Commission on the Limits of the Continental Shelf can play a role in ascertaining the legal nature of the Okinawa Trough in light of Article 76 of the LOS Convention. Note that the main function of the Commission is to make recommendations to coastal States on matters relating to the establishment of the outer limits of their continental shelf where that shelf extends beyond 200 N.M. from the coast.[225] If there is only one coastal State or there are no overlapping claims of the continental shelf in the East China Sea, then surely the recommendations by the Commission on the outer limits of the continental shelf will be final and binding as paragraph 8 of Article 76 provides.[226] But as there are overlapping claims over the continental shelf in the East China Sea, it is doubtful whether the Commission can make recommendations on the legal status of the Okinawa Trough upon the request of a State in the region. It appears from the provisions of Article 9 of Annex II of the LOS Convention that the Commission would not be in a position to make any recommendations on the legal status of the Okinawa Trough at the request of a coastal States in the East China Sea. Because it provides that: "The actions of the Commission shall not prejudice matters relating to the delimitation of boundaries between States with opposite or adjacent coasts". The Rules of Procedure of the Commission further elaborates on this issue. According to the Rules of Procedures, the Commission may examine the submissions in the area under dispute only with prior consent by all States concerned and the recommendations shall not prejudice the position of States that are parties to the dispute.[227]

Therefore, in light of the provisions of the LOS Convention and the Rules of the Procedure, the prior consent by all three coastal States is needed to find out from the Commission whether the Okinawa Trough constitutes a juridical continental margin. And note that even if there is a recommendation by the Commission, it does not affect the positions of the coastal States on the delimitation of the continental shelf. Furthermore, the possibility of Japan giving its prior consent for any submission on the Okinawa Trough to the Commission on the Limits of the Continental Shelf would be very low. Note that Japan has objected to the Chinese marine scientific research in the East China Sea, when the research was conducted in the waters on its side from the hypothetical equidistance line between China and Japan. It was also noted that there were several occasions where the Korean research vessels have been followed and asked to stop the research in the waters on the Japanese side of a hypothetical equidistance line between Korea and Japan.[228] It may mean that the Korean research vessel needs permission for the geomorphological survey of the Okinawa Trough, in order to avoid intervention from the Japanese maritime security agency at sea, although Korea is of the view that the Okinawa Trough is the end of its natural prolongation. It might be that Korea does not wish to apply for Japan's consent for the marine scientific research in the Okinawa Trough, fearing that its application might give the impression that Korea recognises the Japanese position on the delimitation of the maritime boundaries of the continental shelf. Therefore, the best way to solve the disputes on the delimitation of the EEZ and the continental shelf in the East China Sea would be to refer the disputes to the ICJ, ITLOS or arbitration. Note that neither Japan nor Korea nor China has exercised their rights under Article 298 of the LOS Convention to opt out of the compulsory procedures for disputes on the sea boundary delimitation.

3.3.6. *Diaoyutai/Senkaku and Delimitation*

Diaoyutai in Chinese or Senkaku in Japanese is a chain of eight small islands located to the north east of Taiwan and the south west of the Ryukyu islands in the East China Sea.[229]

It seems appropriate to briefly summarise the positions of China (Taiwan) and Japan on the territorial issue over the island, and then to consider the implications the disputed island might have over the delimitation. China (Taiwan) relies for its territorial title on its discovery and her long use of the islands since the Ming Dynasty in the fifteenth century. The Chinese position is that the islets were ceded to Japan by the Simonoseki Treaty of 1895 after the Sino-Japanese War between 1894-1895 along with Taiwan and Japan who renounced its title over the islets along with Taiwan by the San Fransico Peace Treaty of 1951. The Japanese position is that the islets were *terra nullius* when it incorporated the islets in January 1895, and thus the islets were not ceded from China by the Simonoseky Treaty, and the islets have been Japanese territory except in the time of US trusteeship after World War II.[230]

This island appears to sit at the top of the eastern marine of the continental shelf protruding from China and is located to the west of the Okinawa Trough.[231] Therefore China (Taiwan) would have a good argument, if this island is declared to be Chinese (Taiwanese), that the Okinawa Trough should be a boundary delimiting the eastern location of a Chinese (Taiwanese) boundary with Japan; but if this island is declared to be Japanese, then Japan might be in a position to link the maritime zones generated from the Ryukyu islands and Senkaku.[232]

It might not be easy for either Party to argue to disregard the disputed island by simply relying on the provision of Article 121(3) of the LOS

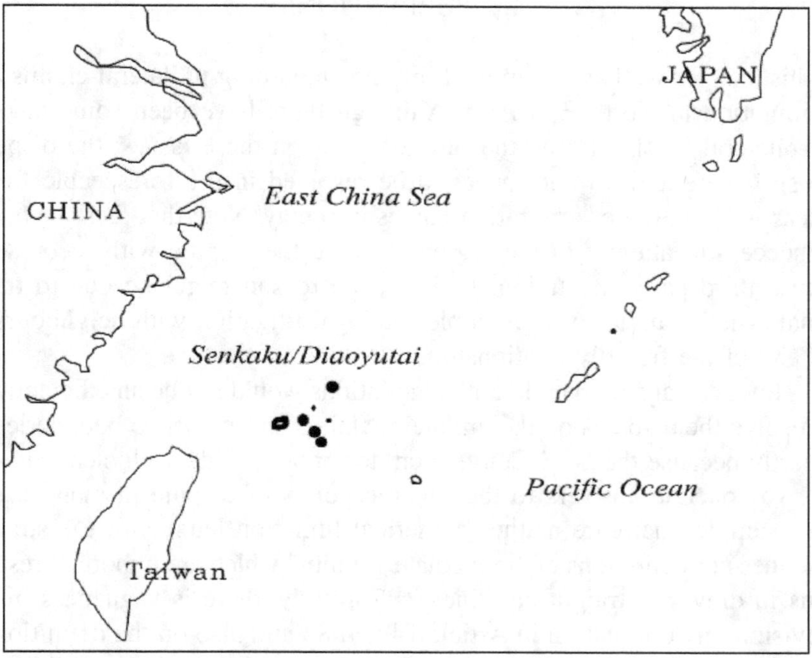

Map 36: Diaoyutai/Senkaku Islands

Convention. It is to be noted that not all of the islands in the group of Diaoyutai /Senkaku appear to be rocks in terms of Article 121(3) of the LOS Convention because some of the islands have soil with a variety of tropical vegetation and the largest one is as large as 3,5 km^2 in size although most of the islets are relatively small.[233] But apart from the provision of Article 121(3) of the LOS Convention, it seems probable that the maritime zone around Diaoyutai/Senkaku would be enclaved or the effect of the island would be very limited for an equitable solution because of its relatively small size and remoteness either from China (Taiwan) or Japan.[234] In this regard, it can be recalled that the Channel Islands, which belong to the UK but are located closer to the French coast than to the British coast, were given only limited effect and thus the continental shelf of the islands were enclaved in the *Anglo-French Arbitration*.[235]

4. OBSERVATIONS

In this chapter we have examined disputes regarding unilateral claims and delimitation in North East Asia. Although there have been some talks at the bilateral level between the littoral States on these issues, the disputes are still there and do not appear to be resolved in the foreseeable future unless ICJ, ITLOS or arbitration comes into play. Note that there has been no successful attempt in the region to solve the dispute with recourse to such a third party institution. Perhaps, the reason might be due to traditional Asiatic culture where people see legal litigation with neighbours as the end of the friendly relationship.

However, again, the bilateral negotiations would not be an efficient way to resolve the disputes on the unilateral claims and maritime boundaries. It is partly because the LOS Convention does not provide such clear rules as to give concrete answers to the specific points of dispute in many cases. For example, there are neither numerical limits on lengths of the straight baselines nor definitions of "immediate vicinity" which are important restrictions in drawing straight baselines.[236] Similarly, there is vagueness in the provision on delimitation in Article 74 and 83 and also on the definition of rocks in paragraph 3 of Article 121. Still, however, there is a possibility of these disputes being brought before a compulsory disputes settlement procedure as none of the coastal States have opted out of the procedure on the sea boundary dispute.

In this situation, the three coastal States in North East Asia concluded three bilateral fisheries pending the ultimate resolutions of disputes on unilateral claims and delimitation reserving their final positions. We will discuss the three provisional fisheries agreements in the next chapter.

NOTES

1. The whole of Korea was under the control of Japan from 1910 to 1945 and there was no distinction between South Korea and North Korea until 1945 when the Korean peninsula was divided along with Latitude 38° North by the Allied

Powers for the purpose of the military operation against the Japanese army in the Korean peninsula.

2. North Korea was placed under the control of the Soviet Union until 1948 when the communist government of the Democratic People's Republic of Korea was established.

3. Official Gazette, USAMGIK Ordinance No. 189, 10 May 1948.

4 The United States adopted a 3-mile territorial sea in 1794 through the Neutrality Act made by Congress on the outbreak of war between Great Britain and France. The Act is understood as the decisive step in the translation of the cannon-shot rule into three-mile rule: see D.P. O'Connel, *The International Law of the Sea*, p. 131. However, the United States extended the extent of the territorial sea to 12 miles in December 1988: United Nations, *The Law of the Sea: National Claims to Maritime Jurisdiction*, 1992, p. 137.

5. The Korean government gave force to laws and regulation made in the reign of the interim government through Article 100 of the first Constitution of the Republic of Korea promulgated on 17 July 1948. It provided that "current laws and regulations continue to be effective as far as they do not conflict with this constitution (translated by the author)". As the provision on the three-mile territorial waters in the Ordinance on the Function of the Korean Coast Guard did not conflict with the first Constitution, the three-mile rule was regarded to be effective. In 1961, however, the Korean government took measure to abolish so-called "old laws and regulations" as of 20 January 1962. Article 1 of the Act on Special Measure regarding Old Laws and Regulations provided that "Old laws and regulations in this Act refer to the laws and regulations which were entered into force before 16 July 1948 and continued to be effective by Article 100 of the Constitution (translated by the author)". Article 2 of the Act provided that "old laws and regulations will be regarded abolished as of 20 January 1962 unless otherwise provided in accordance with Article 1 (translated by the author)". As there was no provision to save the Ordinance on the Functions of the Korean Coast Guard, the Act was considered abolished.

6. The Territorial Sea Act of 1977 went into effect on 30 April 1978 along with its Enforcement Decree of 1978. The title of the Act changed to the "Territorial Sea and Contiguous Zone Act" in 1995 for the purpose of establishing contiguous zone in accordance with relevant provisions of the LOS Convention.

7. The Western Channel of the Korea Strait is situated at an important position linking the East Sea (Sea of Japan) to the north and the East China Sea to the south. The narrowest width of the Channel is only 22,75 miles between the eastern end of Geoje-do on the Korean side and Saozaki on the Tsushima islands of Japan. As both Korea and Japan have adopted three-mile limits for the breadth of their respective territorial waters, there is a high seas corridor in the Western Channel of the Korea Strait: see Hyun-soo Kim, *The Impact of the U.N. Convention on the*

Law of the Sea on the Territorial Sea Regime of Selected East Asian Countries, Ph.D. thesis, University of Wales College of Cardiff (1993), pp. 57-59.

8. Diplomatic notes were sent to the Korean government by the government of the United States of America to show their concern on this matter in 1976. In this regard, Michael and Westerman, referring to *Maritime Claims Reference Manual* (2 vols. 1987) by US Department of Defence, wrote that: "In the Korea Strait, both Japan and South Korea have enacted legislation limiting their otherwise 12-mile territorial seas, leaving a high seas corridor in the western channel of the strait. By setting its baseline for the territorial sea in the manner in Fig. 5.37, however, South Korea has significantly reduced freedom of passage in this corridor. . . . The United States and other international users have refused to recognise South Korea's claim in this area, as well as the attendant provision in South Korean law requiring prior notification for warship entering territorial waters: W. Michael Reisman and Gayl S. Westerman, *Straight Baselines in International Maritime Boundary Delimitation*, Macmillan (1992), p. 178.

9. This provision was derived from the UK proposal. The rationale for the proposal was that "if the strait had a good and wide enough high-seas route down in the middle, it was unnecessary to provide a special right of transit passage since ships and aircraft could navigate on the high sea through the strait". See S.N. Nandan and D.H. Anderson, "Straits Used for International Navigation", 61 *B.Y.I.L.* (1990), pp. 176-178.

10. Note that only ships are entitled to innocent passage. Submarines are required to navigate on the surface showing their flags under the innocent passage, whereas under the transit passage regime both ships and aircraft are entitled to more free passage than under innocent passage. Submarines are under no such obligation as is required under the innocent passage: see Articles 17-20 and Article 39 of the LOS Convention. In this legal framework of the LOS Convention Korea would feel that it is better to keep a high seas/ EEZ corridor in the Korea Strait for the sake of security. Note that there is no part of high seas but exclusive economic zone left in the Korea Strait because both Korea and Japan established exclusive economic zones in 1996. However, as the freedom of the navigation is guaranteed in the waters of the exclusive economic zone like in the high seas and the transit passage regime remains in the Korea Strait as far as there is exclusive economic zone routes left: see Articles 58 and 87 of the LOS Convention.

11. The baselines for measuring the Korea's exclusive fisheries zone were adopted by Exchange Notes between Korea and Japan after consultation between the two countries in accordance with Article 1 of the fisheries agreement. Article 1 of the fisheries agreement provided that: ". . . in case where either Contracting Party uses a straight baseline in establishing its fisheries zone, such straight baseline will be determined upon consultation with the other Party". The Korean government revised its straight baselines on 18 December 2002. The revised straight baselines are, however, basically the same as the old ones as the revision was made in order to

reflect new geodetic system – World Geodetic System, reflect the change of the names of places used for the base points, and minor natural changes in geography in the relevant areas.

12. The issue of drawing straight baselines connecting Cheju island and the Korean peninsula was raised in 1960s in the context of establishing fisheries zone of Korea when Korea and Japan negotiated the fisheries agreement. As a result of compromise, Korea did not draw straight baseline connecting the Korean peninsula and Cheju island but established its fisheries zone between the Korean peninsula and Cheju island.

13. W.M. Reisman and G.S. Westerman, *op. cit.* (1992), p. 175. The Korean peninsula has about 3,500 islands, mostly in the south and the west.

14. W.M. Reisman and G. S. Westerman, *ibid.*, pp. 175-178. They write that: "The southern coast of South Korea is a Norwegian-style coastline, deeply indented and fringed with islands like a *skjaergaard*. The basepoints selected to enclose this area of coast, however, are located on rocks and islands which are not a part of the island fringed and are therefore inappropriate". For a critical examination of the Korea's straight baselines, see US Department of State, *Limits in the Seas No. 121 – Straight Baseline and Territorial Sea Claims: South Korea*, 1998.

15. US Department of State, *Limits in the Sea No. 121 – Straight* Baseline *and Territorial Sea Claims: South Korea*, 1998, p. 6.

16. U.N. Docs. ST/LEG/SER.B/6 (1957) and ST/LEG/SER.B/8 (1959). The text of the proclamation is reproduced by Choon-ho Park, *The Law and Practice Relating to the International Regulation of Fisheries in Asia with Particular Reference to the Korea-Japanese Dispute*, University of Edinburgh, Ph.D. thesis (1971), pp. 398-379.

17. *Statement by President Rhee on February 8, 1952.* The MacArthur Line was imposed by the Occupation Authorities in Japan to restrict the Japanese fishing vessels' fishing activities to the waters around Japan. The MacArthur Line was finally lifted on 25 April 1952, three days before the entry into force of the Peace Treaty between the Allied Powers and Japan.

18. It is notable that a US naval officer expressed sympathy with Korea saying that: "During the years of occupation of Korea by Japan, this area was so badly depleted by the Japanese beam trawlers that the fishing in the area was practically unproductive due to the destruction of bottom feeding grounds and sea grasses . . . Therefore, from the conservation angle, the Korean government has a point in advocating the necessity of controlling fishing in the area". See Burdeck H. Brittin, *International Law for Seagoing Officers*, US Naval Institute, 1956, pp. 78-79.

19. The Japanese Foreign Ministry announced its disapproval of the Korean claims on 25 January 1952 relying on the principle of the freedom of the high seas; see Shigeru Oda, "The Normalization of Relations between Japan and the Republic of Korea", 61 *A.J.I.L.* (1967), pp. 51-52.

20. The Statement by President Rhee on February 8 1952 went on that, ". . . The main objective in establishing a boundary line on the seas is to maintain peace between Korea and Japan, and I believe Japan will naturally comply with this . . .".

21. The question whether or not the Law for Conservation of Fisheries Resources of 1954, based upon the Peace Line, were to be abolished arose within the Korean government when it took measures for establishment of EEZ and enactment of EEZ Fisheries Law. However, no measure was taken to abolish neither the Peace Line nor the Law for Conservation of Fisheries Resources. But, as there is a provision in the Enforcement Decree of the EEZ Fisheries Law of 1996 to the effect that fishing regulations by the Law for Conservation of Fisheries Resources applicable within the Peace Line are superseded by the relevant provisions of the EEZ Fisheries Law, there is no conflict between the EEZ fisheries regime and the Peace Line Fisheries regime: Sang-hoon Cho and Sun Pyo Kim, *Some Critical Issues on the Law of the Sea in the North East Asia between 1995-1997* (written in Korean for internal use within the Foreign Ministry of Korea), 1997.

22. It was estimated that 282 Japanese trawlers and 2,784 fishermen were arrested by the Korean authorities in the period between 1952 and 1964; see S. Oda, *ibid.*, 61 *A.J.I.L.* (1967), p. 51.

23. Article 4 of the Fisheries Agreement made it clear that neither Parties can have jurisdiction in the waters outside the exclusive fisheries zones.

24. Article 1 of the Act provides that: "The purpose of this Act is to contribute to the industrial development by rationally exploiting petroleum and gas, etc. (hereinafter referred to as 'submarine minerals') from among the natural resources existing under the waters adjacent to the coasts of the Korean peninsula and its adjacent islands that are the territory of the Republic of Korea, or in the continental shelf over which the Republic of Korea may exercise all her rights".

25. China did not specifically make its continental shelf claims in 1970s. We will discuss China's claims later in this Chapter.

26. Article 2 of the Exclusive Economic Zone provides that:

> (1) The exclusive economic zone of the Republic of Korea shall be the zone other than the territorial sea from among the sea area stretching from the outer line of the baseline referred to in Article 2 of the Territorial Sea and Contiguous Zone Act in accordance the Convention by 200 nautical miles.
> (2) The delimitation of the exclusive economic zone between States with opposite or adjacent to the Republic of Korea (hereinafter referred to as the "relevant States") shall, notwithstanding the provisions of paragraph (1) be established by agreement with the relevant States on the basis of international law.

27. Paragraph 2 of Article 5 of the Exclusive Economic Zone of the Republic of Korea.

28. Article 5 of the Law on the Exercise of Sovereign Rights with respect to Foreigners' Fishing in the Republic of Korea's EEZ. The law went into force on 7 August 1997.

29. The only exception where foreigners can get access to the Special Prohibition Zones is through a fisheries agreement with Korea as Article 2 of the law makes it clear that provisions of agreements with foreign countries shall override the provisions of the law.

30. P.C. Jessup, *The Law of Territorial Waters and Maritime Jurisdiction*, G.A. Jennings Co., Inc. (1927), pp. 45-46.

31. Mizukami Chiyuki, *Japan and the Law of the Sea* (written in Japanese), Yooshindou Press (1995), pp. 1-5.

32. Article 1 of the Law on the Territorial Sea. The law is reproduced by United Nations, Legislative Series, ST/LEG/SER.B/19, p. 56.

33. Those straits are the Soya Strait, the Tsugaru Strait, the eastern channel of the Tsushima Strait (the Korea Strait), the western channel of the Tsushima Straight and the Osumi Strait: Supplementary provisions of the Law on the Territorial Sea.

34. Article 2 and Annexed Schedule of Enforcement Order of the Law on the Territorial Sea (Cabinet Order No. 210 of June 1977); reproduced by the United Nations, Legislative Series, *ST/LEG/SER.B/19*, p. 57.

35. For an analysis of the Straight Baselines of Japan, see US Department of State, *Limits in the Sea No. 120-Straight Baseline and Territorial Sea Claims: Japan* (1998), p. 3.

36. US Department of State, *Limits in the Sea (No. 120)-Straight Baseline and Territorial Sea Claims: Japan* (1998), p. 2.

37. Jong-hwa Choi, *Modern History of Korea-Japan Fisheries Relations* (written in Korean), Sejong Press (2000), pp. 276-289.

38. The agreement between the Republic of Korea and Japan concerning fisheries is reproduced in English with an illustration map in D.W. Bowett, *The Law of the Sea*, Manchester University Press (1967), pp. 102-109: Paragraph 1 of Article 1 of the agreement provides that:

> The High Contracting Parties mutually recognise that each High Contracting Party has right to establish a sea zone (hereinafter "fishery zone"), extending not more than 12 nautical miles from its respective coastal base line, over which it will have exclusive jurisdiction with respect to fisheries. However, in case either High Contracting Party uses the straight base line shall be determined through consultation with the other High Contracting Party.

39. The view of the Japanese government was conveyed to the Korean government by a diplomatic note dated 19 December 1996 and repeated by Japanese prosecutors in proceedings in the *Japan v. Kim Sun-ki* Case No. 35, 1997 in Japan: See Hee Kwon Park, "Case Reports – Japan v. Kim Sun-Ki Case No. 35. 1997, Matsue District Court, Shimane Prefecture, Japan, August 15, 1997", 92 *A.J.I.L.* (1998), pp. 301-305.

40. Hee Kwon Park, *idem*, 92 *A.J.I.L.* (1998): Masahiro Miyoshi, "Japan's Straight Baseline System", 111 *The Journal of International Affairs*, Aichi University, Japan (1999), pp. 1-17. The Korean vessel, Daedong-ho No. 909, was arrested on charge of illegal fishing in the territorial waters of Japan at a point 19 miles off a nearest Hamada coast on 9 June 1997. The Japanese government tried to apply the "Law on the Regulation of Fisheries by Foreigners of 1967" which prohibits foreign nationals from fishing in the "Japanese waters", except in cases where they are legally resident in Japan and designated as fishermen by the Minister of Agriculture. Here the Japanese waters are understood as indicating internal waters and territorial waters of Japan. However, the appeal court dismissed the ruling of the Matsue District Court and the Supreme Court of Japan upheld the Japanese government position, making distinction between the territorial waters and fisheries zone mentioned in the Agreement of 1965.

41. Article 9 of the agreement provides that:

> 1. Any dispute between the High Contracting Parties concerning the interpretation or implementation of this Agreement shall be settled primarily through diplomatic channels.
> 2. Any dispute which cannot be settled under the provision of paragraph 1 shall be submitted for decision to an arbitration commission of three arbitrators; . . .

42. Korea needs to rely on "the autonomy of the arbitration clause" if it wishes to bring the disputes to the arbitration tribunal because the Fisheries Agreement of 1965 was terminated in 1999. However, it is to be noted that an arbitration clause in a contract is considered as being separated from the contract of which it forms part, thereby ensuring its existence when for all other purposes the contract itself is effectively at an end. See Alan Redfern and Martin Hunter, *Law and Practice of International Commercial Arbitration* (Second Edition), Sweet and Maxwell (1991), p. 174. A. Redfern and M. Hunter pointed out that: "Most claims are brought to arbitration following termination of a contract. It would be a nonsense if for some reason the arbitration clause was held to be terminated. This is the time when it is most needed, as a method of determining the claims and counterclaims of the parties".

43. Mizukami Chiyuki, *Japan and The Law of the Sea* (written in Japanese), Yoosindou Press (1995), p. 112. Its national court indicated Japan's different views on the continental shelf regime on the mineral resources and on the sedentary resources. On 14 March 1984, in *Odeco Nippon Company* Case, Tokyo Higher Court has said that, "Rules embodied in Articles 1, 2 and 3 of the 1958 Continental shelf Convention – excluding the rule on sedentary living resources – had become the international customary law at the latest by the time point of the judgement of 20 February 1969 by ICJ on the *North Sea Continental Shelf Cases*.

44. Choon-ho Park, "Oil under Troubled Waters", 14 *H.I.L.J.* (1973), No. 2, pp. 212-260.

45. Yong-jeon Ma, *Legal Problems of Seabed Boundary Delimitation in the East*

China Sea, Occasional Papers/Reprint Series in Contemporary Asia Studies Inc. (1984), pp. 34-35.

46. Masayuki Takeyama, "UNCLOS and Exclusive Economic Zone in East Asia", A paper presented at the international Conference on the Law of the Sea organised by the SLOC Study Group-Korea, Seoul Korea, 29-30 November, 1995.

47. Law No. 31 of 2 May 1977 on Provisional Measures Relating to the Fishing Zone; reproduced by the United Nations, Legislative Series, *ST/LEG/SEWR.B/19*, pp. 226-240.

48. Mizukami Chiyuki, *op. cit.* (1995), p. 70: The Soviet Union issued "the Decree of the Presidium of the Supreme Soviet for the Provisional Measure Concerning the Conservation of Living Resources and Fisheries Regulation in the Areas of the Oceans Contiguous to the Coasts of the Soviet Union" on 10 December 1976; the US Congress passed the Fisheries Conservation and Management Act in 1976.

49. Article 6 of Enforcement Order of 17 June 1977 of Law No. 31 of 31 May 1977 on Provisional Measures Relating to the Fishing Zone; reproduced by the United Nations, Legislative Series, *ST/LEG/SEWR.B/19*, pp. 226-240.

50. The law is reproduced in English by the United Nations, *Law of the Sea Bulletin* No. 35, 1997.

51. Article 1 of the supplementary provisions of the Law on the Exclusive Economic Zone and the Continental Shelf provides that "This law shall be entered into force on the day when the Convention enters into force *vis-à-vis* Japan". Japan deposited the instrument for the ratification of the LOS Convention with the Secretary-General of the United Nations on 20 June 1996 and thus the Convention went into force on 20 July 1996, the 30th day following the deposit of its instrument of ratification in accordance with Article 308 of the LOS Convention.

52. Article 1 of the Law on the Exclusive Economic Zone and the Continental Shelf provides that: "An exclusive economic zone, in which Japan exercises its sovereign rights and other rights as a coastal State, as prescribed in Part V of the United Nations Convention is hereby established, in accordance with the United Nations Convention on the Law of the Sea". Article 2 of the same law on the continental shelf also refers to the LOS Convention.

53. Note that the phrase within [ ] is this author's own addition and the phrase within ( ) is provided for in the original text. It can be noted here that the identical provision is found in Article 3 of the Law No. 31 of 2 May 1977 on Provisional Measures Relating to the Fishing Zone. The law on fishing zone of 1977 was abolished as the law on the EEZ and the continental shelf went into effect.

54. Japan objected to the adoption of the Geneva Convention on the Continental Shelf in the first United Nations Conference on the Law of the Sea because it feared that the Convention would negatively affect the pearl fisheries and crab fisheries. Mizukami Chiyuki, *op. cit.*, pp. 108-110.

55. For the legislation on the EEZ and the continental shelf by States over the world, see Robert W. Smith, *Exclusive Economic Zone Claims: An Analysis and Primary Document*, Martinus Nijhoff Publishers (1986), pp. 61-499. See also "1. Can a Unilateral Equidistance Line be a Solution", in Chapter Two.

56. The Japanese Maritime Security Agency seized two Korean fishing vessels in the East China Sea where there are no agreed boundaries between China and Japan in February 1999, applying its own equidistance line: see *Chosunilbo*, 27 March 1999, available in English at <www.chosun.com/w21data/html/news>.

57. The Chinese government's initial concept of its territorial sea first appeared in 1864 in its protest against Prussia's seizure of three Danish vessels in the Gulf of Pohai. See Hyun-soo Kim, *op. cit.* (1993/ Ph.D. thesis), pp. 224-225.

58. T. Cheng, "Communist China and the Law of the Sea", 63 *A.J.I.L.* (1969), pp. 54-56.

59. The Law is reproduced in English by Office of Legal Affairs, *The Law of the Sea: National Legislation on the Territorial Sea, the Right of Innocent Passage and the Contiguous Zone*, United Nations (1995), pp. 86-88.

60. Article 2 of the law provides that, ". . . The baseline of the territorial sea of the People's Republic of China is designated with the method of straight baseline, formed by joining the various base points with straight lines".

61. *Window* (A weekly Hong Kong newsmagazine published in English), 24 May 1996, pp. 28-29.

62. The Paracel islands are situated equidistant from the Chinese and Vietnamese coasts and have been in Chinese control since 1974. There was a two-day armed clash between South Vietnam and China in the islands. See Choon-ho Park, *East Asian Executive Reports*, Vol. 3, No. 5 (May 1981), pp. 8-13.

63. The Vietnamese protest was deposited with, and was published by the United Nations. See United Nations, *Law of the Sea Bulletin* No. 32 (1996), p. 91.

64. *Korea Times* (An English daily newspaper published in Korea), 16 May 1996. However, unlike Vietnam, Korea has not deposited its Note Verbales which contain protests with the United Nations. *Window* reported that "The three base points Seoul objects to are located at two lately emerged islands – Macaiheng and Waikejiao – off the sandy coast of Jiangsu Province in the Yellow Sea as well as a rocky island – Haijiao or Dongdao. The objection was on the ground that they were not inhabited": see *Window*, 24 May 1996, p. 28.

65. US Department of State, *Limits in the Seas, No. 117 – Straight Baseline Claim: China* (1996), p. 3.

66. Choon-ho Park, "Oil under Troubled Waters", 14 *H.I.L.J.* (1973), No. 2, pp. 212-260; Jeanette Greenfield, *China's Practice in the Law of the Sea*, Oxford Clarendon Press(1992), p. 117.

67. The statement was reprinted in English in the *Peking Review*, 1 January 1971, p. 28.

68. *Working paper on Sea Area within the Limits of National Jurisdiction Submitted by the Chinese Delegation (A/AC.138/SC.II/L34)*, 16 July 1973: Reproduced by J. Greenfield (1992), pp. 230-232.

69. Ying-jeou Ma, *Legal problems of Seabed Boundary Delimitation in the East China Sea*, Occasional Papers/Reprint Series in Contemporary Asia Studies Inc. (1984), p. 274.

70. The Regulation is reproduced by J. Greenfield, *China's Practice in the Law of the Sea* (1992), pp. 244-249.

71. S.H. Jhe, "Some Delimitation Issues in the Maritime Areas Surrounding the Korean Peninsula", 11 *Hae-yang-jung-chak-yon-ku* (Studies on Marie Policy) No. 1, pp. 109-110. The article was written in English.

72. *Xinhwasha Telegraph*, 15 May 1996.

73. The Act was adopted at the third session of the Standing Committee of the Ninth National People's Congress, 26 June 1998 and entered into force on the same day. For the English translation of the Act, see *Law of the Sea Bulletin No. 38*, United Nations (1998), pp. 28-31.

74. Article 3 of the Act provides that, "in the exclusive economic zone the People's Republic of China shall exercise sovereign rights for the purpose of exploring and exploiting, conserving and managing the natural resources of the waters superjacent to the seabed and its subsoil, and with regard to other activities for the economic exploitation and exploration of the zone, such as the production of energy from the water, current and winds. The People's Republic of China shall have jurisdiction in the exclusive economic zone with regard to the establishment and use of artificial islands, installation and structures: marine scientific research; and the protection and preservation of the marine environment. . .". Article 4 of Act provides that, "The People's Republic of China shall exercise sovereign rights over the continental shelf for the purpose of exploring it and exploiting its natural resources. The People's Republic of China shall have jurisdiction over the continental shelf with regard to the establishment and use of artificial islands, installation and structures; marine scientific research; and the protection and preservation of the marine environment. . . ".

75. Article 2 of the Act on the Exclusive Economic Zone and Continental Shelf.

76. Evans pointed out that the provisions of Articles 74(1) and 83(1) reflect the current status of the customary international law; see M.D. Evans, "Delimitation and Common Maritime Boundary", 64 *B.Y.I.L.* (1993), p. 294.

77. Choon-ho Park, "China and Maritime Jurisdiction", 22 *G.Y.I.L.* (1979), pp. 119-141; J.R.V. Prescott, *Maritime Jurisdiction in East Asian Seas*, East-West Environment and Policy Institute Occasional Paper No. 4 (1987), pp. 41-42. In a

non-governmental fisheries agreement between China and Japan signed in 1955, the Japan-china Fisheries Council of Japan agreed to observe the military zones of China. In 1975 when Japan and China concluded a fisheries agreement between the two governments, Japan stated its intention to guide its fishing vessels to refrain from fishing in the military zones, while reserving its position against the military zones: see Masahiro Miyoshi, "New Japan-China Fisheries Agreement", 41 *J.A.I.L.* (1998), pp. 32-34.

78. R.P. Barston and P.W. Birnie, "The Falkland Islands/ Islas Malvinas conflict: A question of zones", 7 *M.P.* (1983), p. 20.

79. D.P. O'Connell, *The Influence of Law on the Sea Power*, Manchester University Press (1975), p. 167.

80. R.P. Barston and P.W. Birnie, "The Falkland Islands/ Islas Malvinas conflict: A question of zones", 7 *M.P.* (1983), pp. 14-34.

81. F.C. Leiner, "Maritime Security Zones: prohibited yet perpetuated", 24 *V.J.I.L.* (1984), p. 980.

82. The East Sea is about one million sq. km. in size and semi-enclosed by Japan, Korea, and Russia.

83. *Sankei shimbun* (a daily newspaper in Japanese), 29 August 1968.

84. Choon-ho Park, "Report Number 5-12: Agreement between Japan and Republic of Korea Concerning the Establishment of Boundary in the Northern Part of the Continental Shelf Adjacent to two Countries", *International Maritime Boundaries*, pp. 1058-1089.

85. Choon-ho Park, "Report Number 5-12: Japan-South Korea", J.I. Charney and L. Alexander, *International Maritime Boundaries*, p. 1059.

86. Choon-ho Park, *ibid.*, p. 1057.

87. For similar observation, see S. Oda and H. Owada, "Annual Review of Japanese Practice in International Law XV" (1976-77), 28 *J.A.I.L.* (1985), pp. 132-134.

88. Article 1 of the Geneva Convention on the Continental Shelf provides that: "For the purpose of these Articles, the term 'continental shelf' is used as referring (a) to the seabed and subsoil; of the submarine areas adjacent to the coast but outside the area of territorial sea, to a depth of 200 meters or, beyond that limit, to where the depth of the superjacent waters admits of the exploitation of the natural resources of the said areas; (b) to the seabed and subsoil of similar submarine areas adjacent to the coasts of islands".

89. However, Point No. 1 is not a strict equidistant point from Cheju island and Tori-Shima. Point No. 1 is 47.4 miles from Tori-Shima and 47 miles from Cheju island.

90. Korea and Japan agreed to delimit the boundary of northern continental shelf on the basis of the equidistant line in December 1970. In September 1971, Korea

argued and Japan denied that straight baselines should be used for drawing the equidistant line. Later, Korea dropped the position that straight baseline should be used and agreed in principle with Japan that the equidistant line should be drawn from all the relevant basepoints of territories of Korea and Japan. Then, Japan proposed 42 points for the equidistant boundary using all parts of relevant coasts. Later, the 42 points were reduced to 35 points. In the process of negotiation Korea argued to pick up Point No. 35 as it stands now, which was chosen from Korea's domestic legislation.

91. See, for example, the joint proposal on the delimitation of EEZ and the Continental Shelf by Bahamas, Barbados, Canada, Colombia, Cyprus, Democratic Yemen, Denmark, Gambia, Greece, Guyana, Italy, Japan, Kuwait, Malta, Norway, Spain, Sweden, United Arab Emirate, United Kingdom and Yugoslavia, which is reproduced by Platzödor Third United Nations Conference on the Law of the Sea, Vol. IV, p. 468. Also see Japan's proposal on the delimitation of the continental shelf–A/CONF.62/C.2/L.31/Rev.1 (1974) reproduced in Third United Nations Conference on the Law of the Sea Official Records Vol. III, p. 211.

92. US Department of State Bureau of Oceans and International Environmental and Scientific Affairs, *Limits in the Seas* No. 120, 1998, p. 4.

93. In this regard, Churchill and Lowe wrote that: "Having established the situation where the use of straight baselines is permissible, the Territorial Sea and Law of the Sea Convention go on to lay down a number of conditions governing the way in which straight baselines may be drawn". R.R. Churchill and A.V. Lowe, *op. cit.* (1999), p. 35.

94. The Korean government challenged the legality of some of the Japanese straight baselines through diplomatic notes, but it did not circulate the diplomatic notes in the United Nations. Dr. Hee Kwon Park, a former director of international legal affairs division of Foreign Ministry of the Republic of Korea denies legality of some twenty or so straight baselines of Japan in his book, *The Law of the Sea and North East Asia: A Challenge for Co-operation*, Kluwer Law International (2000), at p. 28.

95. For example, there are different views on the maximum permissible length of straight baselines. For the United States Department of State, it is 48 miles. J. Ashley Roach and Robert W. Smith argued for 24 miles. Some attempts were made to find out the maximum length limitation of straight baselines from the *Anglo-Norwegian* Case with different conclusions. Beazley argued for a 45-nautical mile. Hodgson and Alexander suggested the use of a 40-nautical miles: US Department of State, *op. cit.*, p. 15; J. Ashley Roach and Robert W. Smith, *United States Responses to Excessive Maritime Claims*, Martinus Nijhoff Publishers, 1996, p. 64; J. Ashley Roach and Robert W. Smith, "Straight Baselines: The Need for a Universally Applied Norm", 31 *O.D.I.L.*, (2000), p. 50.; P.B. Beazley, *Maritime Limits and Baselines: A Guide to their Delineation*, The Hydrographic Society Special Publication No. 2, 1978, p. 9; R. Hodgson and L. Alexander, "Towards an

Objective Analysis of Special Circumstance of Oceanic Archipelagos and Atolls", *Law of the Sea Institute Occasional Paper*, No. 13, 1971, p. 42.

96. Korea Petroleum Development Corporation, *Strategy for Comprehensive Exploration in the Continental Shelf* (written in Korean, 1996), p. 48.

97. Diplomatic note dated 28 January 1952 sent by the Japanese government to the Korean government (unpublished).

98. Diplomatic note dated 25 September 1954 sent by the Japanese government to the Korean government (unpublished).

99. Hee Kwon Park, *op. cit.* (2000), p. 89.

100. PCIJ, *Ser. A, No. 2*, p. 11.

101. The Exchange of Notes was concluded in Korean and Japanese. The quoted provision here is translated into English by this author.

102. The Exchange of Notes was concluded on 22 June 1965 and entered into force on 18 December 1965.

103. *Southern Bluefin Tuna Arbitration (Australia and New Zealand v. Japan), Award on Jurisdiction and Admissibility*, August 4, 2000, para. 39.

104. *Southern Bluefin Tuna Arbitration* Award, *idem.*

105. Article 16 of the CCSBT provides that:

> 1. If any dispute arises between two or more of the Parties concerning the interpretation or implementation of this Convention, those Parties shall consult among themselves with a view to having the dispute resolved by negotiation, inquiry, mediation, conciliation, arbitration, judicial settlement or other peaceful means of their own choice.
> 2. Any dispute of this character not so resolved shall, with the consent in each case of all parties to the dispute, be referred for settlement to the International Court of Justice or to arbitration; but failure to reach agreement on reference to the International Court of Justice or to arbitration shall not absolve parties to the dispute from the responsibility of continuing to seek to resolve it by any of the various peaceful means referred to in paragraph 1 above.
> 3. In cases where the dispute is referred to arbitration, the arbitration tribunal shall be constituted as provided in the Annex to this Convention. The Annex forms an integral part of this Convention.

106. *Southern Bluefin Tuna Arbitration* Award, para. 57.

107. *Ibid.*

108. *Southern Bluefin Tuna Arbitration* Award, para. 55.

109. Article 298, paragraph 1, subparagraph (a) (i) of the LOS Convention provides that:

> When signing, ratifying or acceding to this Convention or at any time thereafter, a State may, without prejudice to the obligation arising under section 1, declare in writing that it does not accept any one or more of the following categories of disputes:

(a) (i) disputes concerning the interpretation or application of article 15, 74 and 83 relating to sea boundary delimitations, or those involving historic bays or titles, provided that a State having made such a declaration shall, when such a dispute arises subsequent to the entry into force of time is reached in negotiations between the parties, at the request of any party to the dispute, accept submission of the matter to conciliation under Annex V; and provided further that any dispute that necessarily involves the concurrent consideration of any unsettled dispute concerning sovereignty or other rights over continental or insular land territory shall be excluded from such submission.

110. See Article 10 of the Geneva Convention on the Territorial Sea and the Contiguous Zone and Article 1 of the Geneva Convention on the Continental Shelf.

111. B. Gidel, *Le droit international public de la mer* (1934), p. 684; English translation is provided in A. Soons, *Artificial Islands and Installations in International Law*, 17 Law of the Sea Occasional Paper No. 22, (1977).

112. A. Soons, *Artificial Islands and Installations in International Law*, 17 Law of the Sea Occasional Paper No. 22, (1977).

113. U.N. Sea-Bed Committee Doc. A/AC.138/SR.57, p. 167.

114. For example, Malta proposed to distinguish between islands and islets by their size; African States proposed to set up criteria such as size, geological characteristic, population, distance from main land of islands; Romania proposed to use the term "islands and small islands, uninhabited and without economic life;" Turkey proposed to restrict the maritime zones of "islands without economic life", rocks and low-tide elevation; and Colombia proposed to restrict the maritime zones of "islands without a life of their own, without a permanent and settled population:" see M.H. Nordiquist *et al.* eds., *United Nations Convention on the Law of the Sea 1982 A Commentary*, Vol. 3, Kluwer (1985), pp. 328; see also U.N., *The Law of the Sea, Regime of Islands – Legislative History of Part VII (Article 121) of the United Nations Convention on the Law of the Sea* (1988), pp. 22-81.

115. M.H. Nordquist *et al.* eds., *United Nations Convention on the Law of the Sea 1982 A Commentary*, Vol. 3, Kluwer (1985), pp. 328-; U.N., *The Law of the Sea, Regime of Islands – Legislative History of Part VII (Article 121) of the United Nations Convention on the Law of the Sea* (1988), pp. 89-108.

116. U.N., *The Law of the Sea, Regime of Islands – Legislative History of Part VII (Article 121) of the United Nations Convention on the Law of the Sea* (1988), p. 93.

117. B. Kwiatkowska and A.H.A. Soons raised the following questions with regard to the "human habitation or economic life of their own". How large are the groups of people involved?; does economic life need to be of a commercial nature; do fisheries in the waters and mineral exploitation from the sea-bed surrounding the island amount to economic life?; does human habitation and/or economic life depend on the area of the island, existence of potable water, tillable soil, natural resources, biotic life; and does the term "of their own" exclude the possibility of

artificial extension of the area of an island to make it habitable and economically viably? See B. Kwiatkowska and A.H.A. Soons, "Entitlement to maritime areas of rocks which cannot sustain human habitation or economic life of their own", 21 *N.Y.I.L* (1990), p. 165.

118. R.D. Hodgson and R. Smith, "The Informal Single Negotiating Next (Committee II): A Geographical Perspective", 3 *O.D.I.L.* (1976), p. 225.

119. E.D. Brown, *The International Law of the Sea*, Dartmouth (1994), Vol. I, p. 150.

120. B. Kwiatkowska and A.H.A. Soons, *op. cit.*, p. 142.

121. J.I. Charney, "Rocks that cannot sustain human habitation", 93 *A.J.I.L*, 1999, p. 862, and p. 872.

122. J.I. Charney, *idem.*

123. D. Bowett, "Islands, Rocks, Reefs and Low-tide Elevations", *International Maritime Boundaries*, pp. 131-151.

124. D.H. Anderson, "Report Number 9-5: Ireland-United Kingdom", *International Maritime Boundaries*, pp. 1767-1780.

125. J.R. Victor Prescott, "Report Number 6-12: Indonesia-Malaysia-Thailand", *International Maritime Boundaries*, pp. 1443-1454.

126. Kaldone G. Nweihed., "Report Number 2-14; United States (Puerto Rico and the Virgin Islands)-Venezuela", *International Maritime Boundaries*, p. 695.

127. Kaldone G. Nweihed., "Report Number 2-12; The Netherlands (Antilles)-Venezuela", *International Maritime Boundaries*, p. 623.

128. For example, Bowett emphasised, *inter alia*, the impact of the method of delimitation on the role of islands in the delimitation, saying that: "where there is no reliance on equidistance, the relevance of islands diminishes. In other words, where a method other than equidistance is chosen, because it better suits the geographical relationship viewed as a whole, it necessarily follows that islands which have little significance for the totality of the geographical relationship will tend to have effect on the method". He also observed that "Equally, while mere 'rocks' (within the meaning of Article 121(3) of the 1982 Convention) cannot generate a shelf or economic zone, there is no reason why they cannot be used as basepoints", D. Bowett, *op. cit.*, p. 134 and p. 148.

129. J.I. Charney, *idem.* However, note that there are views that the rock provision in the LOS Convention is a rule of customary law; see Churchill and Lowe, *The Law of the Sea* (1990), p. 164; H. Dipla, *Le regime juridique des iles dans dans le droit international de la mer* (1984), pp. 48-49; B. Kwiatkowska and A.H.A. Soons, *op. cit.*, p. 175.

130. Article 309 of the LOS Convention provides that: "No reservation or exceptions may be made to this Convention unless expressly permitted by other articles of this Convention". An article is found in the Convention which expressively permits a reservation.

131. L.D.M. Nelson, "The Delimitation of Maritime Boundaries in the Caribbean", Johnston and Saunders, eds., *Ocean Boundaries Making: Regional Issues and Development* (1988), 167; Nweihed, "EZ (Uneasy) Delimitation in the Semi-enclosed Caribbean Sea", 8 *O.D.I.L* (1980), pp. 20-23.

132. D.H. Anderson, "British Accession to the UN Convention on the Law of the Sea", 46 *I.C.L.Q.* (1997), 778-779: Rockall has been a point of dispute between the UK and other neighbouring countries: Ireland, Denmark and Iceland. The United Kingdom proclaimed a continental shelf in 1964, and a 200 N.M. fisheries zone in 1976. The British government designated some 52,000 square miles of continental shelf in 1974 and used Rockall as a base point under its Fisheries Act 1976. Note that the United Kingdom did not sign the LOS Convention and thus she became a Party to the LOS Convention by way of accession in accordance with Article 307 of the LOS Convention. The UK Foreign and Commonwealth Office wrote that: "Following British accession to the Convention, Rockall remains part of British territory, with a twelve mile territorial sea – it is part of Scotland. But Rockall is not able to sustain human habitation and so upon accession British fishery limits were brought into line with the provisions of the Convention by measuring that part of British fishery limits previously measured from Rockall from St Kilda instead": UK Foreign and Commonwealth Office, "Background Brief: Britain's Accession to the United Nations Convention on the Law of the Sea", London, January 1998, p. 7, available at <www.Faco.gov.uk/news/archive.asp>.

133. For example, a Japanese scholar argued that: "As for the territorial issues, the Takeshima and Senkaku Islands, which are uninhabited under normal conditions, should be disregarded in the drawing of boundaries for the continental shelf as well as for the fisheries zone": Kiyofumi Nakauci, "Problems of Delimitation in the East China Sea and Sea of Japan", 6 *O.D.I.L.* (1979), p. 314.

134. "Syto" in Japanese means islands.

135. *Records of National Assembly* of 21 November 1998. The English translation is made by the author.

136. For the Japanese proposal of 3 May 1978 to delete the paragraph of the ICNT, see Second Committee Informal Suggestion of 3 May 1978, Doc C.2//Informal Meeting/27 (reproduced by Platzöder Vol., V, p. 37); see also *UNCLOS III Official Records* Vol. IX, p. 68.

137. If an equidistance line is to be drawn between Korea's mainland and Japan's mainland in the East Sea then the line passes about 2 miles to the west of Dok-do.

138. The author visited Dok-do on 7 June 2001 and met the Korean paramilitary policemen and lighthouse keepers stationed there.

139. However, there is a view that the seabed around Dok-do is rich in hydrate, which began to be recognised as a new source of energy.

140. In the *Jan Mayen* Conciliation and in the *Jan Mayen* Case in the ICJ, the conciliation committee and the ICJ both mentioned Article 121(3) without attempt to interpret the paragraph. In the *Monte Confurco* Case in the ITLOS, where Sheycell's fishing vessel was arrested on a charge of illegal fishing by a French warship in the French Kergueln's EEZ that was claimed from tiny rocks, Sheichell did not challenge the legality of the zone itself relying on Article 121(3), but simply asked the ITLOS for a reduction of the bond payable and prompt release of the vessel.

141. J.I. Charney, *op. cit.*, 93 *A.J.I.L.*, 1999, p. 868.

142. J.I. Charney, *idem.*

143. J.I. Charney, *ibid.*, p. 870.

144. A.E. Boyle, *Korean Exclusive Economic Zone* (March 1996, written for Ministry of Foreign Affairs of the Republic of Korea), p. 25.

145. 1969 *ICJ Reports*, para. 57.

146. 18 *R.I.A.A.* 1980, reprinted in 18 *I.L.M.* 379 (1979), para. 184.

147. *ibid.*, para. 251.

148. 1982 *ICJ Reports*, para. 129.

149. 1984 *ICJ Reports*, para. 201.

150. 1984 *ICJ Reports*, para. 222.

151. 1985 *ICJ Reports*, para. 64; Faraj Abdullah Ahnish, *The International Law of Maritime Boundaries and the Practice of States in the Mediterranean Sea*, Clarendon Press Oxford (1993). pp. 175-176.

152. 1999 *Eritrea-Yemen Arbitration* Award, paras. 115-119: the award is reproduced in the Permanent Court of Arbitrator's internet homepage, <www.pca-cpa.org>. The sovereignty over these islands were disputed between Yemen and Eritrea along with other islands in the Red Sea, and the arbitral tribunal decided in 1998 that Yemen has sovereignty over the islands. In its award of 1998, the tribunal described the disputed islands as follows: "The disputed islands and islets range from small to tiny, are uniformly unattractive, waterless, and habitable only with great difficulty": 1998 *Eritrea-Yemen Arbitration* Award, para. 93. For the analysis of the two awards, see Barbara Kwiatkowska, "The Eritrea-Yemen *Arbitration*: Landmark Progress in the Acquisition of Territorial Sovereignty and Equitable Maritime Boundary Delimitation", 32 *O.D.I.L.* (2001), pp. 1-25.

153. *Qatar/Bahrain* Case, 2001 *ICJ Reports*, paras. 197 and 219. According to the report of the expert commissioned by Bahrain, Qit'at Jaradah's length and breadth are about 12 by 4 metres at high tide, whereas at low tide they are 600 and 75 metres.

154. J.M. Van Dyke *et al.*, "The Exclusive Economic Zone of the Northern Hawaiian Islands: When do uninhabited islands generate an EEZ", 25 *S.D.L.R.* (1988), p. 449.

155. Isles of Scilly is an archipelago of five inhabited islands with a total population of two thousand and numerous other small rocky islets 28 miles off Lands End – the most South Westerly point of the British Isles; Kerkennah island is 20 km off the coast Sfax and it has been inhabited and Seal island is inhabited: <www. scillyonline. co.uk/home.html>; <www. homestead/ kerkennak/ indexomain.html>; 1984 *ICJ Reports*. para. 222.

156. Various photos and description of these islets are available in internet at <www. fwkc. com/ encyclopedia/ low/ articles/e /e007000245f.html>; <www. seawatchtours. com/ pffins.html>; <www. million1. com/ malta/ 288.Shtml>.

157. A.E. Boyle, *Korean Exclusive Economic Zone*, p. 26.

158. The Yellow Sea is about 400,000 sq. km. in size and is enclosed by Korea on the eastern side and by China on the west side. To its south the Yellow Sea is connected to the East China Sea and on its way to the East China Sea, there is the island of Cheju of Korea on the eastern side.

159. Zou Keyuan, "China's exclusive economic zone and continental shelf: developments, problems and prospects", 25 *M.P.* (2001), pp. 77-78.

160. Records of the First Round of Talks between Korea and China on Delimitation of the EEZ (written in Korean for internal use of the Ministry of Foreign Affairs of Korea), 1997, pp. 12-13.

161. *UN Official Records*, xiv: *Summary Records of Meeting, Third UN Conference on the Law of the Sea*, Geneva (1980), p. 24.

162. Richard T.S. Hsu, "A Rational Approach to Maritime Delimitation", 13 *OD.I.L.* (1983), pp. 108-112; J.R.V. Prescott, *Maritime Jurisdiction in East Asian Seas*, The East-West Environment and Policy Institute, Occasional Paper No. 4, East-West Centre (1987), p. 51.

163. Jonathan I. Charney, "Central East Asian Maritime Boundaries and the Law of the Sea", 89 *A.J.I.L.* (1995), 741.

164. For the discussion on the proportionality, see M.D. Evans, *Relevant Circumstances and Maritime Delimitation*, Oxford Clarendon Press (1989), pp. 224-231.

165. It is to be noted that it took two years for China to legislate the Act on the Exclusive Economic Zone and the Continental Shelf since the Chinese government declared that China had sovereign rights and jurisdiction in its exclusive economic

zone and continental shelf. The main reason for the delay of the legislation of the Act was the worries of the local fishermen and local governments in the coastal area of the Yellow Sea that the introduction of the EEZ regime would facilitate the shaping of new fisheries regime and thus would mean loss of fishing ground to the local fishermen.

166. Ji Guoxing, *Maritime Jurisdiction in the Three China Seas*, University of California Institute on Global Conflict and Co-operation, Policy Paper No. 19 (1995), pp. 9-12.

167. Jeanette Greenfield, *China's Practice in the Law of the Sea*, Clarendon Press Oxford (1992), p. 125.

168. Hee Kwon Park, *The Law of the Sea and North East Asia: A Challenge for Co-operation*, Kluwer Law International (2000), pp. 25-26.

169. Hee Kwon Park, *idem.*

170. US Department of State, Straight Baselines Claim: China, Limits in the Seas No. 117, 9 July 1996, p. 6.

171. Encyclopaedia Britannia, "China Sea", available at <www.britannica.com>.

172. The CCOP/ECAAFE reported that "a high probability exists that the continental shelf between Taiwan and Japan may be one of the most prolific oil reservoirs in the world . . ."; CCOP/ECAFE, "Geological Structure and Some Water Characteristics of the East China Sea and Yellow Sea", 2 *Technical Bulletin* (1969), pp. 39-40; Choon-ho Park, "Oil under Troubled Waters", 14 *A.J.I.L.* (1973), pp. 212-216. Note that ECAFE was renamed as ESCAP (Economic and Social Commission for Asia and Pacific) in 1974.

173. Note that Taiwan, the Republic of China, then represented China in the Security Council of the United Nations and then replaced by the Beijing government, the People's Republic of China in October 1971.

174. Choon-ho Park, *Continental Shelf Issues in the Yellow Sea and the East China Sea*, Law of the Sea Institute Occasional Paper(1972), pp. 20-22.

175. Ji Guoxing, *op. cit.*, p. 10.

176. Choon-ho Park, "The Sino-Japanese-Korean Sea Resources Controversy and the Hypothesis of a 200-Mile Economic Zone", 16 *H.I.L.J.* (1975), pp. 41-42.

177. Zou Keyuan, "China's exclusive economic zone and continental shelf: developments, problems, and prospects", 25 *M.P.* (2001), pp. 77-78.

178. Statement by the Chinese Ministry of Foreign Ministry, 13 June 1977, partly reproduced in English in the *Beijing Review*, 17 June, p. 17.

179. Mark J. Valencia, "North East Asia: Petroleum Potential, Jurisdictional Claims and International Relations", 20 *O.D.I.L.* (1989), p. 47.

180. Zou Keyuan, *op. cit.*, pp. 77-78.

181. As there is a "without-prejudice clause" in the Korea-Japan Fisheries Agreement, it cannot be said that Korea is obliged to adopt distant criteria in future delimitation of EEZ boundaries with Japan.

182. It is to be noted that the ICJ in the *Qatar-Bahrain* Case said that; "the concept of a single maritime boundary does not stem from multilateral treaty law but from State practice, and that ***it finds its explanation in the wish of States*** to establish one uninterrupted boundary line delimiting the various-partially coincident-zones of maritime jurisdiction appertaining to them(emphasis added);" 2001 *ICJ Reports*, para. 173.

183. 1969 *ICJ Reports*, para. 101(d) (3).

184. 1969 *ICJ Reports*, paras. 15 and 18. Note that the Court in this regard distinguishes between apportionment and delimitation, saying that: "its task in the present proceedings relates essentially to the delimitation and not the apportionment of the concerned, . . . Delimitation in an equitable manner is one thing, but not the same thing as awarding a just and equitable share of a previously undelimited area, even though in a number of cases the results may be comparable, or even identical": para. 18.

185. 1982 *ICJ Reports*, para. 131.

186. P. Weil, *The Law of Maritime Delimitation-Reflection*, Cambridge Grotius Publication Limited (1989), p. 238; M.D. Evans, *Relevant Circumstances and Maritime Delimitation*, Oxford Clarendon Press (1989), p. 231.

187. L. Legault and B. Hankey, "Methods, Oppositeness and Adjacency, and Proportionality in Maritime Boundary Delimitation", *International Maritime Boundaries*, p. 219.

188. L. Legault and B. Hankey, *ibid.*, pp. 219-220; D. Anderson, "Report No. 9-2: France-Spain", *International Maritime Boundaries*, pp. 1719-1734.

189. P. Weil, *op. cit.*, p. 238.

190. *St. Pierre and Miquelon Arbitration* Award, para. 33: the award is reproduced in 31 *I.L.M.* (1992).

191. *St. Pierre and Miquelon Arbitration* Award, para. 93.

192. 1993 *ICJ Reports*, Dissenting opinion of Judge *ad hoc* Fischer, para. 13 at 309.

193. 1993 *ICJ Reports*, para. 61.

194. This rough assessment is done on a chart with the scale of 1/1,500,000. In the measurement of the coastal lengths, a number of artificial straight lines were used and the openings in the Ryukyu chain were not included in the Japanese coastline. Also note that the coastlines of China and Japan below Lat. 27° N are excluded

in calculating the coastlines. This rough assessment can be justified for the purpose of this book and note that the arbitral court in the *Anglo-French* Case said that: "the Court does not consider that the course of the boundary between the United Kingdom and the French Republic in that region depends on any nice calculations of proportionality based on conjecture . . .": 1977 Award, para. 27.

195. 1969 *ICJ Reports*, paras. 8 and 15.

196. J.I. Chareney, "International Maritime Boundary Delimitation", 88 *A.J.I.L.* (1994), p. 247.

197. J.I. Chareney, *ibid.*, pp. 247-254.

198. Professor Charney noted that the maximum reach trend is particularly apparent in the Caribbean Sea, the Baltic Sea, Africa and South America. He pointed out several exceptions to the general terms but he asserted that the exceptions appear to be "exceptions to the general practice in maritime boundary agreements": see J.I. Charney, "The Diaoyu/Senkaku Islands Maritime and Territorial Dispute", Taiwan Law Society and Taiwan Institute of International Law eds, *International Law Conference on the Dispute over Diaoyu/Senkaku Islands* (2-3 April 1997, Taiwan), pp. 128-129.

199. The Agreement provided a tri-junction and also provided that the tri-junction was connected to the boundary between Indonesia and Thailand agreed on 17 December 1971, to the boundary between Indonesia and Malaysia agreed on 27 October 1969 and to the boundary between Thailand and Malaysia agreed on 21 December 1971. For the analysis of the Agreement, see Victor Prescott, "Report Number 6-12: Indonesia – Malaysia – Thailand", *International Maritime Boundaries*, pp. 1443-1450.

200. The tri-junction agreement among Sri Lanka, India and the Maldives simply provides for the co-ordinates of the tri-junction and stipulates that the tri-junction is equidistant form the three States. For the analysis of the Agreement, see Victor Prescott, "Report Number 6-9: India-Maldives-Sri Lanka", *International Maritime Boundaries*, pp. 1401-1408.

201. The agreement was signed by the three States on 22 June 1978. The tri-junction is connected to the boundary between India and Indonesia agreed in 1977, to the boundary between Indonesia and Thailand agreed in 1975 and to the boundary between Thailand and India agreed on the same the tri-junction agreement was signed. For the analysis of the Agreement, see Victor Prescott, "Report Number 6-7: India-Indonesia-Thailand", *International Maritime Boundaries*, pp. 1374-1388.

202. The boundary between the United Kingdom agreed in 1966 and the boundary between the United Kingdom and the Netherlands agreed in 1965 meet at a point of Latitude 55° 45° 54° North, Longitude 3° 22° 13° East. Thus the point was a tri-junction before 1971 when the United Kingdom signed three separate agreements with Denmark, Germany and the Netherlands and thus established three

new tri-junctions. For the analysis of these agreements, see Davis H. Anderson, "Report Number 9-10: Denmark-UK"; "Report Number 9-11: Federal Republic of Germany – The Netherlands"; "Report Number 9-12: Federal Republic of Germany-UK;" Report Number 9-13: The Netherlands-UK," *International Maritime Boundaries*, pp. 1825-1869.

203. David H. Anderson, *idem.*

204. In this regard Korea laid a continental shelf claim to the mid-channel of the Okinawa Trough which is about 280 miles for the nearest Korean territory and also argues that EEZ boundaries in the region should be drawn in such a way as not to cut-off the Korea's EEZ from the mid-ocean of the East China sea; Choon-ho Park, *East Asia and the Law of the Sea*, Seoul National University, 2nd edition (1985), p. 11.

205. Shandung Local Government, *Geography of Shandung* (written in Chinese, 1984), pp. 30-33.

206. Maritime Safety Agency of Japan, *Maritime Sea Lanes* (written in Japanese, 1991), vol. 2, pp. 20-100. The Japanese islands are Kotoretto, Takara-shima, Akuseki-shima, Nakano-shima, Kutino-shima, Amaomi-shima, Taira-shima, Yokoate-shima, Tori-shima, Me-shima, Woochimukai-shima, Gazya-shima, Iotori-shima and Kusakaki-shima. Of these islands only the first 8 islands appear to have inhabitants at present. In Kusakaki-shima, there are no inhabitants at the moment, but some seasonal fishing was done in the 19th century.

207. For the legal analysis of the dispute regarding Senkaku/Diaoyutai, see *Dispute over Diayo/Senkaku islands*, edited by Taiwan Law Society and Taiwan International Law Society from the papers and discussions at an international law conference held in Taiwan on 2-3 April 1997.

208. In this regard Japan has been arguing that the tri-equidistance point among Korea's Marado, China's Haijiao and Japan's Tori-shima should be the final tri-junction among Korea and China and Japan. It is noted that Point 6 of the joint development zone between Korea and Japan in 1974 is identical with the tri-equidistance point among Korea's Marado, China's Haijiao and Japan's Tori-shima.

209. According to Keyuan, China recently realised the difficulties of applying the natural prolongation principle to maritime boundary delimitation, and began to consider the principle of proportionality which is favourable to China, however, there is a possibility that China would insists on natural prolongation for the delimitation of the continental shelf: see Zou Keyuan, "China's exclusive economic zones and continental shelf: developments, problems, and prospects", 25 *M.P.* (2000), pp. 77-78.

210. *Qatar-Bahrain* Case, 2001 *ICJ Reports*, para. 173.

211. 1985 *ICJ Reports*, para. 36.

212. 1985 *ICJ Reports*, para. 39.

213. The reason why the Court played down the role of the rift zone might be that the Court was not sure of the existence of the rift zone. The Court said that: "the Court is not satisfied that it would be able to draw any sufficiently cogent conclusions from is as to the existence or not of the fundamental discontinuity on which the Libyan argument relied": see 1985 *ICJ Reports*, para. 41.

214. M.D. Evans, *Relevant Circumstances and Maritime Delimitation*, Clarendon Press, Oxford(1989), p. 101. Evans made it clear that the discussion in his book on this point assumed the inaccuracy of the Court's reasoning in the *Libya/Malta Continental Shelf* Case.

215. 1985 *ICJ Reports*, para. 41.

216. 1985 *ICJ Reports*, para. 39.

217. 1982 *ICJ Reports*, para. 68.

218. 1982 *ICJ Reports*, para. 80.

219. 1982 *ICJ Reports*, para. 80. Note that "it" in the passage means the Tripolitanian Furrow.

220. *Anglo-French Arbitration*, Decision of 1977, paras. 104, 105, and 108.

221. *Anglo-French Arbitration*, Decision of 1977, para. 107. ICJ in the *North Sea Continental Shelf* Cases, noted that "the self area in the North Sea separated from the Norwegian coast by the 80-100 kilometres of the Trough cannot in any physical sense be said to be adjacent to it, nor to be its natural prolongation". However, in the agreement between the UK and Norway in 1965 on the continental shelf boundary, both Parties drew the boundary following the equidistant line sharing the view that the Norwegian Trough did not exclude Norway from the seabed beyond it. D.H. Anderson, "Report Number 9-15: Norway-United Kingdom", *International Maritime Boundaries*, pp. 1881-1882.

222. *Anglo-French Arbitration*, Decision of 1977, para.107.

223. C.H. Park, *Continental shelf issues in the Yellow Sea and the East China Sea*, Law of the Sea Institute Occasional Paper No. 15, University of Rhode Island (1972), p. 2; H. Schulte Nordholt, "Delimitation of the Continental Shelf in the East China Sea", 32 *N.I.L.R.* (1985), p. 136.

224. Kiyofumi Nakauci, "Problems of Delimitation in the East China Sea and Sea of Japan", 6 *O.D.I.L.* (1979), p. 312.

225. See paragraph 9 of Article 76 of the LOS Convention. UK argued with Canada, Australia, and Denmark at the UNCLOS III in favour of a Commission with no more than recommendation competence lest a powerful Commission would erode her rights over the continental margin: see UNCLOS III Official Records, vol. 13.

pp. 17-33; see also "Landlocked and Geographically Disadvantaged States and the Question of the Outer Limit of the Continental Shelf", 59 *B.Y.I.L.* (1988), p. 290. At the Sixth Meeting of States Parties to the United Nations Convention on the Law of the Sea (New York, 10-14 March 1997) 21 members of the Commission were elected for a term of five years on 13 March 1997 and began their term of office on the date of the first meeting of the Commission – 16 June 1997.

226. Paragraph 8 of Article 76 provides that: "Information on the limits of the continental shelf beyond 200 nautical miles from the baselines from which the breadth of the territorial sea is measured shall be submitted by the coastal State to the Commission on the Limits of the Continental Shelf set up under Annex II on the basis of equitable geographical representation. The Commission shall make recommendations to coastal States on matters related to the establishment of the outer limits of their continental shelf. The limits of the shelf established by a coastal State on the basis of these recommendations shall be final and binding".

227. The Rules of Procedure of the Commission on the Limits of the Continental Shelf was adopted in its entirety on 4 September in the 4th Session of the Commission (Document No. CLCS/3/Rev.2) and revised on 5 February in 2001 in the 9th Session (Document No. CLCS/Rev.3). Article 5 of the Annex of the Rules of the Procedure provides that:

> 5. (a) In cases where a land or maritime dispute exists, the Commission shall not examine and qualify a submission made by any of the States concerned in the dispute. However, the Commission may examine one or more submissions in the areas under dispute with prior consent given by all States that are parties to such a dispute.
> (b) The submissions made before the Commission and the recommendations adopted by the Commission thereon shall not prejudice the position of States which are parties to a land or maritime dispute.

228. Korea Working Group on the Continental Margin led by Professor, Yong-ahn Park, a member of the Commission on the Limits of the Continental Shelf, *Internal Working Paper*, June 2000.

229. Kenneth Y. Choy, "Overview of Taiwan's Legal and Jurisdictional Considerations in the Diaoyu/ Senkaku Dispute", in Taiwan Law Society and Taiwan Institute of International Law ed., *International Law Conference on the Dispute over Diaoyu/ Senkaku Islands*, April 1997, p. 119.

230. For the analysis of the territorial dispute see Kenneth Y. Choy, *op. cit.*, pp. 2-11; Yoshiro Matsui, "Legal Bases and Analysis of Japan's Claims to Diaoyu Islands", in *International Law Conference on the Dispute over Diaoyu/Senkaku Islands*, pp. 21-45; J. Greenfield, *China's Practice in the Law of the Sea*, pp. 128-129; Choon-ho Park, "The South China Sea Disputes: Who owns the islands and the natural resources", 5 *O.D.I.L.* (1978), pp. 27-59.

231. C.H. Park, *Continental shelf issues in the Yellow Sea and the East China Sea*, Law of the Sea Institute Occasional Paper No. 15, University of Rhode Island

(1972), p. 2; H. Schulte Nordholt, "Delimitation of the Continental Shelf in the East China Sea", 32 *N.I.L.R.* (1985), p. 136. But there is a differing view that the island is part of the upheaval belt connecting Taiwan and the northern part of the island of Kyushyu in Japan: see Kiyofumi Nakauci, "Problems of Delimitation in the East China Sea and Sea of Japan", 6 *O.D.I.L.* (1979), p. 312.

232. Jonathan I. Charney, "The Diaoyu/Senkaku Islands Maritime and Territorial Dispute", in Taiwan Law Society and Taiwan Institute of International Law ed., *ibid.*, p. 126.

233. Jonathan I. Charney, *ibid.*, p. 120.

234. A.E. Boyle, "UNCLOS, ITLOS and the Settlement of Maritime Boundary Disputes between Taiwan and Japan", in Taiwan Law Society and Taiwan Institute of International Law ed., *ibid.*, pp. 143-144: Jonathan I. Charney, "The Diaoyu/Senkaku Islands Maritime and Territorial Dispute", *ibid.*, p. 126; J. Greenfield, *China's Practice in International Law*, p. 131.

235. Case *concerning the Delimitation of the Continental Shelf between the United Kingdom of Great Britain and Northern Island and the French Republic*, Decision of June 30 1977, 18 *R.I.A.A.*, paras. 184 and 202: the award is reproduced in 18 *I.L.M.* (1979). pp. 397-494.

236. In this regard, the US Department of State has been trying to provide with specific standards for evaluating straight baselines: see US Department of State, *Limits in the Seas No. 106: Developing Standard Guidelines for Evaluating Straight Baselines*, 1987; see also J. Ashley Roach and Robert W. Smith, "Straight Baselines: The Need for a Universally Applied Norm", 31 *O.D.I.L.* (2000), pp. 31-47.

CHAPTER FIVE

Lessons from the Fragile Provisional Regime in North East Asia

1. PROVISIONAL FISHERIES REGIME BETWEEN KOREA AND JAPAN

1.1. *Background*

For 35 years between 1965 and 1999, the fisheries relations have been governed by the Fisheries Agreement of 1965, whereby both countries established their respective 12 N.M. exclusive fisheries zones from the coasts of each State; a joint regulation zone outside Korea's exclusive fisheries zone and a joint resources research zone outside the joint regulation zone.[1]

It is generally understood that the Fisheries Agreement of 1965 between Korea and Japan was negotiated under the influence of UNCLOS I and II where the argument to give the coastal States exclusive or preferential fisheries rights over the waters up to 12 miles had been very strong, although the argument fell short of getting a necessary majority for producing such rules.[2] In the 1960s it was Japan who wanted to have a fisheries agreement concluded with Korea because Korea proclaimed the so-called "Peace line" in a vast area of the sea around the Korean Peninsula in 1952 and enforced the line against the Japanese fishing vessels.[3] By concluding the fisheries agreement, the Japanese government succeeded in rendering the Peace line inapplicable to its fishermen and thus the Japanese fishermen gained access to fisheries resources even within the Peace line as long as they did not fish within the 12-mile fisheries zone of Korea. Although there were some restrictions in the joint regulation zone as to the number and size of fishing vessels, types of fishing gear, time of fishing operations, there had not been serious disputes in the zone because the maximum catch had been set at such a high level so as to satisfy the need for the Japanese fishermen and

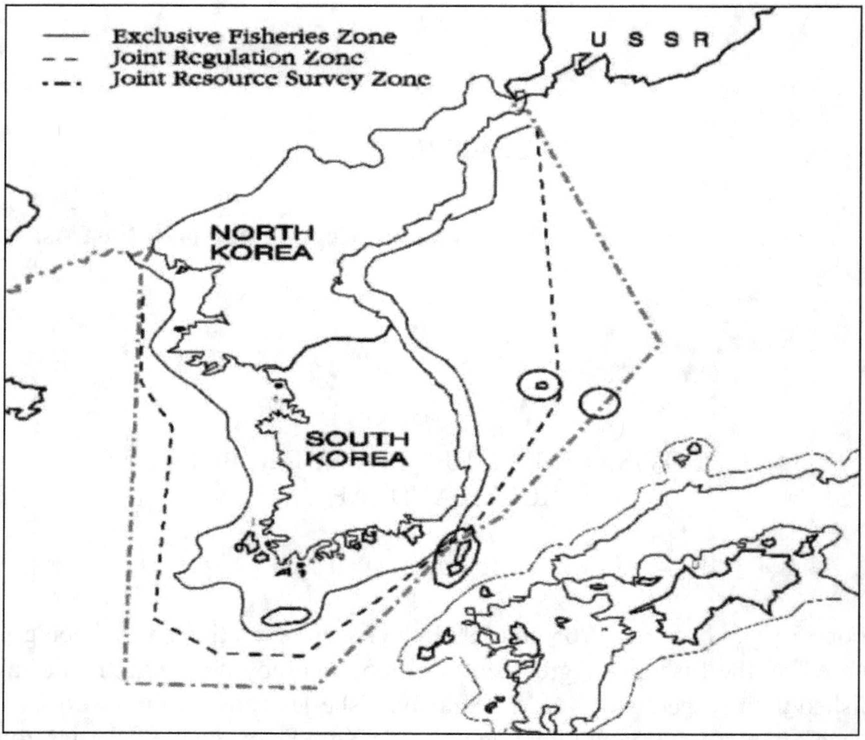

Map 37: Fisheries Agreement between Korea and Japan of 1965
(Source: J.R.V. Prescott, *Maritime Jurisdiction in East China Sea*)

the enforcement over the vessels in the joint regulation was to be exercised solely by the flag State of the vessels.[4]

However, problems began to arise in the late 1970s as the Korean fishing vessels became much more active in the waters off the Japanese coast with improved power and fishing equipment thanks to modern technology. In this situation, the proclamation of the 200-mile fisheries zone by the Soviet Union in 1976 also provided a backdrop against which new fisheries disputes arose anew because the Korean distant water fishing vessels that lost their fishing ground in the waters around the Soviet Unions in Northwest Pacific began to swamp the coastal areas of Japan's Hokkaido to catch Alaska Pollock.[5] Although this situation was not welcomed at all by Japan which also proclaimed its 200-mile fisheries zone in 1977, it could not enforce the zone against Korean fishermen in the waters beyond the 12 miles from its coasts because of the Fisheries Agreement of 1965, which provided for, as Japan had wished, the freedom of fishing under the flag-

State jurisdiction in the area beyond 12 miles from the coasts.[6] Then the Fisheries Agreement of 1965 turned out to benefit Korean fishermen more than Japanese fishermen. Obviously this was not the situation which Japan had expected in the 1960s when it asked Korea to conclude a fisheries agreement.

In this situation Japan raised the need to amend the Fisheries Agreement of 1965 but Korea tried to keep the agreement as it stood because it benefited Korean fishermen. However, after long discussion, Korea and Japan reached a compromise formula in 1980, establishing zones where certain types of fishing regulations were to be applied.[7] The zones were established in the waters around Korea's Cheju island, Japan's Kyushu and Hokkaido.

According to the regulation scheme, Japanese fishermen were to apply fishing regulations in the designated waters around the island of Cheju and Korean fishermen were to apply fishing regulations in the designated waters around Kyushu and Hokkaido of Japan. Although the fishing regulations were agreed between the two governments, this type of fishing regulation was named "autonomous fishing operation regulation measures" because fishing vessels operating in the areas where the measures were to be applied were subject only to the jurisdiction of the State under whose flag the vessels flew.[8] Due to the fact that flag State jurisdiction was applied in the autonomous fishing operation regulation zones, the autonomous fishing operation scheme was not intended to replace the 1965 agreement but to complement it on the basis of Articles 1 and 4 of the agreement, which made it clear that the enforcement jurisdiction in the waters beyond the 12-mile fisheries belonged only to the flag States.

However, the autonomous fishing regulation scheme was not fully successful in preventing fisheries disputes arising between Korean and Japan. This was particularly so with regard to fishing activities by Korean fishermen in the waters around the Japanese island of Hokkaido, as some of the Korean fishing vessels sometimes did not faithfully observe the autonomous fishing operation regulations. There was even antagonism among the Japanese local fishermen of Hokkaido against the Korean fishing vessels because the Japanese fishermen applied their own conservation measures in the waters even beyond 12 miles from the Japanese coasts whereas some of the Korean fishing vessels appeared not to have due regard to the Japanese conservation measures, thus making the Japanese conservation efforts much less effective. Therefore, the local fishermen and politicians from Hokkaido took the initiative for abolishment of the agreement of 1965 in the mid-1990s and establishment of a new fisheries agreement with Korea.[9]

In 1996 when Japan and Korea proclaimed their respective EEZ Japan began to emphasise the need to abolish the Fisheries Agreement of 1965 and conclude a new fisheries agreement on the basis of the EEZ fisheries regime in accordance with the LOS Convention. In March 1996, three Japanese ruling parties jointly declared their position that "a new fisheries agreement between Japan and Korea should be concluded within a year".[10] In the face of the Japanese proposals for a new fisheries agreement, Korea voiced its position that a new fisheries agreement should be established based upon the delimitation of EEZ boundaries, and thus the delimitation of EEZ boundaries should come first. The question whether the delimitation of EEZ or provisional fisheries agreements should come first became a heated issue between Korea and Japan even in the summit talks between the two countries. In the summit meeting of 25 January 1997, the Japanese Prime Minister, Ryutaro Hashimoto, asked the Korean President, Youngsam Kim, for co-operation from the Korean side for attainment of an objective of concluding a new fisheries agreement within the year 1997, arguing that the delimitation of EEZ boundaries should not be a prerequisite for concluding a new fisheries agreement.[11]

In March 1997 the Korean side presented a proposal of an outline of a new fisheries agreement, which provided that a new fisheries agreement should come into effect after Korea and Japan agreed on the delimitation of EEZ boundaries and that the level of current fishing catch should be respected for five years after the coming into effect of the new fisheries agreement. Various proposals were exchanged since March 1997 between Korea and Japan. At this time, Korea appears to have tried hard to get an agreement on an EEZ boundary line between Korea and Japan because the momentum for the delimitation of an EEZ boundary would be significantly lessened if the EEZ boundary were not drawn now before the conclusion of a new fisheries agreement. When the delimitation of an EEZ boundary was proven difficult to do before the conclusion of a new fisheries agreement, the Korean government proposed in September 1997 to draw a provisional fisheries boundary following the equidistance line between Korea's Ulleung-Do and Japan's Oki-syto. But the Japanese side did not accept this proposal.[12] Since November 1997 Korea and Japan have begun to discuss in detail the shape and extent of joint fishing zones and the zones where EEZ fisheries regime is to be applied. In January 1998 as the Japanese government notified its intention to terminate the Fisheries Agreement of 1965, the agreement of 1965 was doomed to be terminated on 22 January 1999, a year after the notification.[13] After hard negotiations conducted at various levels, Korea and Japan reached an agreement on a new fisheries agreement in September 1998 on the basis of modified 35-mile EEZ fisheries zones of each country and establishment of joint fishing zones outside the

EEZ fisheries zones.[14] The new fisheries agreement was signed on 28 November 1998 in Kagoshima, Japan and the instruments of ratification were exchanged on 22 January 1999 in Seoul, Korea bringing the new agreement into effect on the same day.[15]

1.2. *Main Features of the New Fisheries Agreement between Korea and Japan*

1.2.1. *Basic Structure of the Agreement*

The Agreement of 1999 is agreed not upon the ultimate EEZ boundaries but upon the provisional joint fishing zones and EEZ fisheries zones of a limited extent which are to be considered as each Party's EEZ for the purpose of implementation of the agreement.[16] As such, there is no doubt that the agreement is a provisional arrangement of a practical nature envisaged in Paragraph 3 of Article 74 of the LOS Convention pending the ultimate agreement on boundaries of EEZ. Thus the agreement purports to regulate fisheries relations between Korea and Japan pending the ultimate delimitation of EEZ boundaries between the two countries. Related to this, the agreement obliges the two countries "to continue to negotiate in good faith for an earlier delimitation of the exclusive economic zone".[17] The "without-prejudice clause" of the agreement provides a quite comprehensive protection of the positions of each Party, providing that: "Nothing in this Agreement shall be deemed to prejudice the position of each Contracting Party relating to issues on international law other than matters on fisheries".[18]

Here it will be appropriate to examine the basic structure of the new agreement. As the new agreement is based on the EEZ fisheries regime, Article 1 of the agreement provides that: "This Agreement applies to the exclusive economic zone of the Republic of Korea and the exclusive economic zone of Japan". Then, Articles 2 to 6 of the agreement provide detailed provisions on fishing access of fishing vessels of one Party to the EEZ of the other Party. Each Party shall permit nationals and fishing vessels of the other Party to harvest in its exclusive economic zone pursuant to the agreement and its relevant laws and regulations (Article 2). Each Party may take necessary measures in accordance with international law in its exclusive economic zone to ensure that nationals and fishing vessels of the other Party engaged in fishing activities in its exclusive economic zone comply with the fisheries agreement, its relevant law and regulations and specific conditions imposed upon them (Article 5).

However, since there is no EEZ boundary between Korea and Japan the agreement deals with the question of what are the geographical limits of the EEZ of each Party in applying the provisions on EEZ fishing access. In this regard, the Parties agreed to use the Northern Continental Shelf Boundary of 1974 as a fisheries boundary for the purpose of the application of the agreement (Paragraph 2 of Article 7). Thus the continental shelf boundary is now also a fisheries boundary even if it is a provisional one. And for the waters where there is no continental shelf boundary, Korea and Japan agreed to set up two joint fishing zones; one is in the East Sea connected to the northern terminus of the Northern Continental Shelf Boundary and the other is in the East China Sea connected to the southern terminus of the Shelf Boundary (Article 9).

The agreement makes it clear that the provisions of Articles 2 to 6 on EEZ fishing access do not apply to the joint fishing zones (Article 8). The lines constituting the outer limits of the joint fishing zones are to be considered as geographical limits for the exercise of sovereign rights for the purpose of application of the agreement by each State (Annex II). Therefore each Party is to exercise EEZ fisheries rights against the other Party under the agreement in the area situated between the outer limits of its territorial sea and the outer limits of the joint fishing zones on its side. This is a sort of presumption on the fisheries boundaries using the outer limits of the joint fishing zones. However, one exception to this assumption is provided for the area where the fisheries relations between North Korea and Japan are concerned (Paragraph 3 of Annex).[19]

Article 9 does not attempt to name the two joint fishing zones, and thus there are no official names for the two zones. However, the Korean government favours the term "middle zones" or "intermediate zones", whereas the Japanese government prefers the term "provisional zones".[20] Detailed provisions are laid down on the management of marine living resources in the two joint fishing zones (Annex I).

Korea and Japan set up "Korea-Japan Fisheries Committee" to consult and render recommendations and binding decisions to both governments on the various matters relating to the implementation of the agreement (Article 12). The Committee can make recommendations or decisions only through an agreement between the representatives of the Parties participating in the Committee (Paragraph 2 of Article 12).[21] The range of issues that the Committee may deal with is very broad. The Committee may consult and render recommendations, *inter alia*, on matters relating to EEZ fishing access such as the species of fish allowed to be caught, quotas of catch, areas of fishing and specific conditions on fishing operations (Articles 3 and Article 12). The Committee can consult also on the matters relating to the State of marine living resources in the two joint fishing zones. However, a distinction was deliberately made between the joint fishing zone in the East Sea and the joint fishing zone in the East China Sea in this regard. That is to say that the Committee can make only recommendations with regard to conservation and management of marine living resources in the East Sea, whereas it can even make decisions with regard to the same matters in the joint fishing zone in the East China Sea. This distinction is crucial to the Korean government because it wishes to avoid the impression that Korea and Japan take "joint" measures with regard to the waters around Dok-do, which is "an inherent part of Korean territory", and is "geographically" located in the joint fishing zone in the East Sea.[22] Note that the presence of Dok-do in the middle of the "middle zone" in the East Sea appears to be intolerable to some Korean people.[23] Therefore, on the basis of this distinction the government of the Republic of Korea explained to the Korean people that the middle zone in the East Sea is not "jointly" managed by Korea and Japan, whereas the middle zone in the East China Sea is "jointly" managed by Korea and Japan.[24]

The dispute settlement procedures are also provided in the agreement mentioning arbitration procedure in detail. However, unlike the Fisheries Agreement of 1965, the arbitration envisaged in the new fisheries agreement is not compulsory; only through the consent of both Parties can disputes be brought to arbitration (Article 13). These provisions on dispute settlements can have a significant impact when a dispute arises and the dispute relates to the provisions of the LOS Convention as well as the new fisheries agreement. In the case where such a dispute arises and either Party wishes to bring the case before an arbitral court constituted under the LOS Convention, then it is highly probable that the other Party can successfully challenge the jurisdiction of the arbitral court as did Japan in the *Southern Bluefin Tuna Arbitration* of 2000.[25] It was Japan who preferred to provide for consensual arbitration, whereas Korea wished to reproduce the compulsory arbitration clause in the Fisheries Agreement of 1965.[26]

1.2.2. *Shaping of the Joint Fishing Zones (Intermediate Zones)*

Korea and Japan needed to establish the two joint fishing zones in the East Sea and in the East China Sea because they did not reach an agreement on delimitation of their overlapping EEZ claims. In other words, if Korea and Japan succeeded in drawing EEZ boundaries then the fisheries agreement could have been simpler than the one they have now. It took more than two years of intensive negotiations for Korea and Japan to reach an agreement in shaping the outer limits of two joint fishing zones.[27] In shaping the joint fishing zone in the East Sea, the following elements were considered:

35-mile lines from the coasts, in particular from Ulleung-do and Oki-syto, the line of Lat. 38° 37.0' E. being the northern limits line between South Korea and North Korea, the line of Lat. 135.5° E. as the eastern limits of the joint fishing zone, the location of Yamato Tai which is an important squid fishing ground, and the basepoints which were used in making the northern terminus (Point No. 35) of the Northern Continental Shelf Boundary.[28] However, the most important consideration in shaping the joint fishing zones was the equitable division of the zones which are regarded as EEZ of each Party for the purpose of application of the fisheries agreement.[29]

With regard to the shaping and location of the Intermediate Zone in the East Sea, a serious question arose as to whether the Intermediate Zone had an effect on the territorial issue of Dok-do. However, it seems absurd to assume that the provisional fisheries agreement between Korea and Japan based upon Article 74(3) of the LOS Convention affects a territorial issue, since the "without-prejudice clause" clearly provides that: "Nothing in this Agreement shall be deemed to prejudice the position of each Contracting Party relating to issues on international law other then matters on fisheries".[30]

It is to be recalled that in the *Minquiers and Ecrehos* Case where the French government tried to reinforce its position by using a bilateral fisheries agreement of 1839, arguing that the Minquires and Ecrehos groups were included in the common fisheries zone,[31] the ICJ held that:

> The Court does not consider it necessary, for the purpose of deciding the present case, to determine whether the waters of the Ecrehos and Minquires groups are inside or outside the common fishery zone established by Article 3. Even if it be held that these groups lie within this common fishery zone, the Court cannot admit that such an agreed common fishery zone in these waters would involve a regime of common use of the land territory of the islets and rocks, since the Articles relied on refer to fishery only and not to any kind of use of land.[32]

Here, however, another associated question lingers as to whether the joint fishing zone in the East Sea can be seen as a precedent where Dok-do is excluded from being used as a basepoint. It can be said that Dok-do was not used as Korea's basepoint in shaping the joint fishing zone in the East Sea, in the sense that the joint fishing zone could have been established to the east of Dok-do if it had been used as Korea's basepoint. But at the same time by the same token, it can be said that Dok-do was not used as a Japanese basepoint in shaping the joint fishing zone. If Dok-do had been used as a Japanese basepoint then the joint fishing zone could have been situated somewhere between Ulleung-do and Dok-do. Let us approach the

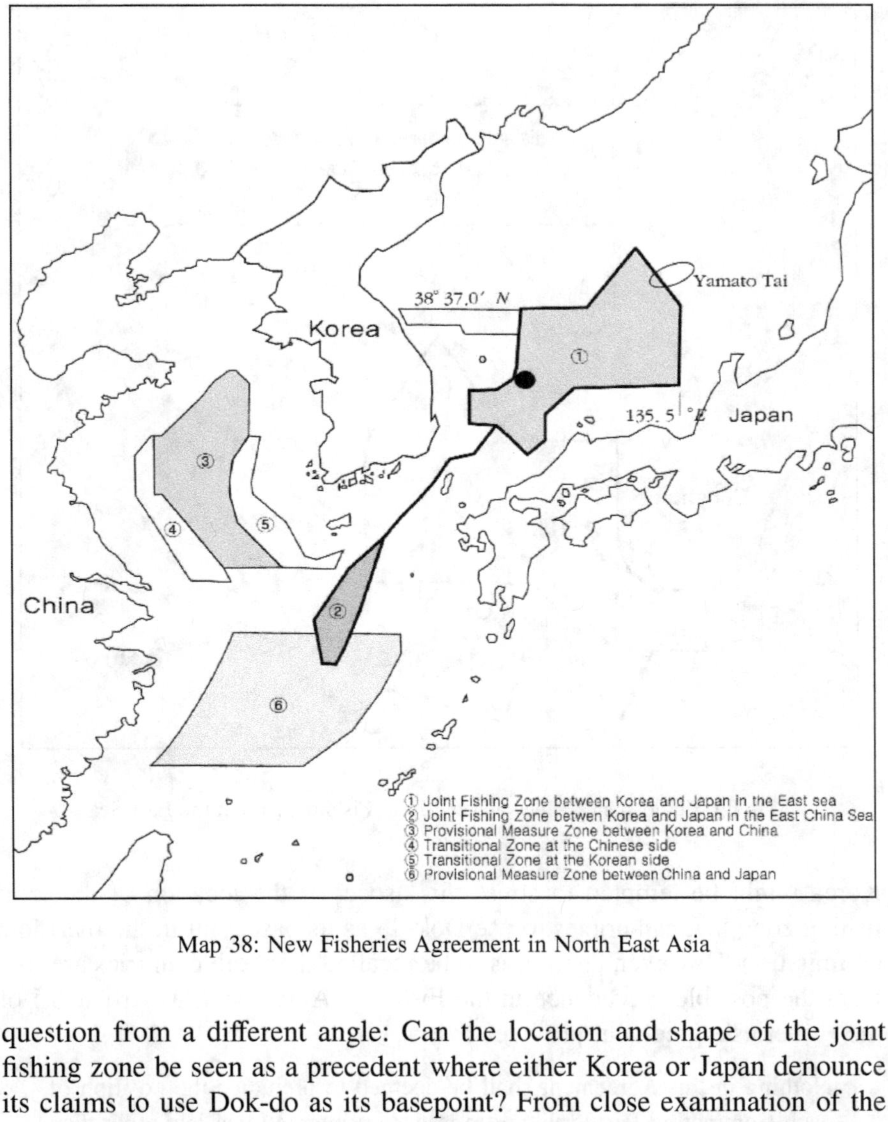

Map 38: New Fisheries Agreement in North East Asia

question from a different angle: Can the location and shape of the joint fishing zone be seen as a precedent where either Korea or Japan denounce its claims to use Dok-do as its basepoint? From close examination of the map, it can be said that the shaping of the joint fishing zone is a precedent where Japan's claim to use Dok-do was denounced whereas Korea's claim is not. Note that some part of the median line between Ulleung-do and Dok-do, which could have been argued by Japan in the delimitation negotiation is located to the west of the joint fishing zone, i.e., in Korea's EEZ, whereas the median line between Dok-do and Oki-syto is well within the joint fishing zone.

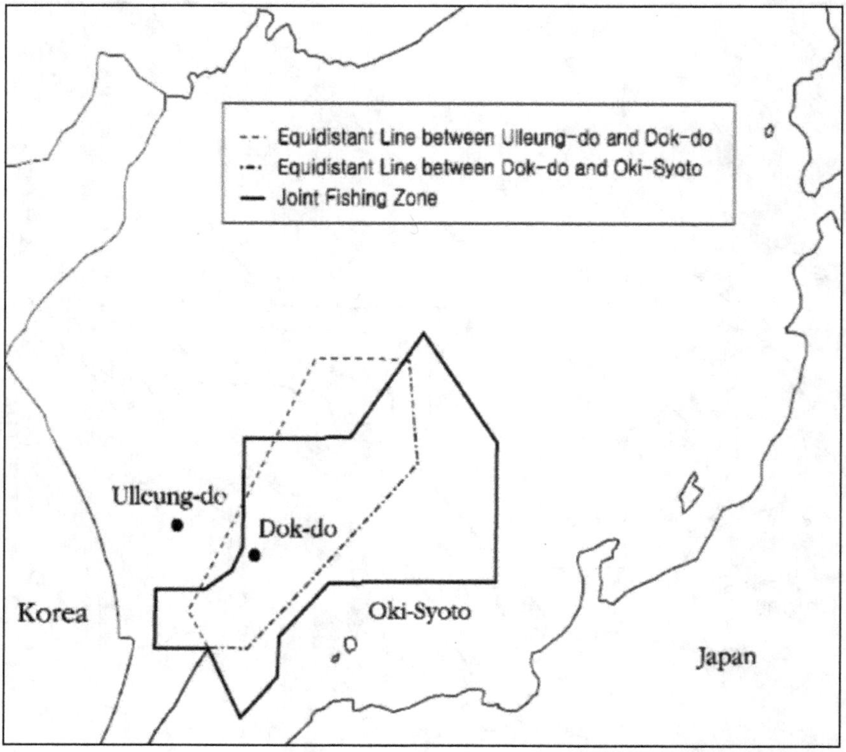

Map 39: Equidistance Lines and the Joint Fishing Zone in the East Sea

Korea might be tempted to stress this aspect of the location of the joint fishing zone if Japan argues to use Dok-do as its basepoint in the maritime delimitation. However, again it is to be recalled that both countries are free from the possible precedence in the Fisheries Agreement, as Article 15 of the Fisheries Agreement provides that:

> Nothing in this Agreement shall be deemed to prejudice the position of each Contracting Party relating to issues on international law other than matters on fisheries.

Now let us turn to the joint fishing zone in the East China Sea. The outer limits of the zone are adopted in such a way as to include the two following hypothetical water column boundaries between Korea and Japan: one is the equidistance line drawn by using all the nearest features in the coasts of Korea and Japan in the region whether they are rocks or islands, and the other is a line drawn by excluding rocks in terms of Paragraph 3 of Article

121 of the LOS Convention. Even after having agreed to include two such lines in the joint fishing zone, Korea and Japan had further difficulty in agreeing on the co-ordinates of the southern limits of the joint fishing zone in the East China Sea, because there were different views on the southern-most point of the water column boundaries between Korea and Japan. Korea's basic position in this regard is that the water column boundary in the East China Sea between Korea and Japan should be drawn in such a way as not to produce a cut-off effect for Korea. Thus the boundary should reach the 200 mile line from Korea's southernmost island Mara-do and the southern limits of the joint fishing zone should be identical with the 200-mile line from Mara-do.[33] For Japan it was also difficult to agree with Korea's position particularly because Japan and China had already concluded a fisheries agreement in November 1997 thereby establishing a joint fishing zone in the East China Sea which encroaches upon the 200-mile line from Mara-do.[34] As they could not agree on the specific co-ordinates of the southern limits of the joint fishing zone in the East China Sea, they agreed to adopt the term "the southern-most parallel of the exclusive economic zone of the Republic of Korea" to describe the southern limits of the joint fishing zone in the East China Sea with a mutual understanding that different views on the southern limits of the joint fishing zone in the East China Sea exist between Korea and Japan.[35] Thus for Korea the southern limits line of the joint fishing zone in the East China Sea is at the parallel of Lat. 29° 43' N. in line with the 200-mile line from Mara-do, whereas for Japan the line would be at the parallel somewhere of Lat. 30° 40' N. following the latitude of a hypothetical tri-equidistant point among Korea's Mara-do, Japan's Tori-shima and China's Haijiao. Note that China's position on the delimitation of maritime boundaries is not for the equidistance principle but for equitable principles particularly with regard to the East China Sea and it did not participate in the negotiation on the shaping of the Korea-Japan joint fishing zone in the area. In this context, China expressed its position that it does not recognise the Korea-Japan joint fishing zone in the East China.[36]

1.2.3. *Jurisdiction in the Joint Fishing Zones*

The separate provisions on the two joint fishing zones with respect to the jurisdiction therein and to the conservation and management of the marine living resources thereof are almost identical except that the Korea-Japan Fisheries Committee can make decisions with regard to the joint fishing zone in the East China Sea whereas it can only make recommendations with regard to the joint fishing zone in the East China Sea.[37] It will be appropriate to examine the exceptional difference between the two zones and then to discuss the common aspects of the two zones with respect to jurisdiction in the zones and conservation and management of the marine living resources of the zones.

As mentioned earlier the difference is deliberately made by, and meaningful to, Korea. Korea wished to avoid any implication that the joint fishing zone in the East Sea where Dok-do is geographically situated is jointly managed by Korea and Japan. The Korean government's position on this point is that the joint fishing zone in the East Sea is not a zone jointly managed by Japan and Korea, because it would be either Korea alone or Japan alone, which decides any conservation and management measures in the joint fishing zone in the East Sea and thus the Korea-Japan Fisheries Committee, which is an international juridical body, does not have any prescriptive competence with regard to the joint fishing zone in the East China Sea.[38] Here, the Korean government appears to presuppose that the joint fishing zone might affect the territorial issue if it is jointly managed by the joint committee. An important question arises here: if the joint fisheries committee has the competence to prescribe any conservation measures with regard to the zone in the East Sea and thus the zone is to be called a "joint management" zone, then does the competence of the joint committee affect territorial status of Dok-do? It seems unreasonable to suggest that a joint fishing zone affects sovereignty over an island if its fisheries resources are jointly managed by the fisheries committee and that the zone does not affect a territorial issue if the committee has no power to decide on the conservation of fisheries resources in the zone. Interestingly, however, a few Korean scholars have argued that the new fisheries agreement between Korea and Japan affects negatively Korea's territorial sovereignty of Dok-do.[39] However, it would be absurd to assume that a provisional fisheries agreement such as the one between Korea and Japan affects the issue of territorial sovereignty. As we discussed elsewhere, the clear distinction should be made between a joint fishing zone or joint development zone and condominium, since the former is related to sovereign rights and the latter is related to sovereignty. We need to recall the provisions of Paragraph 3 of Article 74, the without-prejudice clause in the agreement and the *dicta* of the ICJ in the *Minquiers and Ecrehos* Case. If a provisional fisheries agreement can affect territorial issues against the clear intentions of the Parties notwithstanding all the safeguards, then no country in the world would venture to conclude a provisional fisheries agreement fearing that the provisional agreement would affect its position on the delimitation of EEZ boundaries, even when there is no territorial dispute.

Now let us turn to the common aspects of the two joint fishing zones. The fundamental common aspect is that the flag-State jurisdiction is applied in the two joint fishing zones. It is provided in identical terms that: "Each Contracting party shall not apply its relevant laws and regulations on fisheries to the nationals and fishing vessels of the other Contracting Party in the zone".[40] And even if there is a conservation measure adopted by a decision by the Korea-Japan Fisheries Committee, it is up to each Party to take any necessary measures upon its nationals and fishing vessels.[41] However, an indirect form of co-operation in enforcement can be made in these joint fishing zones if there is a measure adopted by the Committee or a measure adopted by each State in accordance with recommendations by the Committee: in the case where a Party finds a fishing vessel of the other Party engaged in fishing in violation of such a measure, the Party notifies the violation to the other Party, and then the other Party should take necessary measures with regard to the violation by its fishing vessels and report back to the Party of the measures it has taken.[42] However, such indirect co-operation has not happened thus far because the Korea-Japan Fisheries Committee has not yet made any recommendation or decision. The reason why the Committee has not made any recommendation with regard to the conservation and management of the living resources in the joint fishing zone in the East Sea, is that Korea has been reluctant to agree on any measure in the face of a domestic criticism that the status of Dok-do is impaired because of the establishment of the joint fishing zone in the East Sea.[43] And it appears that there has been no serious discussion on the conservation and management of the joint fishing zone in the East China Sea.[44]

It seems appropriate to discuss further the issue of the conservation and management of the living resources in the joint fishing zone in the East Sea because this issue has been the most controversial issue in the discussion of the Korea-Japan joint fisheries committee. In the zone in the East Sea, the main species caught are squids, Spanish mackerel and large crab.[45] Korean fishermen increased their catch in the joint fishing zone in the East Sea after the Korea-Japan fisheries agreement came into effect because their catch in the waters which was part of the high seas and turned into Japan's EEZ fisheries zone decreased significantly.[46] Against this backdrop, Japan asked Korea that the joint committee should adopt a number of conservation measures and some form of enforcement co-operation with regard to the joint fishing zone in the East Sea. However, Korea refused to adopt any conservation measure or even placing a ceiling on the number of fishing vessels per mode in the zone, which is an example of a conservation measure in the fisheries agreement.[47] In the negotiations of 1999 between

Korea and Japan for the fishing quotas and fishing conditions in each other's EEZ in 2000, Japan argued that unless the effective conservation measure is agreed with regard to the middle zone in the East Sea, then the negotiations on fishing quotas in the EEZ cannot be made. But surprisingly Korea responded to Japan saying that it would not agree to adopt any conservation measure through recommendation of the joint committee with regard to the East Sea joint fishing zone, because it might affect its territorial ownership over Dok-Do, even if it gave up its fishing quota in Japan's EEZ. In this dramatic confrontation, both countries agreed in 1999 that each Party should implement its own conservation measures under the principle of flag-State jurisdiction in the joint fishing zones and that autonomous fishing regulations of non-governmental nature may be agreed and implemented through the consultation between their fishermen's organisations.[48] It is not reasonably expected that Korea would agree to adopt any conservation measures in the joint committee with regard to the joint fishing zone in the East Sea because of its unyielding position on this issue influenced by the public worries about the possible implications of conservation measures on Dok-do.[49]

In this regard, a question is raised in this author's mind as to whether Japan is really interested in proper management of the marine living resources in the Korea-Japan joint fishing zone in the East Sea. Because notwithstanding Japan's strong demands for the adoption of conservation measures through the joint committee, Japan agreed with China that it would allow Chinese fishermen to fish up to 7,000 tons of squid annually in the middle of the Korea-Japan joint fishing zone in the East Sea without consultation with Korea.[50] Also note that Japan has never asked to adopt a conservation measure in the Korea-Japan fishing zone in the East China Sea. Now a further question can arise: is it Japan's hidden intention, as the Korean public worries, to take some advantage in territorial issues over Dok-do from the fisheries agreement? The answer is unknown to this author. But one thing is very clear. The provisional fisheries agreement cannot affect territorial questions at all as we have discussed fully thus far.

1.2.4. *The Korea-Japan Fisheries Agreement and the Fishing Order in the East China Sea*

As the "Provisional Measure Zone" between Japan and China overlaps with the Korean version of the joint fishing zone in the East China Sea, there arises a need to address this problem in the negotiations between Korea and Japan. If this problem is not to be solved in one way or another, then jurisdictional conflicts could occur in the East China Sea. Suppose that a

Korean fishing vessel is engaged in the overlapped zone on the presumption that it is only subject to the jurisdiction of the Korean authorities, since it is part of the Korea-Japan joint fishing zone in accordance with the provisions of the Korea-Japan fisheries agreement. Suppose also that a military ship or coast guard ship of China or Japan finds the Korean vessel engaged in fishing in the zone and tries to enforce its fisheries laws on the Korean vessel. In this situation, a jurisdictional conflict could take place when a ship from the Korean Maritime Police Agency comes along and argues that it has the sole jurisdiction over the Korean fishing vessel under the principle of the flag-State jurisdiction, provided in the Korea-Japan fisheries agreement. To avoid such jurisdictional conflicts, Korea and Japan tried to address the problem by signing an Agreed Minutes.[51] In the Agreed Minutes, the Korean government expressed its "intention to co-operate with the government of Japan in order not to damage fisheries relations established by Japan and a third country in ***a certain area*** of the East China Sea (emphasis added)".[52] In this passage the third State seems to imply China because China is the only country relevant in this context.[53] However, it is not clear from its ordinary meaning what the term "a certain area of the East China Sea" indicates. However, in the light of the fact that the Agreed Minutes were intended to deal mainly with the problem of overlap between the Korean version of Korea-Japanese joint fishing zone in the East China Sea and the Sino-Japanese provisional measure zone, it can be reasonably assumed that the term "a certain area" indicates the part of the Korean version of the Korea-Japanese joint fishing zone which is located within the Sino-Japanese provisional measure zone.[54]

Thus interpreted, Korea's intention in the Agreed Minutes can be seen as a compromise from its position with regard to the location of the southern limits line of the Korea-Japanese joint fishing zone and also to its challenge of the legality of the northern limit line of the Sino-Japanese provisional measure zone.[55] Therefore, immediately after this undertaking is made it is provided that: "However, this shall not be deemed to prejudice the position of the Republic of Korea on the fisheries agreement concluded by Japan with such a third country".[56]

In the same document, the Japanese government also made an important *quid pro quo* non-binding undertaking, saying that it has "an intention to co-operate with the Government of such a third country in order to render possible certain fishing activities by nationals and fishing vessels of the Republic of Korea in ***the other certain area*** of the East China Sea under the fisheries relations established by Japan and the third country (emphasis added)".[57] As the Agreed Minutes do not show where "the other certain

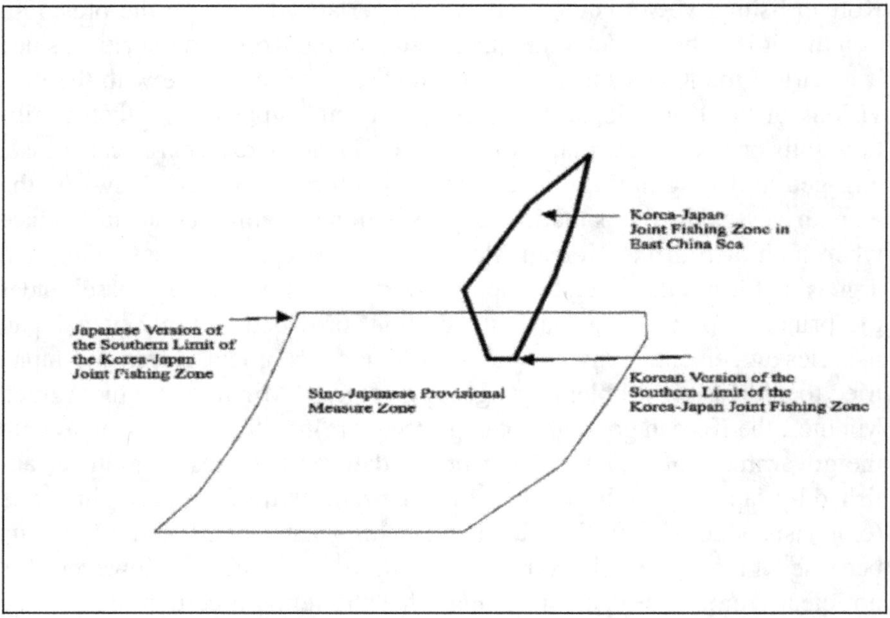

Map 40: Korea-Japan Joint Fishing Zone in the East China Sea

area" in the East China Sea is located, the need arises to figure out where it is. Firstly, it is clear that "the other certain area" is an area that does not belong to "a certain area" where Korea is willing to co-operate with Japan in order not to damage fisheries relations between Japan and China. Thus, the part of the Korean version of the Korea-Japanese joint fishing zone which is located in the Sino-Japanese provisional measure zone is excluded from consideration. Secondly, it should be noted that the Japan's undertaking is related to co-operation with a third State (China) for allowing the fishing activities by Korean fishermen. Thus it can be assumed that the other certain area is an area where both China and Japan have some rights over fisheries and where Japan cannot exercise the rights alone.[58] Therefore it can be assumed that "the other certain area" is located inside the Sino-Japanese provisional measure zone but outside the aforementioned "a certain area".[59]

Korea and Japan jointly expressed in the Agreed Minutes their intention "to negotiate specific ways for maintenance of good fisheries order in the East China Sea" through the three separate joint fisheries committees between Korea, Japan and China. Therefore, the role of the three joint fisheries committees between Korea, China and Japan is expected to maintain a good fishing order in the East China Sea where the three coastal States made overlapping claims of EEZ.

2. PROVISIONAL FISHERIES REGIME BETWEEN KOREA AND CHINA

2.1. *Background*

As there were no diplomatic relations between the Republic of Korea and the People's Republic of China during the Cold War, talks for a fisheries agreement between Korea and China at government level were not possible until 1992 when the two countries established diplomatic relations.[60] However, from time to time incidents of arrest of Korean fishing vessels by the Chinese authorities occurred in the area within the East China Motor Trawl Prohibition Line (the so-called Mao Tse-teung line) as claimed by China in the early 1950s. The Korean government had to advise the Korean fishing vessels not to approach the Mao Tse-teung line for the sake of safety.[61] Thus, in 1975 the Korean government drew a fishing operation restriction line along and off the Mao Tse-teung line for keeping Korean fishing vessels from approaching the line.[62]

Since the mid-1980s, the fishing activities by the Chinese fishing vessels became very active and thus disputes at sea between the fishing vessels of Korea and China frequently arose. The fishermen's organisations of the two countries concluded an "agreement for dealing with disputes between their fishing vessels at sea" in 1989.[63]

The two governments began to negotiate for a fisheries agreement in 1993, a year after they established diplomatic relations. In the early stages of the negotiations, China argued that the whole area between the outer limits of the territorial waters of the two countries should be a joint fishing zone until final delimitation of EEZ boundaries were drawn, whereas Korea argued that EEZ fisheries regimes should be applied in an area as large as possible and a joint fishing zone should be as small as possible.[64] While the negotiations between Korea and China made little progress and massive fishing activities were conducted by the Chinese fishermen off the Korean coasts, a difficult question arose within the Korean government as to whether it should enforce its EEZ Law of 1996 and its EEZ Fisheries Law of 1997 against the Chinese fishermen applying a hypothetical line as a fisheries boundary.[65] The issues involved in this question are (i) whether a unilateral line can be justified in international law, (ii) whether the enforcement activities by one coastal State can be justified under international law when the negotiations for a fisheries agreement is under way with its neighbouring States, and (iii) whether a coastal State can drive out completely foreigners' fishing activities without any reasonable ground based on a total allowable catch (TAC) system.[66] After serious consideration, Korea decided not to apply its EEZ fisheries regime against the Chinese fishermen by inserting a clause in the Enforcement Decree of its EEZ Fisheries Law.[67]

In September 1997, four years since the first round of talks, Korea and China agreed in principle to establish joint fishing zones in the middle of the Yellow Sea.[68] From December 1997, various proposals on how to shape a joint fishing zone in the Yellow Sea were exchanged between the two countries.[69] As Korea wished to set up a joint fishing zone of a narrow width whereas China wanted a joint fishing zone of a wide width, it was difficult for the two countries to agree on the shape. From this situation, the idea of "transitional zones" was introduced to settle the dilemma in the negotiations on the shape of the joint fishing zone in September 1998. The concept is as follows: there are to be established two transitional zones, one is situated between the joint fishing zone and Korea's zone which is regarded as Korea's EEZ for the purpose of application of the fisheries agreement and the other transitional zone which is situated between the joint fishing zone and China's zone, and fishermen from both Korea and China can fish in the transitional zones until the fourth year after the entry of the fisheries agreement, and then the transitional zones turn into part of the zones where each State can exercise its sovereign rights on fisheries for the purpose of the fisheries agreement.[70] After settling the different views on the shape of the joint fishing zone and transitional zones, the two countries were able to initialise the fisheries agreement in November 1998.[71] However, it took another two years since the initialling of the two governments to finally sign the fisheries agreement because there arose different views on the interpretation of the provisions of the initialled text on the fishing orders in the waters off the Yangste river, as we will see later. The Parties agreed to bring the fisheries agreement into force on 30 June 2001.[72]

2.2. *Main Features of the Fisheries Agreement between Korea and China*

2.2.1. *Basic Structure of the Agreement*

The Fisheries Agreement between Korea and China shares some of the fundamental aspects with the Fisheries Agreement between Korea and Japan:

it is a fisheries agreement based upon the EEZ fisheries regime of the LOS Convention and it is a provisional fisheries agreement pending the ultimate delimitation of EEZ boundaries. As it is a fisheries agreement based upon the EEZ regime of Korea and China, Article 1 of the agreement makes it clear that it applies to Korea's EEZ and China's EEZ. And then Article 2 of the agreement provided that: "Each Contracting Party shall permit, in accordance with the provisions of this Agreement and its relevant laws and regulations, the fishing activities by the nationals and fishing vessels of the other Party in its exclusive economic zone".[73] And then detailed provisions on mutual fishing access follow in Articles 2 to 5 of the agreement.[74]

However, as there is no EEZ boundary drawn between Korea and China, joint fishing zones were established under two different nomenclatures: one is the so-called "Provisional Measure Zone" which is in the middle of the Yellow Sea and the others are the so-called "Transitional Zones", one of which is situated near Korea's coast and the other is situated near Chinese coast.[75] We will discuss in detail the characteristics of the joint fishing zones at a later stage.

Since the Korean-Chinese Fisheries Agreement is a provisional arrangement of a practical nature envisaged in Paragraph 3 of Article 74 of the LOS Convention, it shall be without prejudice to the final delimitation. For the sake of a double safeguard, the "without-prejudice clause" is adopted in the Fisheries Agreement, providing that: "No provision in this Agreement shall be interpreted in such a way as to prejudice the position of either Contracting Party on issues in the law of the sea".[76]

A Korean-Chinese Joint Fisheries Committee is to be established in order to facilitate the implementation of the agreement.[77] The Committee has the power to decide matters regarding the conservation and management measures in the Provisional Measure Zone and Transitional Zones, and it can recommend to the government of each Contracting Party measures relating to fishing access to EEZ such as the species of fish allowed to be caught, fishing quotas, and fishing operation conditions.[78] The Committee needs consent between the representatives of both Parties in order to make any decision or recommendation.[79] The governments of Korea and China should respect recommendations made by the Committee and should take necessary measures following the Committee's decision.[80]

It is notable that the agreement does not contain provisions on dispute settlement, because China did not want to have them in there.[81] China's position might have been derived from its wishful thinking that a dispute would not arise, and a dispute, if it did arise by any chance, should be resolved amicably through diplomatic negotiations. However, due to the very fact that there is no provision on the dispute settlement procedure in the fisheries agreement, a Party can take a fisheries dispute, if it arises, to the dispute settlement procedure under the LOS Convention without restrictions, which would have been imposed otherwise by the fisheries agreement, when the dispute relates to the interpretation and implementation of the LOS Convention as well as to the fisheries agreement.[82] It is to be recalled that Japan successfully challenged the jurisdiction of an arbitration tribunal, constituted under the LOS Convention relying on the dispute settlement procedure in the CCSBT (Convention for Conservation of Southern Bluefin Tuna) in a dispute regarding its unilateral experimental fishing of Southern Bluefin Tuna with Australia and New Zealand.[83] The agreement is to be effective initially for five years and can be terminated by a one-year prior notice.[84]

2.2.2. *Provisional Measure Zone and Transitional Zones*

The shape of the Provisional Measure Zone is not a modified form of overlapping claims of EEZ by both Parties. In negotiating the shape of the Provisional Measure Zone, negotiators of both Parties bore in mind a hypothetical line which equally divides the Yellow Sea in size, in order to check whether the shaping of the Provisional Measure Zone and the zones where each State exercises its sovereign rights on fisheries *vis-à-vis* the nationals of the other State, are equitable.[85] Note also that the size of the two Transitional Zones are almost the same.[86]

In the Provisional Measure Zone the fishing vessels are only subject to the enforcement jurisdiction of the State under whose flag the vessels are flying.[87] In the Transitional Zones, as in the Provisional Measure Zone, enforcement authorities of one Party cannot exercise its enforcement jurisdiction against the fishing vessels of the other Party.[88]

However, note that joint monitoring and surveillance measures can take place in Transitional Zones.[89] Even if joint monitoring and surveillance are applied, a law enforcement officer of one Party would not be able to take an initiative role for its own enforcement purpose when he/she is onboard a fishing vessel of the other Party with an officer of the other Party, because the agreement clearly Stated that "each Contracting Party may not apply management or other measures to nationals and fishing vessels of the other Party" in the joint fishing zones,[90] and there is no provision which alludes to this strict application of flag-State jurisdiction being relaxed in a case of joint monitoring and surveillance.

However, an indirect form of co-operation in enforcement is possible in addition to the joint monitoring and surveillance scheme: in the case

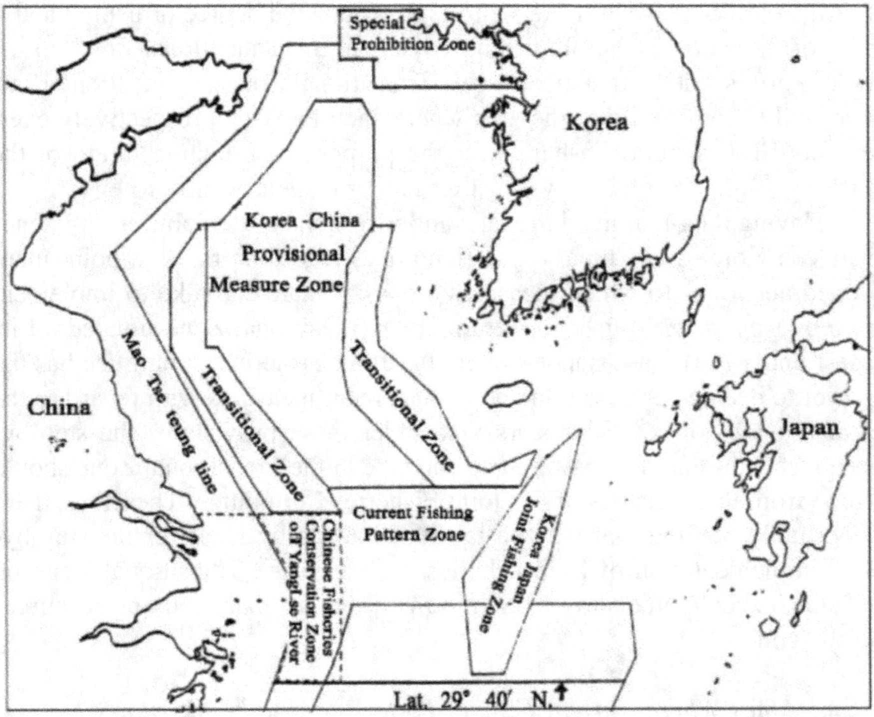

Map 41: Korea-China Fisheries Agreement of 2000

where one Party finds a fishing vessel of the other Party engaged in fishing, in violation of decisions by the Korea-Chinese Joint Fisheries Committee in either Provisional Measure Zones or Transitional Zones, then the former Party can draw attention of the fishing vessel to the violation at sea, and notify the fact to the latter Party together with relevant circumstances, and the latter Party should take necessary measures with regard to the vessel in question and report back to the former Party about the measures it has taken.[91]

Although the Transitional Zones are a sort of joint fishing zone like the Provisional Measure Zone, these zones have their own distinctive characteristics with respect to conservation and management of the fisheries in comparison with that of the Provisional Measure Zone. Firstly, each coastal State is expected to take necessary measures to implement EEZ fisheries regimes step by step on the Transitional Zones situated on its side; secondly, each Party should notify the list of its nationals and its fishing vessels fishing in the Transitional Zone situated on the other Party's side;

thirdly, each State should exert efforts to reduce the degree of fishing activities of its nationals and its fishing vessels in the Transitional Zone on the other Party's side; and lastly, the two Transitional Zones, as mentioned earlier, will become parts of the area where each Party can respectively exercise its EEZ rights on fisheries for the purpose of implementation of the Fisheries Agreement, four years after the agreement went into effect.[92]

Having thus examined the relevant provisions of the joint fishing zones between Korea and China, a question arises regarding the first point mentioned above, as to which measures a coastal State can take to implement step by step the EEZ fisheries regime in a Transitional Zone situated off its coast under the circumstances where the Joint Fisheries Committee has the power to decide conservation and management measures with regard to the Transitional Zones.[93] It appears that under these provisions, the step-by-step measures that the coastal State can take in the Transitional Zone should come from the decisions of the Joint Fisheries Committee. Therefore, there is virtually nothing that the coastal State can do by itself for the step-by-step implementation of EEZ fisheries regime in the Transitional Zone on its side, except some sort of scientific research on marine living resources in the zone.

2.2.3. *Zones Where Current Fishing Patterns are to be Maintained*

Besides the EEZ Fisheries Regime Zones, Provisional Measure Zones and Transitional Zones, there are areas where the current fishing patterns will be maintained. It appears that the term "current fishing pattern" was intended to indicate "free fishing activities" which the nationals and fishing vessels of the Parties have enjoyed thus far in the most parts of the Sea between them.[94] In the Current Fishing Patterns Zone, the two Parties are not to apply their laws and regulations on fisheries against nationals and fishing vessels of the other State, unless agreed otherwise between the two Parties.[95]

Where are the areas where free fishing activities can take place in the Yellow Sea and in the East China Sea? According to the complicated language of Article 9 of the agreement, the Current Fishing Pattern Zones are "certain areas", one of which is situated to the north of the Provisional Measure Zone and the other of which is situated to the south of the Provisional Measure Zone and Transitional Zones.[96] In other words, fishermen of Korea and China can engage in free fishing activities in "certain areas" situated to the north of the Provisional Measure Zone and also in "certain areas" situated to the south of the Provisional Measure Zone and the Transitional Zones, unless there is a special agreement to the contrary. As the term "certain" implies, that not all the areas situated to the north of the Provisional

Measure Zone and not all the areas to the south of the Provisional Measure Zone and Transitional Zones are areas where free-fishing activities are guaranteed. Notice that the whole area of the East China Sea is situated to the south of the Provisional Measure Zone and Transitional Zones.

In order to find out the "certain areas" where the current fishing patterns are to be maintained, we first need to examine the Memorandum of Understanding which shows that there are zones where the free fishing patterns are not to be maintained, although these zones are situated to the north of the Provisional Measure Zone or to the south of the Provisional Measure Zone and the Transitional Zones. These zones are again indicated as certain areas: one is situated in the Korean side to the north of the Provisional Measure Zone and the other is situated in the Chinese side to the south of the Provisional Measure Zone and Transitional Zones. In these areas, the two Parties shall "respect the laws and regulations which the coastal States are now enforcing and the two Parties should take necessary measures to ensure that their respective nationals and fishing vessels observe such laws and regulations", as these zones are not part of the areas where the current fishing patterns are to be maintained. With regard to this, Korea and China have a common understanding that the zone where Korea can enforce its laws and regulations, on the Korean side of the north of the Provisional Measure Zone, is Korea's Special Prohibition Zone. This is situated near the Northern Limits Line between South Korea and North Korea and here foreign nationals and fishing vessels are not permitted to fish at all. However, different views arose after they initiallised the text of the Fisheries Agreement and the Memorandum of Understanding with regard to the location of the other zone where China can enforce its fisheries laws and regulations on Korean fishermen, which is situated on the Chinese side to the south of the Provisional Measure Zone and Transitional Zones. Korea argued that the zone is a small box off the Chinese Yangtse river, of which the eastern limit is basically Mao Tse-teung line. However, China revealed its intention to enforce its fisheries laws and regulations in a wider area for tighter conservation than Korea thought.[97]

The reason was that China adopted a new conservation measure in March 1999, which is larger in its geographical scope and is more prohibitive in conservation terms than the previous measures in the waters off the Yangtse river: the new measure prohibits all the fishing activities in the area except by Chinese local fishermen in the area.[98]

It took one year and three months to settle this dispute regarding the new Chinese conservation measure. In August 2000, the two Parties managed to settle their differences through the following compromise formula: the fishing activities by the Korean fishermen are to be allowed in the area where China has adopted the above-mentioned new conservation measure under the gradual phase-out scheme within two years after the entry into force of the fisheries agreement, on the condition that Korean fishermen can be allowed to fish again in the area if the fisheries stocks in the area improve in the future.[99]

However, even after Korea and China had settled this problem, they had to have further negotiations to settle the different views on how far south the southern limits of the Current Fishing Pattern Zones go down.[100] In the negotiations, Korea argued that the southern limits line of Current Fishing Patterns Zone between Korea and China should be drawn at Lat. 29° 43' N. where the 200-mile line reaches from Korea's southernmost island of Mara-do. Note that the parallel of Lat. 29° 43' N. is far south of the parallel of Lat. 30° 40' N. at which the northern limits line of the Sino-Japanese Provisional Measure Zone is drawn. Thus, in the argument Korea challenged the legality of the northern limits line of the Provisional Measure Zone between Japan and China stressing that the Sino-Japanese line was drawn without consultation with Korea whereas Korea's EEZ claims go further south beyond the line. Korea and China at last agreed to set the southern limits of the Current Fishing Pattern Zone between Korea and China at Lat. 29° 40' N. under the condition that Korean fishermen should observe seasonal fishing restrictions in the areas like Chinese fishermen are observing.[101]

3. PROVISIONAL FISHERIES REGIME BETWEEN CHINA AND JAPAN

3.1. *Background*

In the early 1950s China established the so-called "Mao Tse-teung line" and began to enforce the line against the Japanese fishermen.[102] As Japanese fishermen were much more active with their advanced technology than the Chinese fishermen in the 1950s until the mid-1980s in the waters in North East Asia, Japan wished to conclude a fisheries agreement in order to have more and safer access to the resources in the waters off the Chinese coasts.[103] However, as there were no diplomatic relations between the Peoples' Republic of China and Japan until 1972, the Japan-China Fisheries Council on the Japanese side negotiated with the Chinese counterpart, the China Fisheries Council, for a fisheries agreement of a non-governmental nature.

The two Councils succeeded in concluding a fisheries agreement in 1955. But in the agreement, Japan did not succeed in repudiating any of the Chinese claims such as the Mao Tse-teung line, fishing restriction zones in the East China and Yellow seas, and several military zones, but instead Japan had to acknowledge these Chinese claims.[104] However, Japan did not lose in the negotiations completely. Japan was successful in safeguarding Japanese fishing vessels from seizure by the Chinese authorities in the waters beyond the territorial sea of China through provisions on flag-State jurisdiction in the agreement.[105] From 1955 onwards, "the agreement of 1955" has been a solid basis for the fisheries relations between China and Japan, although the agreement has been revised several times thereafter and sometimes there was no valid agreement between China and Japan.[106] Even in 1975 when the two governments concluded a fisheries agreement, there was no substantial change to the fisheries relations between the two countries as the agreement of 1975 was predicated upon the basis of the agreement of 1955 and its subsequent revisions. Since the mid-1980s, as the fishing activities by the Chinese fishing vessels became much more active even in the waters off the Japanese coasts, the fisheries dispute began to arise yet again albeit on a different footing.[107]

In 1996 when the LOS Convention was brought into effect in respect of China and Japan, the Japanese fishing industry voiced its strong wish for a new fisheries agreement with China on the basis of the EEZ fisheries regime. In this situation, China and Japan began talks for a new fisheries agreement in April 1996. After hard negotiations, China and Japan signed a new fisheries agreement in November 1997, agreeing to shelve the delimitation issue and to establish a joint fishing zone in the East China Sea, the so-called "Provisional Measure Zone". However, it was not until February 2000 that China and Japan reached an agreement on fishing quotas and fishing conditions in each other's EEZ, and on the fishing order in the waters to the north of the Provisional Measure Zone. In February 2000, China and Japan also agreed to bring their new fisheries agreement into force on 1 June 2000.[108]

3.2. *Main Features of the Sino-Japanese Fisheries Agreement of 1997*

3.2.1. *Basic Structure of the Agreement*

The new fisheries agreement between China and Japan is a provisional fisheries agreement for regulating fisheries relations between them on the basis of an EEZ fisheries regime pending the ultimate delimitation of EEZ ultimate boundaries between China and Japan. Thus the two governments expressed "their intentions to continue the consultations on the delimitation of the exclusive economic zones and the continental shelves of both countries", and "to make efforts so that a mutually acceptable agreement be reached".[109] Both Parties agreed that: "Nothing in this Agreement shall be deemed to prejudice the position of either Contracting Party in regard to any question on the law of the sea".[110] Also more specifically, both Parties confirmed in the Agreed Minute that the Provisional Measure Zone "should not be deemed to prejudice the position of either Contracting Party in regard to the delimitation of the exclusive economic zones and the continental shelves".[111]

Although the agreement is a provisional arrangement, it is based on the EEZ fisheries regime. Article 1 of the agreement makes it clear on this point, providing that: "The area to which this Agreement is applied shall consist of the exclusive economic zones of both Japan and of the Peoples' Republic China". And Articles 2 to 5 of the agreement provide detailed provisions on fishing access of fishing vessels of one Party to the EEZ of the other Party on a mutual basis. Each Party shall permit nationals and fishing vessels of the other Party to conduct fishing operations within its exclusive economic zone in accordance with the agreement and its relevant laws and regulation (Article 2). Each Party may take necessary measures, subject to international law, to ensure that nationals and fishing vessels of the other Party engaged in fishing activities in its exclusive economic zone, comply with the conservation measures of marine living resources and other terms and conditions established in its laws and regulations (Article 5).

Having set out detailed provisions on fishing access to EEZ, the agreement provides for zones where fishing can be conducted by fishermen of both Parties without a need to get fishing permits from each other, which will be discussed in detail later.

A Japan-China Joint Fisheries Committee was established for the purpose of achieving the objectives of the fisheries agreement.[112] The Committee has the power to recommend the government of each Party on matters relating to fishing access to EEZ such as species allowed to catch, fishing quotas, fishing operation conditions, and it can determine matters relating to the conservation and management measures in the Provisional Measure Zone.[113] All the recommendations and determinations by the Committee shall be reached only through consent between the members of both Parties in the Committee.[114] The governments of Japan and China should respect recommendations by the Committee and should take necessary measures in accordance with the determinations.[115]

The agreement does not contain a provision on a dispute settlement mechanism like the Fisheries Agreement between Korea and China, and thus the dispute settlement procedures of the LOS Convention might be applied without restrictions which would have been imposed otherwise by the agreement if a dispute arose relating to the interpretation and implementation of the LOS Convention as well as to the fisheries agreement between them.[116] The Fisheries Agreement between China and Japan has been in force since 1 June 2000 for an initial period of five years and it can be terminated on or after the date of expiration of the initial 5 years, by a six month prior written notice of one Party.[117]

3.2.2. *Provisional Measure Zone*

Japan and China established the Provisional Measure Zone in the East China Sea where the EEZ claims of both countries overlap. The Provisional Measure Zone is situated between the parallels of Lat. 27° N and Lat. 30° 40'. Generally, the western and eastern limits of the Provisional Measure Zone are situated at the distance of 52 miles from the nearest coasts. Thus in the area between the parallels of Lat. 27° N and Lat. 30° 40' in the East China Sea, China and Japan can exercise their respective EEZ fisheries rights against each other up to broadly 52 miles from its respective coasts. Korea challenged the legality of the northern limit of the Sino-Japanese Provisional Measure Zone arguing that Korea's EEZ claims reach far south beyond the northern limit of the Sino-Japanese Provisional Zone and China and Japan did not consult with Korea in setting the northern limit.[118]

China and Japan undertook to take proper conservation measures in the Provisional Measure Zone in accordance with the determinations by the China-Japan Joint Fisheries Committee in order to ensure that the maintenance of marine living resources is not endangered by over-exploitation.[119] The vessels of the Parties fishing in the Provisional Measure Zone are subject only to the enforcement jurisdiction of the Party of which flag they fly. However, an indirect form of co-operation in enforcement can take place between the two Parties: in the case where one Party finds a fishing vessel of the other Party engaged in fishing in violation of conservation measures determined by the China-Japan Joint Fisheries Committee, then the former Party can draw the violation to the attention of the fishing vessel and notify the fact to the latter Party together with relevant circumstances, and the latter Party should take necessary measures with regard to the vessel in question and report back to the former Party about the measures it has taken.[120]

3.2.3. *Two Untitled Joint Fishing Zones*

There are two more joint fishing zones agreed between China and Japan besides the Provisional Measure Zone, which are situated in the south of the Provisional Measure Zone and in the north of the Provisional Measure

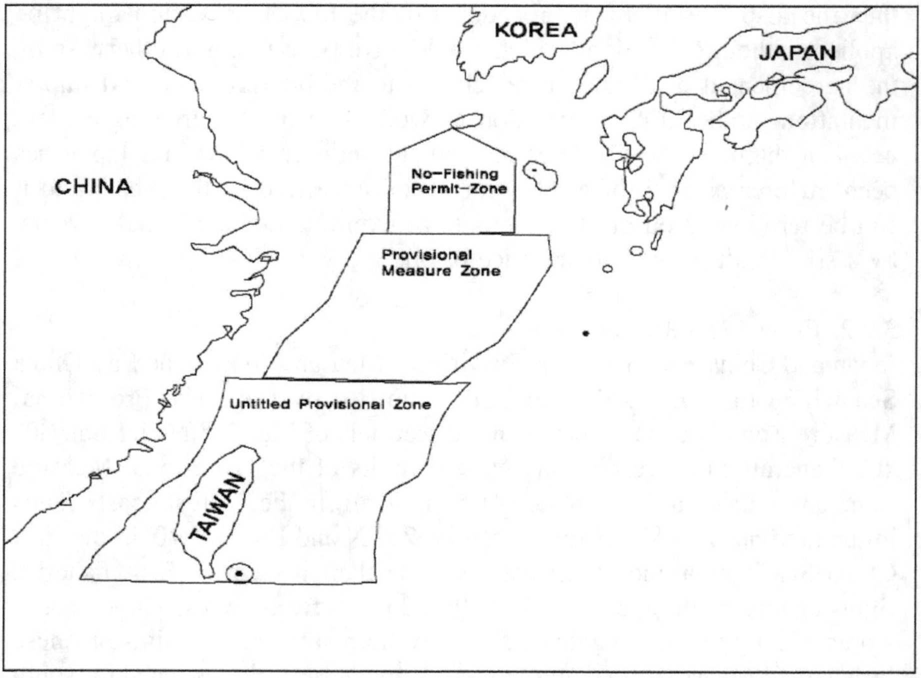

Map 42: Two Untitled Fishing Zones between Japan and China

Zone. The former is located to the south of the parallel of 27° in the East China Sea and to the west of line of longitude 125° 30' E excluding the South China Sea.[121] This joint fishing zone is the area concerning the disputed Senkaku/Diyaoyutai and Taiwan. In this sensitive area those provisions on fishing access to each other's EEZ does not apply, but the China-Japan Joint Fisheries Committee can recommend to each Party on the State of the marine living resources and fishing order in the area. Letters were exchanged between China and Japan regarding the fishing order in the area. The letters expressed the intentions of the two governments not to apply their laws and regulations on fisheries against each other's fishermen in the area.[122]

The joint fishing zone in the north of the Provisional Measure Zone is provided for in the Agreed Minute, through which both Parties expressed their intentions "to respect current fishing activities in a certain area" to the north of northern limit of the China-Japan Provisional Measure Zone. It took two years and three months for Japan and China to fix the certain area, as Japan wanted a narrow zone and China wanted to designate a large area.

In February 2000, China and Japan agreed to set up an area, so-called "No-Fishing-Permits-are-Required-Zone" (hereinafter "No Fishing Permit Zone") where fishermen of both Parties could fish without permits from each other and thus where the current fishing activities could be maintained. The No-Fishing-Permit-Zone is set up in the waters to the north of the China-Japan Provisional Measure Zone between Lat. 124° 45' E and Lat. 127° 30'.[123] Interestingly, the No-Fishing-Permit-Zone designated by China and Japan is closer to the Korean coasts than to the Chinese and Japanese coasts and it comprises mostly parts of the Korea-Japan joint fishing zone in the East China Sea and also a bit of the Transitional Zones between Korea and China.

It will be appropriate to consider here what the implications of the designation of the area by Japan and China will be on Korea's sovereign rights on fisheries in this area and on the management of the living resources in the Korea-Japan Provisional Measure Zone and in the China-Korea Transitional Zones. As Korea is a third Party to the agreement between China and Japan, Korea would not be bound by the agreement, as the maxim *pacta tertiis nec nocent nec prosunt* indicates. Also Article 34 of the Vienna Convention on the Law of Treaties will be applied, which States that: "a treaty does not create either obligations or rights for a third State without its consent". Thus Korea will not be bound by the agreement by China and Japan on the zone where fishing permits are not required. Furthermore, Korea is in a position to require Japanese fishermen to get fishing permit from it in accordance with the provisions of the Korea-Japan Fisheries Agreement, in order to fish in the waters which is to be regarded as Korea's EEZ for the implementation of the Korea-Japan Fisheries Agreement. Note that the waters on the Korean side to the Korea-Japan joint fishing zone in the East China Sea is to be regarded as part of Korean EEZ by the provisions.[124]

However, the agreement between China and Japan can be seen as an expression of their intention that they are not interested in conservation and management of living resources in the area: China would argue that the area included in the No-Fishing-Permit-Zone between China and Japan should also be included into the Current Fishing Pattern Zone between Korea and China; and Japan would be very reluctant to adopt any effective conservation measures with regard to the living resources in the Korea-Japan joint fishing zone in the East China Sea.[125] If this is what Japan and China wish to pursue, for the benefit of their fishermen at the cost of the conservation of marine living resources, in the waters where Korea has the most share to the EEZ in this area, then Korea has to do something for the sustainable maintenance of fisheries resources in the area.

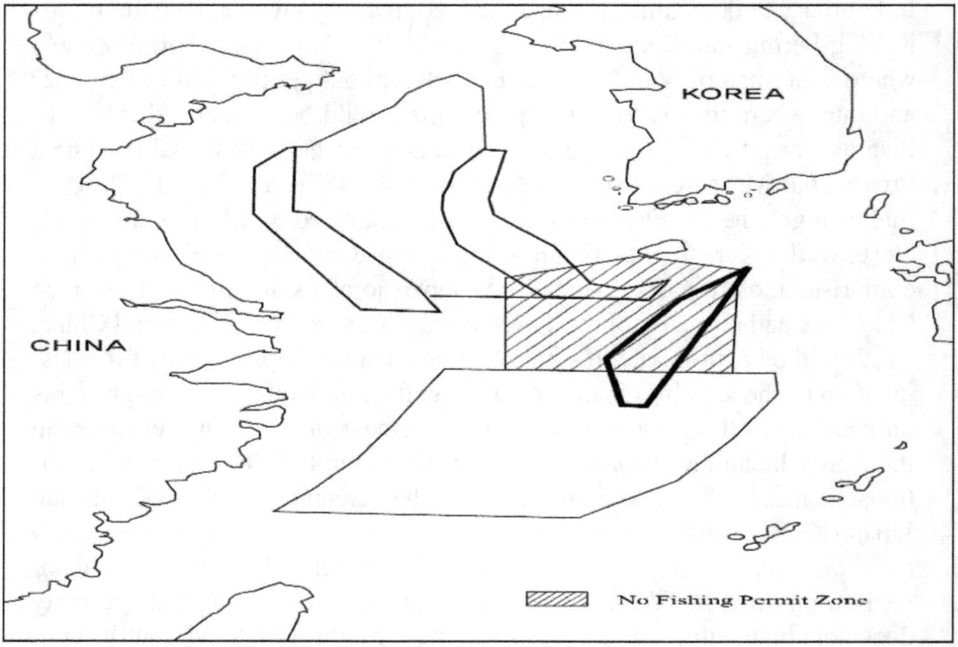

Map 43: No Fishing Permit Zone between China and Japan

4. CONSERVATION OF THE TRANSBOUNDARY FISH STOCKS IN NORTH EAST ASIA

Although China, Japan and Korea are major fishing countries in the world and the living resources in the seas in North East Asia have been over-exploited,[126] there is no regional fisheries organisation in the region. The absence of any regional fisheries organisations in the region might reflect the fact that the littoral States in the region have been interested in catching more but paying little attention to the need for the conservation of fish stocks. Note that until very recently a vast area in North East Asia was part of the high seas without a proper conservation scheme in terms of the LOS Convention, and the littoral States in North East Asia have concluded fisheries agreements on the EEZ regime very recently. This situation does not seem to be in line with the global trend which has been restricting the freedom of fishing, even in the high seas, noting the problem of the over-exploitation of the marine living resources in the seas. It is to be recalled that Agenda 21 noted the general problem in fisheries that:

> There are problems of unregulated fishing, overcapitalisation, excessive fleet size, vessel reflagging to escape controls, insufficiently selective gear, unreliable databases and lack of sufficient co-operation between States. Action by States whose nationals and vessels fish on the high seas, as well as co-operation at the bilateral, subregional, regional and global levels, is essential particularly for highly migratory species and straddling stocks.[127]

On the continuation of this conception manifested in Agenda 21 above, the international community has been developing international instruments to enhance sustainable fisheries. These include the Agreement to Promote Compliance with International Conservation and Management Measures by Fishing Vessels on the High Seas of 1993 (hereinafter the "Compliance Agreement");[128] the FAO Code of Conduct for Responsible Fisheries of 1995 (hereinafter "the Code of Conduct");[129] the Agreement for the Implementation of the Provisions of the United Nations Convention on the Law of the Sea of 10 December 1982 relating to the Conservation and Management of Straddling Fish Stocks and Highly Migratory Fish Stocks of 1995 (hereinafter the "UN Fish Stocks Agreement");[130] and the International Plan of Actions to Prevent, Deter and Eliminate Illegal, Unreported and Unregulated Fishing of 2001 (hereinafter the "IPOA for IUU Fishing").[131]

The Compliance Agreement is trying to discourage flag-of-convenience and re-flagging by the fishing vessels, through reinforcing the responsibility of the flag-States and exchange of information between States on the high seas fishing activities.[132] The Code of Conduct is not a binding legal instrument but voluntary, although certain parts of the Code are based on relevant rules of international law and the Compliance Agreement is an integral part of the Code.[133] The Code of Conduct provides principles and standards applicable to conservation, management and development of all fisheries.[134] The UN Fish Stocks Agreement restricts the time-honoured principles of flag-State jurisdiction in the high seas in the sense that it enables a member of a regional fisheries organisation to board and inspect fishing vessels flying the flag of another State Party to the UN Fish Stocks Agreement engaged in fishing activities on the high seas which is covered by the regional agreement even if the latter State is not a party to the regional fisheries organisation.[135] The IPOA for IUU Fishing is not a binding instrument but a voluntary one.[136] It is to be noted that some provisions of the IPOA might turn into binding domestic rules by those countries that try to combat the IUU fishing at their own jurisdiction through legislation of the provisions in the IPOA. Note that the IPOA strengthens the power of the port State with regard to IUU fishing vessels. Under the IPOA the port States should require fishing vessels seeking permission to enter their ports to provide a copy of their authorisation to fish, details of their fishing trip and other information.[137] Also under the IPOA, the port States can inspect the vessels.[138] The IPOA also recognises the rights of the members of a regional fisheries organisation to adopt measures against non-members of the regional fisheries organisation, if the non-member State fails to ensure that fishing vessels flying its flag do not engage in IUU fishing activities.[139] Note that neither Korea, Japan nor China joins the UN Fish Stocks Agreement and only Japan joins the Compliance Agreement whereas Korea and China do not. It appears that the littoral States do not really welcome the new developments in the law of fisheries, which places emphasis on the conservation of fish stocks and restricts the freedom of fishing even in the high seas.

Now as there is no part of the high seas left in North East Asia, most of the UN Fish Stocks Agreement does not seem to be directly applicable to the conservation of the marine living resources in the region. Notice that the UN Fish Stocks Agreement of 1995 purports to deal with those fish stocks which occur both within the exclusive economic zone and in the high seas and highly migratory fish stocks.[140] However, there is still a need for multilateral co-operation between the littoral States in the region because a number of important fish stocks such as mackerels (*Scomber japonicus Houttuyn*), squids (*Todarodes pacificus*) and hairtails (*Trichiurus lepturus Linnaeus*) occur in economic zones of more than two States in the region.[141]

In this regard, it seems appropriate for the three riparian States to seek trilateral co-operation for conservation of these fish stocks in accordance with Article 63(1) of the LOS Convention, which provides that:

> Where the same stocks or stocks of associated species occur within the exclusive economic zones of two or more coastal States, these States shall seek either directly or through appropriate subregional or regional organisations, to agree upon the measures necessary to co-ordinate and ensure the conservation and development of such stocks without prejudice to the other provisions of this Part.

It is to be noted that the above-mentioned provision obliges States sharing the transboundary fish stocks to negotiate to agree upon the necessary conservation measures and thus the three coastal States in the region are under this *pactum de negotiando* for the proper conservation and management of the shared stocks.[142]

It would be undesirable to try to establish trilateral joint fishing zones among the three littoral States for their co-operation scheme, because it would take a lot of unnecessary time and energy as the transboundary fish

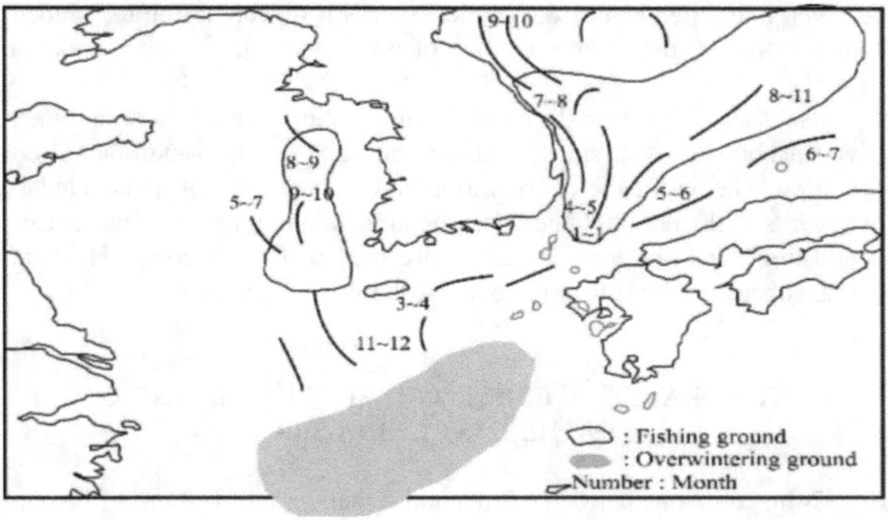

Map 44: Migration Route of Squid

stocks know no manmade boundaries or joint fishing zones. It is to be noted that the tribunal in the *Eritrea-Yemen Arbitration*, decided to keep the traditional fishing in the disputed areas even after delimitation of maritime boundaries without making specific limits of joint fishing zones.[143]

For the conservation of the shared stocks, simply setting up a trilateral fisheries committee and identifying shared fish stocks would be an appropriate start. In fact, a number of multilateral co-operation schemes have been in place for shared stocks without having a joint fishing zone or a fisheries boundary. As we have seen in Chapter 3, the trilateral fisheries agreement between Denmark (Greenland), Iceland, and Norway (Jan Mayen) for the Capelin stocks is a good example.[144]

In addition, the International Baltic Fisheries Commission, which was established in 1973 by the Baltic coastal States for the co-operation on preserving the living resources of the Baltic Sea and the Belts, is also a good example of multilateral agreement for conservation of transboundary fish stocks without making any boundary or joint fishing zone.[145] The participating Parties in the Commission are the European Community, Finland, Germany, Poland, Sweden, Estonia, Latvia and Lithuania. The Commission works by means of making recommendations to the Parties with regard to measures establishing total allowable catch or fishing efforts according to species, stocks, areas and fishing periods including total allowable catches for areas under the fisheries jurisdiction of Contracting States.[146] Although the Convention uses the term "recommendation", the recommendation should be given effect by each of the Parties, unless it objects to the recommendation within 90 days from the date of notification of a recommendation to it.[147]

It seems advisable for the coastal States in North East Asia to first identify transboundary fish stocks in the region and set up a multilateral co-operation scheme for the conservation and management of transboundary fish stocks in the region. Otherwise, the three bilateral fisheries agreements would turn out to be ineffective because of their limited competence and overlaps between joint fishing zones of the three agreements.

5. EXPLORATION AND EXPLOITATION OF OIL AND GAS IN THE EAST CHINA SEA

In 1974 negotiations between Korea and Japan on delimitation of the continental shelf in the East China Sea hardly moved forward mainly due to the fundamentally different positions of the two countries on the applicable delimitation principle. Japan argued for a strict equidistance method for the delimitation, whereas Korea based its position on natural prolongation relying on the existence of the Okinawa Trough in the East China Sea, along which Korea established the southern limit of its submarine mineral exploitation area by the Enforcement Decree of the Submarine Mineral Resources Development Act of 1970. The ICJ judgement of 1969 on the *North Sea Continental Shelf* Cases, which was based on the natural prolongation theory reinforced the position of Korea in the negotiations.[148] It is to be recalled that the ICJ recognised the natural prolongation as a basis of title to the continental shelf[149] and also as an important principle in the delimitation, saying that:

> Delimitation is to be effected . . . in such a way as to leave as much as possible to each Party all those parts of the continental shelf that constitute a natural prolongation of its land territory into and under the sea, without encroachment on the natural prolongation of the land territory of the other.[150]

However, the confrontation between the two countries on the applicable principles did not appear to be resolved by negotiation.[151] So the Japanese government then proposed to bring the issue to the ICJ, but the Korean government rejected the proposal.[152] Then in 1974, both countries agreed to establish a joint development zone accommodating both sides' positions by signing the "Agreement between the Republic of Korea and Japan con-

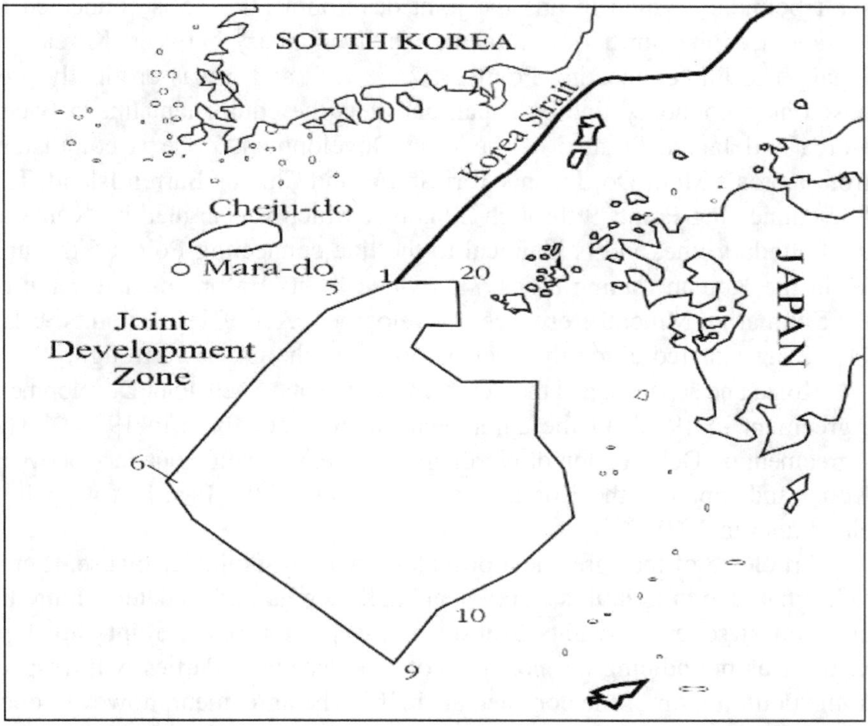

Map 45: Joint Development Zone between Korea and Japan

cerning Joint Development of the Southern Part of Continental Shelf Adjacent to the two Countries" (hereinafter referred to as "the Southern Continental Shelf Joint Development Agreement").

In shaping the joint development zone in the East China Sea, both Korea's and Japan's arguments were all accommodated. The joint development zone encompasses about 24,101 square miles. And the joint development zone was divided into 9 sub-zones.

The equidistant line which had been argued by Japan and the line along the Okinawa Trough which had been argued by Korea constituted the limits of the joint development zone. Since China is also a coastal State, the question of drawing a western boundary of the joint development zone was a difficult task as China did not participate in the negotiations for political reasons. Korea and Japan used a hypothetical equidistance line between China, Korea and Japan for drawing the western limits of the joint development zone.[153]

The joint development zone is comprised of 35 points. Point 1 of the joint development zone is identical to point 1 of the northern continental shelf boundary which means the joint development zone is connected to the south of the northern continental shelf boundary between Korea and Japan. The line connecting Points 1, 2, 3, 4, 5, and 6 was originally proposed as a boundary line by Japan and it is the equidistant line between Korea and Japan. Point 6 of the Joint Development Zone is equidistant from Korea's Mara-Do, Japan's Tori-Shima and China's Barren Island. The line connecting Points 9 through 20 is that which was argued by Korea as the boundary line, and is identical to the line connecting Points 5 through 16 in the Korean mining block 7 provided by its Enforcement Decree of the Submarine Mineral Resources Development Act of 1970. Points 9, 10, 11, 12 are situated along the deepest points in the Okinawa trough.

Korea and Japan signed the Southern Continental Shelf Joint Development Agreement in 1974 but the agreement entered into force in 1978.[154] The Agreement on Delimitation of Northern Continental Shelf Boundary between Korea and Japan in the Korea Strait and south of the East Sea went into effect also in 1978.[155]

Article 28 of the agreement provides a without-prejudice clause. It provides that "Nothing in this Agreement shall be regarded as determining the question of sovereign rights over all or any portion of the Joint Development or as prejudicing the positions of the respective Parties with respect to the delimitation of the continental shelf". The agreement, however, does not oblige the parties to negotiate for ultimate maritime boundaries. The Agreement was to be in force 50 years from 22 June 1978. And one of the parties can terminate the agreement at the end of the initial 50 years or at any time with a three-year prior written notice.

Each Party was to authorise concessionaires with respect to each subzone and the concessionaires of both Parties shall enter into an operating agreement to carry out joint exploitation and exploration of natural resources. However, so far any production of oil has not been made.

After the agreement was signed, the Chinese government immediately protested against it.[156] A Statement of the Chinese Foreign Ministry read that:

> The Chinese Government holds that according to the principles that the continental shelf is the natural extension of the continent, it stands to reason that the question of how to divide those parts of the continental shelf in the East China Sea involving other countries should be decided by China and the related countries through consultation[157]

In this regard, the Japanese government tried to justify the joint development zone agreement relying on the fact that the joint development zone had been restricted within the Korean and Japanese side of the hypothetical equidistance lines between Korea, Japan and China.[158]

Although Korea and Japan have conducted some exploration research in the joint development zone, they were not successful in finding any oil reservoirs in the zone.[159] Ironically, however, it was China who was successful in finding and exploiting some sizeable deposits of oil and gas in the East China Sea. China embarked on its own exploration activities in the East China Sea in 1974. After ten years of exploration in the East China Sea, China succeeded in finding a sizeable amount of oil and gas in the so-called "Pinghoo No. 1 well" in 1983. China also found a good amount of oil and gas in the so-called "Chunxiao No. 1 well" in 1995, and in the so-called "Chunxiao No. 3 well" in 2000.[160] China set up an oil platform in the Pinghoo No. 1 well and installed two 400 km long pipelines for transporting oil and gas from the well to Shanghai in 1998.[161] Those wells developed by China are within the Chinese side of the hypothetical equidistance lines between Korea and China and between China and Japan and thus are not in the joint development zone between Korea and Japan.

However, encouraged by its successful exploration and exploitation in the East China Sea, China begun to conduct some sort of scientific research and exploration activities beyond the hypothetical equidistance line between China and Japan in the late 1990s.[162] Most of the exploration activities conducted by the Chinese vessels are seismic surveys, and analysis of temperature and salinity by putting instruments into the sea. However, an actual drilling was reported to have taken place by Chinese vessels.[163] Japan regarded these exploration activities by the Chinese vessels on the Japanese side of the hypothetical equidistance line as an infringement on its rights and thus the Japanese foreign minister, Yohei Gono, conveyed Japan's view when he paid a visit to China on 28 August 2000.[164] Then the two foreign ministers of China and Japan agreed in principle to set up through negotiations at the working level, a prior notification scheme regarding marine scientific research in the East China Sea between the two countries under the premise that such a regime does not affect the delimitation of maritime boundaries.[165] However, it appears that the negotiators of the two governments at working level encountered a problem of designating an area where the mutual prior notification scheme is to be applied and thus agreed on a prior notification scheme without a specific geographical scope. Here arises a question of whether such a scheme can be effective or valid, because Korea, one of the littoral States in the East China Sea, has not been invited to take part in the scheme.

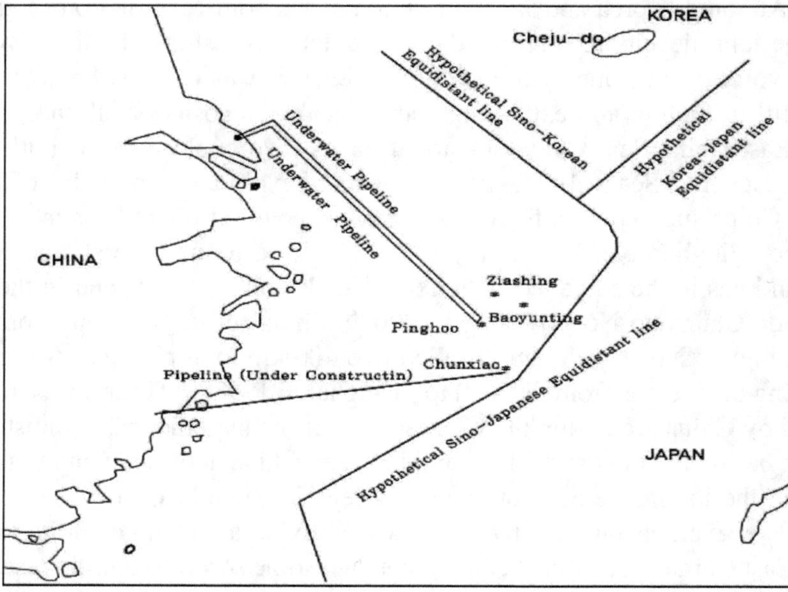

Map 46: Chinese Exploitation Activities in the East China Sea

6. PROTECTION AND PRESERVATION OF THE MARINE ENVIRONMENT IN NORTH EAST ASIA

Although there is a possibility of jurisdictional conflicts between the neighbouring coastal States with regard to the protection and preservation of the marine environment when there is no boundary, it does not mean that co-operation cannot take place due to the absence of boundaries. The neighbouring coastal States can co-operate with each other at a bilateral plane or a multilateral plane for better management of the marine environment, which they share, regardless of whether there is a maritime boundary in place or not. The vehicles for the co-operation need not be a provisional arrangement envisaged in Article 74(3) or Article 83(3) of the LOS Convention. As there have been attempts to co-operate between the littoral States in the region for the protection and preservation of the marine environment in the region, it will be useful to briefly examine those attempts.

6.1. *Bilateral Agreements for Co-operation on Environmental Matters*

Since the East Sea, Yellow Sea and the East China Sea in North East Asia are semi-enclosed seas,[166] they are more vulnerable to pollution than the seas which are not semi-enclosed or enclosed. There have been worries about the dumping of radioactive waste such as nuclear submarines, waste from nuclear power plants, and nuclear containers into the East Sea by the Soviet Union (later Russia) over the three decades between 1959 and 1993.[167] The Yellow Sea is also said to be seriously contaminated by industrial water waste and oil spills from oil and gas explorations.[168] It is to be noted that under the LOS Convention, the coastal States bordering a semi-enclosed sea are obliged to co-operate with each other in the exercise of their rights and in the performance of their duties with respect to the protection and preservation of the marine environment of the semi-enclosed sea.[169]

There have been attempts to improve the situation through co-operation both at the bilateral level and at the multilateral level among the littoral States in the region. Let us first look at the efforts made at the bilateral level. The following bilateral agreements were concluded for co-operation on the environmental issues: the agreement between Korea and Japan of 1993, the agreement between China and Korea of 1993 and the agreement between Korea and Russia of 1997.[170] The form and substance of the agreements are almost identical to each other: they are relatively brief; they are intended to deal with the environment as a whole and thus are not restricted to the marine environmental issue in their scope; and, they do not contain provisions on the issues of the exercise of jurisdiction by the coastal States in the disputed areas. However, they do contain several important principles and vehicles for co-operation between the neighbouring States on the environmental issues on a broad basis of regionalism.

Regionalism in the Agreements

These three agreements are all based upon an environmental "regionalism" in North East Asia. This can be illustrated by the preamble to the Korea-Russia Agreement, which States: "Believing that the co-operation between the Parties in the field of environment is of mutual advantage for addressing the environmental challenges and is essential ***for the protection and improvement of the regional and global environment*** . . . (emphasis added)".[171] Similarly, the other two agreements recognise "the urgent need of regional efforts" to prevent environmental deterioration.[172]

However, as there is no explanation of "region" in the agreements we need to consider on what basis the term "region" was mentioned. Some scholars have tried to identify the basis on which the regions have been established in the marine environmental matters. Professor Boyle tried to distinguish the following three senses in which the term "region" has been used in a maritime context: formal dimension focusing on physical and geographical character of a marine region; functional definition focusing on patterns of use of the marine region; and political dimension which is essentially defined by the decisions of a group of States to co-operate.[173] Alexander identified two kinds of basis for regional groupings: first, ideological or political identity of interest, and second, mutual needs and interests of States regardless of ideological consideration. Alexander named the grouping based upon the first kind "common policy grouping" and the second kind "complementary-use regions".[174] It seems to this author that Alexander's first type of grouping is rather irrelevant considering the wide range of ideological backgrounds in North East Asia. Okid also accepted that geopolitical factors can be the banner for some regional groupings, but he was more inclined to the "problem sheds" type which seems basically the same as Alexander's "complementary-use" regions.[175] "Problem sheds" is a region within which the levels of pollution are relatively or completely independent of the discharge of pollutants elsewhere, and thus the States in the "problem sheds" will have common interests in working together. In North East Asia, several "problem sheds" can be identified. For example, in terms of the air mass, the whole North East Asia is an environmental community which can be distinguishable from the rest of the global environment.[176] In terms of the marine environment which we are now discussing, each of the Yellow Sea and the East Sea (the Sea of Japan) and the East China Sea can constitute an independent region.

However, as the agreements deal not only with the marine environment nor the air mass, but the environment as a whole, the word "regional" in the agreements appears to have been used in a rather broad and flexible way so that the co-operation between the Parties can take place to target at various problem-sheds in North East Asia. In terms of the LOS Convention the regionalism in the three agreements can be associated with the co-operation provided in Article 194, which provides that: "***States shall take***, individually or ***jointly*** as appropriate, ***all measures*** consistent with this Convention that are necessary ***to prevent, reduce and control pollution*** of the marine environment from any source, using for this purpose the best practicable means at their disposal and in accordance with their capabilities, and ***they shall endeavour to harmonise their policies in this connection*** (emphasis added)".[177]

Preventive Approach and Co-operation Mechanism in the Agreements

The three agreements are based upon the preventive approach in the sense that its main objective is to protect the environment through co-operation before damages occur. For instance, the Korea-China Agreement provides that "preventive measures should serve as important elements in the co-operative activities of the Parties to minimise the possible adverse effects of environmental damage".[178]

The preventive approach is well justified from economic, scientific, and legal points of view. Economically, it will be much more costly to clean up the damage than to prevent it. The cost might even be prohibitive. In a worse case scenario, the damage would be irreversible involving important synergism in the environment. Moreover, it is difficult or even impossible sometimes to calculate damage accurately, to provide adequate compensation and to apportion liability. According to the preventive approach, lack of full scientific certainty shall not be used as a reason for postponing cost-effective measures to prevent environmental degradation.[179] It is to be noted that many international instruments adopt the preventive approach, and various methods are being developed for a preventive approach such as an early-warning system, risk assessment, a monitoring-system etc.[180] Principles are emerging that would strengthen procedural requirements such as notification, consultation, access to information or environmental impact assessment, that would have to be fulfilled before States could engage in activities that could significantly harm the environment outside their jurisdiction.[181]

On the basis of the preventive approach, the three agreements provide for environmental impact assessment, and exchange of information, science and technology on the environmental protection.[182] As the agreements are primarily intended to provide better opportunities to exchange information, technology, and experience related to the environment including the marine environment, it is expected that the agreements will advance "the level of the best practicable means" at the disposal of the three Parties, and therefore the duties of the three Parties under Article 194 of the LOS Convention "to prevent, reduce, and control pollution of the marine environment" will be more substantive and more pressing.[183]

Environmental Impact Assessment

Article 2(4) of the Korea-Chinese Agreement provides for the "joint assessment of environmental impact" as one of the co-operative activities between the two countries.[184] Environmental impact assessment can be said to be a "preliminary method of investigation aimed at determining not just the viability of a proposed project but also, and more significantly, the effect of the proposed activity" on the environment.[185]

Environmental impact assessment is developed, in a sense, from the recognised need to integrate environmental concerns into development planning process. As the World Commission on Environment and Development put it: "to make development sustainable – to ensure that development meets the needs of the present without compromising the ability of future generations to meet their own needs".[186]

To encourage the incorporation of assessment information in planning and decision-making for activities that risk environmental impacts, the UNEP Governing Council in 1982 requested that appropriate guidelines, standard and model legislation be drawn up in the field of environmental impact assessment. The resulting 13 principles were approved by the UNEP Governing Council for use as a basis for preparing appropriate national measures, including legislation and international co-operation.[187] Principle 17 of the Rio Declaration strongly recommends environmental impact assessment for planned activities that are likely to have a significant adverse impact on the environment.

However, as the environmental impact is usually related to the possible impact of a planned project – in other words, future activities, environmental impact assessment would not be sufficient in preventing environmental damage unless it goes along with the monitoring of current activities. Although the three agreements in the region, however, do not provide the monitoring programme, the possibility is not excluded that the monitoring programme will be agreed upon as one of "other forms of co-operation as may be mutually agreed upon".[188] Under the Korea-Chinese Agreements, Korea and China conducted the joint monitoring of the marine environment in the Yellow Sea in 1997 and in 1998.[189]

Exchange of Information and Co-operation on Environmental Science of Technology

The main objective of the three agreements appears to be "to provide better opportunities to exchange information, technology and experience related to the environmental protection".[190] It is well recognised that the availability of scientific and technological information on environmentally sound technologies are essential for environmental protection.[191] It is recalled that the LOS Convention encourages States, for the protection and preservation, to co-operate in scientific research and the exchange of information and data about marine pollution.[192]

Technology co-operation involves joint efforts by private enterprises and government, of suppliers and recipients. Therefore such co-operation entails an interactive process involving government, the private sector and research and development facilities to ensure the best possible results from technological co-operation.[193] As a large body of useful technological knowledge lies in the public domain, there is a need to improve access to the information about such technologies which are not covered by patents.[194]

In such cases, a clearing house of information will be instrumental in promoting the flow of technology. However, as many of the State-of-the-art technologies are owned by the private sector and the royalties for the use of the technology is usually high, there is a need for technological co-operation on preferential terms.[195] For example, the Convention on Biological Diversity provides that: "Access to and transfer of technology . . . to developing countries shall be provided and/or facilitated under fair and most favourable terms, including on confessional and preferential terms where mutually agreed. . . ". In North East Asia, as the levels of technologies and economic development of the countries are different, preferential terms for co-operation on environmental science and technology would be useful in facilitating the co-operation. Technological co-operation on preferential terms seems difficult to take place under the agreements as the three agreements emphasise "the basis of equality and mutual benefit".[196]

6.2. *Co-operation at the Multilateral Level on the Marine Environmental Issues*

As the three environmental agreements recognise, there is "the urgent need of regional efforts" to prevent environmental deterioration in North East Asia. It is to be recalled that many regions where there are semi-enclosed seas, regional co-operation regimes are in place. It appears that the littoral States in the Mediterranean Sea are the States which have adopted the largest number of international instruments for combating specific types of pollution as well as a comprehensive convention on a regional basis. These are the Convention for the Protection of the Marine Environment and the Coastal Region of the Mediterranean adopted in 1976 and amended in 1995;[197] the Protocol for the Prevention and Elimination of Pollution of the Mediterranean Sea by Dumping from Ships and Aircraft or Incineration at Sea adopted in 1976 and amended in 1995 of 1976; the Protocol Concerning Co-operation in Combating Pollution from Ships and Aircraft of the Mediterranean Sea by Oil and Other Harmful Substances in Cases of Emergency of 1976; the Protocol Concerning Mediterranean Sea Against Pollution from Land-Based Sources and Activities adopted in 1980 and amended in 1996; Protocol for the Protection of the Mediterranean Sea against Pollution resulting from Exploration and Exploitation of the Continental shelf and the Seabed and its Subsoil of 1994; Protocol Concerning Specially Protected Areas and Biological Diversity in the Mediterranean Sea of 1995; and Protocol; on the Prevention of Pollution of the Mediterranean Sea by Transboundary Movements of Hazardous Wastes and Their Disposal of 1996.[198]

Although there is no multilateral regional environmental convention in place in North East Asia, some environmental forums have been developed since the 1990s for co-ordinated action for better management of the marine environment among the States in the region. The North East Pacific Action Plan, the so-called NOWPAP was launched as a part of UNEP Regional Seas Programme, for protection of the marine environment in North East Asia in Seoul in September 1994.[199] The geographical scope of the Action Plans is the East Sea and the Yellow Sea.[200] The participating States are Korea, Japan, China and Russia.[201] The Action Plan is not a binding international legal instrument, although there is a possibility that a convention will be negotiated and adopted in the future.[202]

The implementation of the Action Plan is divided into five stages: the establishment of a comprehensive database and information management; a survey of national legislation, objectives, strategies and policies; the establishment of a collaborative regional monitoring programme; the development of effective measures for regional co-operation in marine pollution preparedness and response; and public awareness of the marine, coastal, and associated freshwater environment.[203] The participating States agreed in 1997 to raise a trust fund with a total annual deposit of half a million US dollars and are now planning to set up its regional co-ordinating unit in the near future.[204] Although the NOWPAP is not a fully-fledged body and it has yet to produce a tangible achievement,[205] the prospects of the NOWPAP are good.[206]

There are other regional forums for co-operation on the environment in North East Asia such as "the Meeting of Senior Officials on Environmental Co-operation in North East Asia", and "the Environment Congress for Asia and the Pacific" (the so-called "Eco-Asia)".[207] These forums have been focusing on the issues of air pollution, acid rain, pollution from the energy plants and the conservation of forests, and it seems that no work has been done on the marine environment so far by these forums.[208]

Having seen that a multi-lateral co-operation for the protection and preservation of the marine environment is in the initial stage in North East Asia, it seems necessary to begin to develop regimes for dealing with specific pollution in the region, like the regime in the Mediterranean Sea. NOWPAP can be a base for the development of such a regional environmental regime. The littoral States can establish a comprehensive convention and then develop a pollution-specific protocol like in the Mediterranean Sea and Persian Gulf. Whatever the case might be, urgent measures seem to be required for preventing the degradation of the Yellow Sea caused by the land based sources, and nuclear waste in the East Sea.

7. OBSERVATIONS

Those agreements which the littoral States have produced for maritime delimitation and provisional arrangements thus far, are as follows: the partial continental shelf boundary between Korea and Japan; a joint continental shelf development agreement between Korea and Japan; and three provisional fisheries agreements between Korea, China and Japan.

The fisheries agreements are all provisional agreements pending the ultimate delimitation of EEZ boundaries between them. All these fisheries agreements purport to establish joint fishing zones under various names or without names. As we have discussed in Chapter 3, there are already a number of joint fishing zones in place around the globe: some of these are classified into so-called "Grey Zones", some of these are into "White Zones" and some of these are somewhere between the two categories.

We also saw that in this chapter that there are potential conflicts between the three littoral States because some of the zones overlap with each other. These overlaps are between the Sino-Japanese Provisional Measure Zone and Korea-Japan joint fishing zone in the East China Sea; the Sino-Korean Current Fishing Pattern Zone and the Sino-Japanese Provisional Measure Zone; the Sino-Japanese No-Fishing-Permit-Zone and the Korean-Japanese joint fishing zone in the East China Sea; and the Sino-Japanese No-Fishing-Permit-Zone and the Sino-Korean Transitional Zone at the Korean side.

In the overlapping areas, proper conservation of the living resources would be very hard to achieve because agreement between the two Parties on conservation measures is difficult to achieve. These phenomena may be the result of the fact that third parties in each do not recognise provisional zones between two others and thus deliberately try to break the provisional zones. Note that Korea does not recognise the northern limits of the Sino-Japanese Provision Measure Zone and it succeeded in breaking it and making a current fishing Pattern Zone with China. Similarly, China broke the Korea-Japan joint fishing zone in the East China Sea, establishing a No-Fishing-Permit-Zone with Japan there. Japan who has been insisting on adopting conservation measures through the joint fisheries committee agree with China to allow Chinese fishing vessels to fish in the joint fishing zone in the East Sea.

In this very complicated situation, confused further by conflicts between one exclusive bilateralism against another, the proper management of fisheries resources in the areas would be difficult to achieve. In this regard, we need to note that in the three fisheries agreements, there is no provision on the exercise of jurisdiction over the fishing activities in the various joint fishing zones by the vessels of third countries. Thus each joint committee should consider the issues on the allocation of fishing quotas to, issuing of fishing permits to, and enforcement jurisdiction over the vessels of third countries.[209] Apart from the issues of the challenges to the agreements and overlaps between the joint fishing zones, these provisional fisheries agreements would not be effective for the proper conservation of certain important species which straddle the EEZs of the three States. We have seen that there is an obligation on the coastal States which share transboundary fish stocks to negotiate to enter into an agreement to properly conserve the transboundary fish stocks and this obligation is applicable to the three States in North East Asia. Besides this obligation, there is an urgent need for the coastal States in the region to set up co-operation schemes for the conservation of transboundary fish stocks to supplement the three fisheries agreements between Korea, China and Japan.

Let us turn to the northern continental shelf boundary of 1974 between Korea and Japan. Its geographical scope is limited to the southern part of the East Sea and it does not delimit the water column. It is to be recalled that Korea and Japan have not succeeded in finding oil in the joint development zone under the complicated procedures of the joint development agreement of 1974, which is often cited as a good example of a joint development zone agreement.[210] It seems that sufficient exploration has not been made partly due to the complicated consensual procedures that should be applied at almost every stage of exploration and exploitation, and due to the Chinese challenge to the development zone.

It also seems desirable for the littoral States in North East Asia to the enter into provisional arrangements on the marine scientific research, making sure that the research conducted in the disputed areas by a coastal State does not reinforce its position on the delimitation and also the third parties enjoy freedom of the marine scientific research in the disputed areas.

NOTES

1. For detailed analysis of the agreement, see Choon-ho Park, *The Law and Practice Relating to the International Regulation of Fisheries in Asia with Particular*

Reference to the Korea-Japanese Dispute, Ph.D. thesis at the University of Edinburgh (1971); Hideo Takabayashi, "Normalisation of Relations between Japan and the Republic of Korea", 10 *J.A.I.L.* (1966), pp. 16-23.

2. Although there was an attempt to distinguish inner six and outer six in the Geneva Conferences, there was no such distinction in the 12-mile fisheries zone envisaged in the agreement of 1965. Choon-ho Park gave 14 examples of twelve-mile fishery zone agreements which were concluded in a various forms in 1960s under the influence of the two Geneva Conferences; Choon-ho Park, *ibid.*, pp. 303-304.

3. Shigeru Oda, "The Normalisation of Relations between Japan and the Republic of Korea", 61 *A.J.I.L.* (1967), pp. 51-51.

4. The total maximum catch of mackerel in the joint regulation zone was set at 150,000 tons per year with a 10 percent fluctuation according to the status of the stocks: see Article 6 of the agreement of 1965.

5. Ministry of Foreign Affairs of Korea, *Issues of Fishing in the Waters off Cheju-island and Hokkaido* (written in Korean), 1981, pp. 7-10.

6. It is also to be noted that the joint regulation was set up only in the waters outside Korea's exclusive fisheries zone and there was no such zone around the Japanese coasts.

7. Ministry of Foreign Affairs of Korea, *Negotiation History of the Autonomous Fishing Operation Regulation Measures between Korea and Japan* (written in Korean for internal use with the Ministry), 1982. The Autonomous Fishing Operation Regulation Measure was revised five times until it was abolished along with the fisheries agreement of 1965 in 1999.

8. Mark J. Valencia, *A Maritime Regime for North East Asia*, Oxford University Press (1996), p. 255.

9. Tsuneo Akaha, "Fishery Relations in North East Asia", a Paper presented at the International Conference on the UN Convention on the Law of the Sea and East Asia, held in Seoul on 29-30 November, 1995, p. 9. He argued that "The 1965 pact has proved woefully inadequate, from the Japanese perspective, in controlling South Korean fishing in Japanese coastal areas and this has prompted coastal fishery interests in the northern and north-western parts of Japan to demand the expansion of Japanese jurisdiction uniformly to a distance of 200 miles from their coasts".

10. *On dealing with the Issues of Sino-Japanese and Korean-Japanese fisheries agreements on the occasion of ratification of the LOS Convention*, issued in Japanese jointly by representatives from Liberal Democrats, Social Democrats and New *Sakigake*, Party of Japan, 22 March 1996.

11. For the history of the negotiations between Korea and Japan on new fisheries agreement, see Jong-hwa Choi, *The History of Modern Fisheries Relations between Korea and Japan* (written in Korean), Sejong Press (2000), pp. 55-90. The Korean

President said that Korea will co-operate with Japan for settlement of fisheries issues but the delimitation of EEZ boundaries should come first before the conclusion of a fisheries agreement.

12. S.P. Kim and S.G. Hong, *A Study on the Provisional Fisheries Zones between Korea and Japan* (written in Korean), Korea Maritime Institute (1999), pp. 18-22.

13. On the same day when Japan notified its intention to terminate the Fisheries Agreement of 1965, Korea notified, as a counter measure to Japan's notification, its intention that it would suspend the implementation of the autonomous fishing operation regulation measures for an indefinite period, which Korea has been keeping in the waters off the Japanese coasts since 1980. This counter measure was unexpected by Japan and was in fact effective in reminding Japan that mutual co-operation is important for the successful management of a good fisheries regime between Korea and Japan. One of the reasons why the Korean government took this kind of counter measure was that it had been asking the Japanese government not to terminate the Fisheries Agreement in the middle of the negotiations to set the deadline for the conclusion of a new fisheries agreement unilaterally. The Korean government resumed the implementation of autonomous fishing operation regulation measures on 1 July 1998 taking account of Japan's request conveyed by a Japanese Member of Parliament, Koukou Sato, who was the Chairman of the International Fisheries Committee in the ruling Liberal Democrat Party, to the Korean President, Foreign Minister and Minister of Maritime Affairs and Fisheries on June 25 1998.

14. *The Korea Herald*, 26 September 1998.

15. *The Korea Times*, 23 January 1999.

16. The text of the Agreement is reproduced in the Appendix to this book.

17. Paragraph 1 of Annex I of the Agreement.

18. Article 15 of the Agreement.

19. Paragraph 3 of Annex II provides for specific three points chosen from the points which constitute the joint fishing zone in the East Sea and provides that the abovementioned assumption is not applied to a certain part of the areas situated to the north west of the line formed by the straight lines connecting those point. This exception was originally proposed by Japan and Korea accepted it with the insertion of "certain". It provides thus: "The provisions of paragraphs 1 and 2 shall not be applied to a certain part of the Agreement Zone situated to the north west of the line formed by the straight lines joining the following points in sequence". Note that the "Agreement Zone" refers to the exclusive economic zone of the Republic of Korea and the exclusive economic zone of Japan (Article 1). There appears to be room for different views on the meaning of "certain part of the Agreement Zone" between Korea and Japan.

20. S.P. Kim and S.G. Hong, *op. cit.*, pp. 1-6; Hee Kwon Park, *op. cit.*, pp. 60-61; Sinsuge Sukiyama, "The Meaning of the Conclusion of the New Fisheries Agreement between Korea and Japan" (written in Japanese), Jurist (1999), No. 1151, pp. 98-104. Note that Sukiyama was a director at the Treaties Division of the Japanese Foreign Ministry when the fisheries agreement was negotiated. The reason why the Korean government prefers the term "middle zones" or "intermediate zones" is that it wishes to avoid the implications that the zones are set up because the delimitation negotiations was unsuccessful.

21. The Committee is composed of one Representative and one Member appointed respectively by Governments of the Contracting Parties. Usually the Representatives are the Assistant Minister in the Ministry of Maritime Affairs and Fisheries of Korea and the Minister of Fisheries Office of Japan.

22. Ministry of Foreign Affairs and Trade of Korea, "Korea-Japan Fisheries Agreement", available at <www.mofat.go.kr/main/top.Html>. The Korean government used the term "geographically" to emphasise that Dok-do is not part of the joint fishing zone.

23. For example, *Kwangbokhoi* (an old Korean Association for Independence from Japan), made a Statement on its positions that the new Korea-Japan fisheries agreement should not be ratified because Dok-do is included in the joint fishing zone in the East Sea: the Statement was published in *Dongah Ilbo* (a daily newspaper), 22 December 1998.

24. S. Hong, "The Contribution of the New Korea-Japan Fisheries Agreement to the Development of the Korea-Japan Relation (written in Korean)", at <www.mofat.go.kr /main/top.html>. Note that Mr. Hong was Foreign Minister at the time of his writing the article. See also Hai-ung Jung, "The EEZ Regime and the Korea-Japan Fisheries Agreement (written in Korean)", 6 *Seoul International Law Study* (1999), pp. 24-27: Seo-Hang Lee, "The Negotiations on and Main Contents of the Korea-Japan Fisheries Agreement (written in Korean)", presented at Maritime Policy Seminar held in Seoul on 23 October 1999.

25. *Southern Bluefin Tuna arbitration (Australia and New Zealand v. Japan), Award on Jurisdiction on Jurisdiction and Admissibility*, August 4, 2000, para. 57.

26. It is not clear why Japan wanted to avoid the compulsory arbitration provisions in the new Fisheries Agreement. But it might be due to the fact that Japan experienced the situation where an arbitration tribunal could have been set up at the request of Korea with regard to the disputes on the arrests of Korean fishing vessels by the Japanese authorities applying its new straight baselines in late 1990s and Japan does not want such a situation arising again. Or it might be that Japan foresaw that the Japanese authorities would seize Korean fishing vessels more than the Korean authorities would seize the Japanese fishing vessels and did not want to stand before an arbitration tribunal on the charge of arresting Korean fishing vessels.

27. "Negotiations on Korea-Japan Fisheries Agreement", *Shin-Dong-Ah* (a monthly magazine published in Korean), January 1999, pp. 67-88.

28. Hai Uung Jung, *op. cit.*, p. 13. The Northern Limits Line (NLL) or the Northern Boundary Line (NBL) in the East Sea and in the Yellow Sea was drawn by the United Nations Command in Korea immediately following the conclusion of the Armistice Agreement in 1953 since the Armistice Agreement contains no provisions on maritime boundary between the South and North Korea. The NLLs are generally understood as *de facto* maritime boundaries between the two Koreas. However, North Korea has challenged the NLL in the Yellow Sea arguing that the NLL is drawn in such a way to cut into the 12-mile line of the North Korea's territorial waters: see Charn-kyu Kim, "Northern Limit Line as Part of the Armistice System", *Korea Focus on Current Topics*, Korea Foundation, Vol. 7, No. 4, July-August 1999, pp. 100-105.

29. S.P. Kim and S.G. Hong, *op. cit.*, pp. 18-20.

30. Article 15 of the Korean-Japanese Agreement.

31. The *Minquiers and Ecrehos* Case, 1953 *ICJ Reports*, p. 58.

32. 1953 *ICJ Reports, idem.*

33. For the discussion of the cut-off effect in the delimitation, see "3.3.3. Proportionality and Cut-off Effect in the East China Sea" in Chapter Four.

34. China and Japan agreed to establish a joint fishing zone, the so-called "Provisional Measure Zone" and the latitude of the northern limits of the zone is Lat. 30° 40'. We will discuss the new China-Japan Fisheries Agreement later in this Chapter.

35. Paragraph 2 of Article 9 of the Fisheries Agreement.

36. *Records of Negotiations on Sino-Korean Fisheries Talks* (recorded in Korean for internal use of the Korean Foreign Ministry, unpublished), 1998.

37. Paragraph 2 of Annex I of the Agreement is about the joint fishing zone in the East Sea and Paragraph 3 of the same Annex is about the joint fishing zone in the East China Sea.

38. Ministry of Foreign Affairs and Trade of Korea, "Korea-Japan Fisheries Agreement", <www.mofat.go.kr/main/top.html>. For discussion on legislative jurisdiction, see D.W. Bowett, "Jurisdiction: Changing Patterns of Authority over Activities and Resources", 52 *B.Y.I.L.* (1983), pp. 12-22.

39. See, for instance, Myung-ki Kim, "New Korea-Japan Fisheries Agreement and Territorial Ownership of Dok-Do" (written in Korean), a paper delivered at the Great Debate on the Territorial Ownership of Dok-Do, held in the Seoul Press Centre, Seoul, 8 September 2000, pp. 7-9; Young-koo Kim, "Territorial Ownership and New Korea-Japan Fisheries Agreement in the Light of International Law" (written in Korean), a paper delivered at the Great Debate on Dok-Do, held in Seoul

Press Centre, Seoul, 21 October 1998; Sang-myun Lee, "Issues of Dok-do and its Adjacent Waters under the New Korea-Japanese Fisheries Agreement" (written in Korean), 40 *Seoul Law Journal* (1999), pp. 109-132, Sang-myun Lee, "The Problems of the Intermediate Zones in the New Korea-Japan Fisheries Agreement", (written in Korean), *The Korean Journal of International Law*, Vol. 43, No. 2 December 1998, p. 151. The main argument of those scholars is that the Dok-do and its territorial water should have been expressly excluded from the joint fishing zone in the East Sea. In response to their argument, the Korean government explained that the *ratione loci* of the fisheries agreement is exclusive economic zones of Korea and Japan as Article 1 provides and thus it has nothing to do with territory or its territorial waters.

40. Paragraph 2(a) and Paragraph 3(a) of Annex I of the Agreement.

41. Paragraph 3(b) of Annex I of the Agreement.

42. Paragraph 2(e) and Paragraph 3(e) of Annex I of the Agreement. Note here that the Party who finds a fishing vessel of the other Party engaged in fishing in such violation may not draw attention of the fishing vessel to the fact of such violation. Japan proposed in the negotiations a provision which would enable the enforcement authorities of a Party to draw the attention of the other Party's fishing vessels to their violation of the fishing regulation in the joint fishing zone in the East Sea if they are found engaged in such fishing activities, but Korea refused. This restriction in calling to attention fishing vessels of the other Party is the only one but the most important difference from the indirect form of co-operation provided in the China-Japan Fisheries agreement and the China-Korea fisheries agreement as we will see later.

43. *Pusanilbo* (a newspaper published in Korean), 21 September 1999.

44. Records of the first, second and third meetings of the Committee (in Korean, unpublished).

45. According to statistics by Korea's National Fisheries Promotion Institute in 1998, about 700 Korean fishing vessels fish in the joint fishing zone in the East Sea, catching 55,000 tonnes of fish every year.

46. In 1999, Korean fishermen were allowed to catch 149,218 tonnes of fish in the Japanese EEZ. However, at the end of the 1999, it turned out that they fished only 24,358 tonnes in the Japanese EEZ; in 2000, Korea was allowed to catch 130,197 tonnes in Japan's EEZ, but caught only 31,422 tonnes; in 2001, Korea was allowed to catch 109,773 tonnes. Japanese fishermen have been in a similar situation, with their catch having fallen far short of their fishing quota in the Korea's EEZ; Japanese fishermen were allowed to fish 93,773 tonnes in Korea's EEZ in 2000 and in 2001: *The Korea Herald*, 23 December 2000; Ministry of Maritime Affairs and Fisheries of Korea, "Agreement on Fishing Quotas between Korea and Japan for 2001"(written in Korean), at <www.momaf.go.kr>.

47. Paragraph 2(b) of Annex I to the Agreement.

48. Ministry of Maritime Affairs and Fisheries of Korea and Korea Maritime Institute, *A Study on the Legal Status, and Management of the Intermediate Zones of the Fisheries Agreements between Korea, Japan and China*, (written in Korean for internal use), June 2000, pp. 8-9.

49. In the agreement between Korea and Japan on fishing quotas for 2001, no recommendation was adopted with regard to the conservation and management of fisheries resources in the East Sea joint fishing zone; see Ministry of Maritime Affairs and Fisheries of Korea, "Agreement on Fishing Quotas between Korea and Japan for 2001" (written in Korean), at <www.momaf.go.kr>.

50. A letter dated 11 November 1997 was sent to the Chinese ambassador in Tokyo by the Japanese Foreign Minister to show that the Japanese government took note of the Chinese request to allow the Chinese fishermen to fish for squid in the East Sea and the Pacific Ocean without paying a fee to Japan for 5 years after the entry into force of the Sino-Japanese fisheries agreement. In 18 May 2000, the Japanese government agreed to allow the Chinese fishermen to fish for squid up to 7,000 tonnes in the middle of the Korea-Japan joint fishing zone in East Sea except the Yamatotai area. The letter is reproduced in the Appendix to this book.

51. The Agreed Minute was signed on 28 November 1998. The Agreed Minute is reproduced in the Appendix to this book.

52. Paragraph 2 of the Agreed Minute.

53. Hai-ung Jung, "EEZ Regime and Korea-Japan Fisheries Agreement" (written in Korean), 6 *Seoul International Law Study* (1999), reproduced at <www.mofat.go.kr/main/top.html>.

54. However, a Japanese official in the Treaties Bureau of the Japanese Foreign Ministry told this author that in his view the "a certain area" indicates the area corresponding to the Japanese version of the Korea-Japanese joint fishing zone in the East China Sea. His explanation has a point in the light of the fact that Japan and China agreed in their Agreed Minutes signed in November 1997 along with the New Sino-Japanese Agreement, to try to respect current fishing activities in an area north of the Sino-Japanese provisional measure zone and thus Japan need to get an understanding from Korea in this regard. But there was no such understanding between Korea and Japan in drafting the Agreed Minutes. The understanding was that both Korea and Japan needed to address the problems which might arise from the different views on the southern limits of the Korea-Japan joint fishing zone in the East China Sea.

55. The position of the Korean government is that the northern limit line of the Sino-Japanese provisional measure zone is not valid under international law because the line is chosen by Japan and China without consultation with Korea which is

also a coastal State and has EEZ entitlement. We will discuss this issue when we discuss the fisheries agreement between Japan and China.

56. Second sentence of Paragraph 2 of the Agreed Minutes.

57. Paragraph 3 of the Agreed Minutes.

58. Hai-ung Jung, *op. cit.*, reproduced at <www.mofat.go.kr/main/top.html>.

59. Ministry of Maritime Affairs and Fisheries of Korea and Korea Maritime Institute, *A Study on the Legal Status, and Management of the Intermediate Zones of the Fisheries Agreements between Korea, Japan and China,* (written in Korean for internal use), June 2000, pp. 95-97. However, a Japanese official in the Treaties Bureau of the Foreign Ministry told this author that in his view "the other certain area" indicates the whole of the Korea-Japanese joint fishing zone.

60. The Republic of Korea had diplomatic relations with the Republic of China (Taiwan) until 1992. In the Korean War between 1950-1953, the People's Republic of China's volunteer military troops participated in the war to assist North Korea. After the cold war, the Republic of Korea established diplomatic relations with the People's Republic of China cutting off the diplomatic relations with the Republic of China. Since 1992, China has maintained diplomatic relations both with the Republic of Korea and the People's Republic of Korea (North Korea).

61. The East China Motor Trawl Prohibition Line was drawn on 16 December 1950 by the Fisheries Management Bureau of the East China Military Administration Committee. The line is 30-70 miles from the Chinese shore; see Choon-ho Park, "Fishing under Troubled Waters: The North East Asia Fisheries Controversy", 2 *O.D.I.L.* (1974), pp. 114-115.

62. Rules on Safety of Vessel Operation, announced in 1968 amended thereafter several times.

63. Since the late 1990s, the number of Chinese fishing vessels found fishing in the territorial waters of Korea increased rapidly. The number was 249 vessels in 1991, but it was 472 in 1995. According to statistics by the Korean Ministry of Maritime Affairs and Fisheries, the number of Chinese fishing vessels seized by the Agency on charge of illegal fishing in the territorial waters of Korea numbered 15 in 1992; 17 in 1993 and 1994; 45 in 1995 and 1996; 39 in 1997; 31 in 1998; 60 in 1999; and 24 in 2000.

64. Korea also proposed that two countries adopt a vertical line, as a fisheries boundary, which equally divides the Yellow Sea in size if the delimitation takes some time: Ministry of Foreign Affairs and Trade of Korea and The Ministry of Maritime Affairs and Fisheries, *Exposition on Korea-China Fisheries Agreement* (written in Korean, unpublished), April 1999, pp. 8-9. The Ministry of Maritime Affairs and Fisheries, "Exposition Material on Korea-China Fisheries Agreement (written in Korean)", February 2001, at <www.momaf.go.kr>, p. 6.

65. In the negotiations Korea emphasised that it is was going to enforce its EEZ fisheries law against China if a fisheries agreement was not concluded in the near future. In fact, various plans for such enforcement were examined by the Korean government. Note, however, that such a question did not arise with regard to the fisheries relations between Korea and Japan because the fisheries relations between Korea and Japan were regulated by the Fisheries Agreement of 1965, which guaranteed freedom of fishing in the areas beyond 12 miles from the coasts.

66. Korea enacted a regulation on TAC Management in April 1998 after revising the Fisheries Law in 1995 and the Presidential Decree on Preservation of Fisheries Resources in 1996 for introduction of TAC in its EEZ and then implemented a trial TAC on a limited species between 1999-2000 and planed to embark upon a real TAC in 2001. See Ministry of Maritime Affairs and Fisheries of Korea, *Basic Management Plan of TAC System and Implementation Plan in 2001* (written in Korean), at <www.momaf.go.kr>.

67. As the Enforcement Decree promulgated on 6 August 1997 provided for an exception of application of its EEZ Fisheries Law against the Chinese fishermen for one year, the exception expired on 5 August 1998, when Korea and China still negotiated for a fisheries agreement. However, thereafter a tacit mutual understanding was reached between Korea and China to reserve the application of their EEZ fisheries regime against each other's fishermen until the entry into force on the fisheries agreement.

68. Ministry of Foreign Affairs and Trade of Korea and The Ministry of Maritime Affairs and Fisheries, *Exposition on Korea-China Fisheries Agreement* (written in Korean, unpublished), April 1999, pp. 7-9.

69. Ministry of Foreign Affairs and Trade of Korea and The Ministry of Maritime Affairs and Fisheries, *Exposition on Korea-China Fisheries Agreement* (written in Korean, unpublished), April 1999, pp. 8-9.

70. Article of the Fisheries Agreement between Korea and China. The text of the Fisheries Agreement is reproduced in English in the Appendix to this book.

71. *The Korea Herald*, 12 November 1999.

72. Korean Ministry of Maritime Affairs and Fisheries, *Press Release*, 5 April 2001, available at <www.momaf.go.kr>.

73. Paragraph 1 of Article 2 of the Korean-Chinese Fisheries Agreement.

74. It is estimated that the annual fishing quotas for the Korean fishermen in the China's EEZ will be 60,000 tonnes, and the Korean government is trying to set the fishing quota for the Chinese fishermen in Korea's EEZ within 120,000 whereas China argued that its traditional annual fishing catch, 400,000 tonnes should be respected in setting its quotas. Source: Ministry of Maritime Affairs and Fisheries, "Situations of the Korea-China Fisheries Negotiations and Plans" (written in Korean), March 2001, at <www.momaf.go.kr>, pp. 5-7.

75. Articles 7 and 8 of the Korean-Chinese Fisheries Agreement.

76. Article 14 of the Korean-Chinese Agreement. A question arises here why the agreement adopted the term "issues in the law of the sea" rather than the term "issues in the law of the sea other than fisheries". Because obviously the fisheries agreement can have an effect of modifying the rights and obligations of the LOS Convention as between the Parties with regard to fisheries matters between the two Parties, and thus the agreement affects each Party's position on the issues of fisheries though it is a provisional arrangement. Note that the fisheries agreement between China and Japan also adopts the same language in its without-prejudice clause.

77. Paragraph 1 of Article 13 of the Korean-Chinese Fisheries Agreement.

78. Paragraph 2 of Article 13 of the Korean-Chinese Fisheries Agreement.

79. Paragraph 3 of Article 13 of the Korean-Chinese Fisheries Agreement.

80. Paragraph 4 of Article 13 of the Korean-Chinese Fisheries Agreement.

81. In the negotiations China proposed not to have a dispute settlement clause in the Fisheries Agreement saying that the Sino-Japanese Fisheries Agreement signed in November 1997 also does not have provisions on dispute settlement procedure: Ministry of Foreign Affairs and Trade of Korea, "Records of Negotiations on Sino-Korean Fisheries Talks" (recorded in Korean for internal use, unpublished), 1998.

82. See Paragraph 1 of Article 281 of the LOS Convention.

83. *Southern Bluefin Tuna arbitration (Australia and New Zealand v. Japan), Award on Jurisdiction and Admissibility*, 4 August 2000, para. 39.

84. Article 16 of the Korean-Chinese Fisheries Agreement.

85. *Records of Negotiations on Sino-Korean Fisheries Talks* (recorded in Korean for internal use, unpublished), 1998.

86. Transitional Zone in the Korean side is 28.716 sq. km and the Transitional Zone in the Chinese side is 26,192 sq. km; calculated by using a software programme "Delmar".

87. Paragraph 2 and 3 of Article 7 the Korean-Chinese Fisheries Agreement.

88. Paragraph 3 of Article 8 of the Korean-Chinese Fisheries Agreement.

89. *Ibid.*

90. Paragraph 3 of Article 7 and Paragraph 2 of Article 8 of the Korean-Chinese Fisheries Agreement.

91. Paragraph 3 of Article 8 of the Korean-Chinese Fisheries Agreement.

92. Paragraphs 2, 3 and 5 of Article 8 of the Korean-Chinese Agreement.

93. Paragraph 2 of Article 7 and Paragraph 3 of Article 8 of the Korea-Chinese Agreement.

94. The term "current fishing pattern" might seem inappropriate for a term to indicate "free fishing" because both Korea and China already proclaimed their respective EEZ when they initialled the fishing agreement in 1998. It appears that negotiators of both countries had begun to use the term "current fishing pattern" before both countries proclaimed their respective EEZ.

95. Article 9 of the Agreement of the Korean-Chinese Fisheries Agreement.

96. *Ibid.*

97. The area where the new conservation measure is to be applied is the area bound by the lines connecting the following points: Point at Lat. 34° N, Log. 122° 30', Point at Lat. 32° N, Log. 122° 30', Point at Lat. 32° N, Log. 124° and Point at Lat. 29° 30' N, Log. 124°.

98. As the Chinese government adopted the new conservation measure after it had initialled the Agreement and the Memorandum of Understanding, Korea argued that the new measure cannot be applicable to Korean fishermen because the measures were not known when the Agreement and Memorandum of Understanding were initialled. However, it was difficult for Korea to base its argument on the law of treaties because the Chinese new conservation measure was adopted when the Agreement and the Memorandum of Understanding were not yet formally signed and initialling would not have any legal effect for preventing China from adopting a new conservation measure. It appears the new conservation measure by China in the waters off the Yangtse river was adopted under its new fisheries policy adopted in 1999 which aims at 0 % growth of catch in the marine capture fisheries sector for the purpose of sustainable development of fisheries, and thus it would be inappropriate to say that the new conservation measure adopted by China after it initialled the Agreement was to cheat Korea. For the new fisheries policy of China, see Ministry of Agriculture of China, Fisheries Bureau, *Analysis of Fisheries Economics of China in 1999 and Plans for Fisheries Policy in 2000* (written in Chinese), Policy Paper 28 January 2000, at <www.agri.gov.cn/news/2000/yuyezhln/jibenqk.htm>.

99. Ministry of Maritime Affairs and Fisheries, "Situations of the Korea-China Fisheries Negotiations and Plans" (written in Korean), March 2001, *at* <www.momaf.go.kr>, p. 3.

100. No such differences arose between Korea and China with regard to the area situated in the north of the Provisional Measure Zone once they reached the understanding that Chinese fishermen are not allowed to fish in Korea's Special Prohibition Zone, because the fishing activities by Korean fishermen are not active elsewhere other than in the Special Prohibition Zone in the area to the north of the Provisional Measure Zone and the fishing activities by Chinese fishermen in the area are not matters of concern to Korea.

101. Korean Ministry of Maritime Affairs and Fisheries, "Situations of the Korea-China Fisheries Negotiations and Plans" (written in Korean), March 2001 and Press

Release of 5 April 2001, available *at* <www.momaf.go.kr>, p. 6. According to the seasonal restriction, trawl and purse seine vessels are not allowed to fish in the area between 16 June and 19 September.

102. The number of the Japanese fishing vessels seized by the Chinese authorities between 1950-1954 was 154 and the number of Japanese crew seized was 1909. Z. Ohira and T. Kuwahara, "Fisheries Problems between Japan and People's Republic of China", 3 *Japanese Annual of International Law* (1959), p. 109.

103. Masahiro Miyoshi, "New Japan-China Fisheries Agreement", 41 *Japanese Annual of International Law* (1998), pp. 31-33; Hee Kwon Park, *The Law of the Sea and North East Asia: A Challenge for Co-operation,* Kluwer Law International Law (2000), p. 51.

104. Masahiro Miyoshi, *ibid.*, p. 31.

105. Article 3 of the Fisheries Agreement of 1955.

106. The Fisheries Agreement of 1955 was concluded for a period of one year, and it was renewed in 1956 and in 1957. In 1958 the Chinese side refused to renew the agreement asserting that Japanese fishing vessels did not faithfully observe the fishing regulations by the agreement and thus there was no fisheries agreement from 1958 to 1962. But there were no substantial disputes between them in this period as Japanese fishing vessels followed the fishing regulations provided for in the agreement of 1955. In 1963 Japan and China concluded a new fisheries agreement of a non-governmental nature with some modification to the Mao Tse-teung Line, which was valid for 2 years. In 1965 a third agreement of a non-governmental nature was renewed six times until 1975 when the governments of China and Japan concluded a fisheries agreement, on the basis of the fisheries agreements of non-governmental nature: see Choon-ho Park, "Fishing under Troubled Waters: The North East Asia Controversy", 2 *O.D.I.L.* (1974), pp. 116-118.

107. The fisheries production of China increased dramatically from 1980. It was 5.16 million tonnes in 1980, 14.30 millions tonnes in 1990, and 41.22 million tonnes in 1999. This dramatic increase is not only due to the development of captured fisheries but also the development of aqua-culture fisheries. Source: "Basic Situation of the Chinese Fisheries Economy and System of the Fisheries Product Quality Control"(written in Chinese) Fisheries Bureau of the Chinese Ministry of Agriculture, at <www.argi.gov.cn/nres/2000/yuyezhln/jibenqk.htm>.

108. The two countries also agreed on the annual fishing quotas in each other's EEZ on 18 May 2001. Japanese fishermen are allowed to fish up to 78,000 tonnes in the China's EEZ and Chinese fishermen are allowed to fish up to 70,000 tonnes in the Japan's EEZ: Press releases by Japanese Fisheries Agency (written in Japanese), 27 February 2000, and 18 May 2000.

109. First sentence of Paragraph 1 of the Agreed Minute. The Agreed Minute is reproduced together with Fisheries Agreement in the appendix to this book.

110. Article 12 of the Fisheries Agreement of China-Japan Fisheries Agreement.

111. Second sentence of Paragraph 1 of the Agreed Minutes.

112. Paragraph 1 of Article 11 of the China-Japan Fisheries Agreement.

113. Paragraph 2 of Article 11 of the China-Japan Fisheries Agreement.

114. Paragraph 3 of Article 11 of the China-Japan Fisheries Agreement.

115. Paragraph 4 of Article 11 of the China-Japan Fisheries Agreement.

116. *Southern Bluefin Tuna Arbitration* Case *(Australia and New Zealand v. Japan), Award on Jurisdiction and Admissibility*, 4 August 2000, para. 39.

117. Paragraph 1 of Article 14 of the China-Japan Fisheries Agreement.

118. Hee Kwon Park, *op. cit.*, p. 57.

119. Paragraph 2 of Article 7 of the China-Japan Fisheries Agreement.

120. Paragraph 3 of Article 7 of the China-Japan Fisheries Agreement.

121. Article 6(b) of the China-Japan Fisheries Agreement.

122 The letters are reproduced in the Appendix to this book along with the China-Japan Fisheries Agreement.

123. *Mainich Shimbun* (a Japanese daily newspaper), 27 February 2000. The Japanese newspapers used the term "intermediate zone" for the No Fishing Permit Zone.

124. Annex I to the Korea-Japan Fisheries Agreement of 1998.

125. This area that is included into the Fishing-Permit-Is-Not-Required between Japan and China is the area where the Japanese fishermen is the most active among the waters around the Korean peninsula. Chinese fishermen are also very active in this area.

126. FAO, *The State of World Fisheries and Aquaculture* (2000), pp. 7-10.

127. A/CONF.151/26 (Vol. II): Report of the United Nations Conference on Environment and Development (Rio de Janeiro, 3-14 June 1992) Chapter 17/45.

128. The Agreement was adopted in November 1993 was adopted by the FAO Conference.

129. The Code was unanimously adopted on 31 October 1995 by the FAO Conference.

130. A/CONF.164/37 (8 September 1995).

131. Cofi/2001/7Add (February 2001, FAO).

132. G. Moore, "The Food and Agriculture Organisation of the United Nations Compliance", 10 *I.J.M.C.L.* (1995), pp. 412-416.

133. 1.1 of the Code of Conduct.

134. FAO, Introduction to the Code of Conduct for Responsible Fisheries.

135. Article 21 of the UN Fish Stocks Agreement. Peter G.G. Davis and Catherine Redgwell, "The International Legal Regulation of Straddling Fish Stocks", 74 *B.Y.I.L.* (1997), pp. 265-272. D.H. Anderson, "The Straddling Fish Stocks Agreement of 1995-an initial assessment", 45 *I.C.L.Q.* (1996), pp. 463-475.

136. Paragraph 4 of the IPOA for IUU Fishing.

137. Paragraph 45 of the IPOA for IUU Fishing.

138. Paragraph 46 of the IPOA for IUU Fishing.

139. Paragraph 75 of the IPOA for IUU Fishing.

140. Article 3 (1) of the UN Fish Stocks Agreement provides that: "Unless otherwise provided, this Agreement applies to the conservation and management of straddling fish stocks and highly migratory fish stocks beyond areas under national jurisdiction, except that Articles 6 and 7 apply also to the conservation and management of such stocks within areas under national jurisdiction, subject to the different legal regimes that apply within areas under national jurisdiction and in areas beyond national jurisdiction as provided for in the Convention".

141. Korean Ministry of Maritime Affairs and Fisheries, *Studies on the TAC-based Fisheries Management System and Quota Allocations for Jointly Exploited Fisheries Resources under the EEZ Regime* (written in Korean, 2000), pp. 391-399.

142. Moritaka Hayashi, "The Management of Transboundary Fish Stocks under the LOS Convention", 8 *I.J.M.C.L.* (1993), p. 249.

143. Barbara Kwiatkowska, "The Eritrea-Yemen *Arbitration*: Landmark Progress in the Acquisition of Territorial Sovereignty and Equitable Maritime Boundary Delimitation", 32 *O.D.I.L.* (2001), p. 13. *Eritrea-Yemen Arbitration Award* (1999), para. 63; the award is reproduced in the Permanent Court of Arbitrator's internet home page at <www.pca-cpa.org>. For the first stage of the arbitration on territorial disputes on island with map, see Nuno Sergio Marques Anunes, "The Eritrea-Yemen *Arbitration*: First Stage-The Law of Title to Territory Re-Averred", 48 *I.C.L.Q.* (1999), pp. 362-386.

144. As we have seen in Chapter Three, the Convention for the Halibut Fisheries of the North Pacific and Bearing Sea of 1953 as amended by a Protocol of 1979 between US and Canada is be an example of bilateral agreement for conservation and management of transboundary fish stocks without setting-up of a joint fishing, but focusing on a specific fish stock.

145. Article 5 of the Convention on Fishing and Conservation of the Living Resources in the Baltic Sea and the Belts (the Gdansk Convention), which was signed on 13 September 1973. Note that the waters to which the Gdansk Convention

applies as an area which consists wholly of territorial sea and fisheries zone or exclusive economic zone of the participating States. Information on the Convention is available at the IBSFC internet home page at <www.insfc.org>.

146. Article 10 of the Convention on Fishing and Conservation of the Living Resources in the Baltic Sea and the Belts. Article 10 of the Convention provides that:

> Measures relating to the purposes of this Convention which the Commission may consider and in regard of which it may take recommendations to the Contracting States are:
>
> a) any measures for the regulation of fishing gear, appliances and catching methods,
> b) any measures regulating the size limits of fish that may be retained on board vessels or landed, exposed or offered for sale,
> c) any measures establishing closed seasons,
> d) any measures establishing closed areas,
> e) any measures improving and increasing the living marine resources, official reproduction and transplantation of fish and other organisms,
> f) any measures improving and increasing the living marine resources, official reproduction and transplantation of fish and other organisms,
> g) any other measures related to the conservation and rational exploitation of the living marine resources.

147. Article 11 of the Convention on Fishing and Conservation of the Living Resources in the Baltic Sea and the Belts.

148. M. Takeyama, "Japan's Foreign Negotiations over Offshore Petroleum Development", in R.L. Friedheim *et al.*, eds., *Japan and New Ocean Regime*, 1984, p. 287.

149. The ICJ said that: ". . . the rights of the coastal State in respect of the area of continental shelf that constitutes a natural prolongation of its territory into and under the sea exist *ipso facto* and *ab initio*, by virtue of its sovereignty over the land . . .": 1969 *ICJ Reports*, para. 19.

150. 1969 *ICJ Reports*, para, 101.

151. Official Records of the House of Representatives Standing Committee on Foreign Affairs (Japan), 22 April 1977, p. 23; partly translated unto English in 28 *J.A.I.L.* (1985), pp. 134-137.

152. *J.A.I.L., idem.*

153. Immediately after the conclusion of the agreement, the Chinese-government announced its position that it did not recognise the joint development zone.

154. The Agreement between Japan and the Republic of Korea Concerning Joint Development of the Southern Part of the Continental Shelf Adjacent to the Two Countries. The text is reproduced in Choon-ho Park, "Report Number 5-12: Japan-South-Korea", *International Maritime Boundaries*, pp. 1073-1089.

155. The Agreement between Japan and the Republic of Korea Concerning the Establishment of Boundary in the Northern Part of the Continental Shelf Adjacent to the Two Countries. The text is reproduced in Choon-ho Park, *ibid.*, pp. 1063-1065.

156. Hee Kwon Park, *op. cit.* (2000), p. 39.

157. *Xinhua Monthly*, Beijing, 11 February 1974.

158. *J.A.I.L.* (1986), p. 129.

159. An unpublished paper by the Korea Petroleum Development Company which was established by the Korean government. However, there is still a possibility of finding sizeable amounts of oil and gas in the joint development zone, because the exploration in the zone was not sufficiently done due to the China's protests and foreign companies' caution. See P.A. Soeparjadi *et al.*, "World Oil Development: Far East", 70 *AAPG Bulletin* (1985), pp. 1479-1485.

160. S. Hiramotsu, "China's Oil Wells Development in the East China Sea" (written in Japanese), *Towoa Monthly* (June 2000).

161. S. Hiramotsu, *idem.*

162. *Ibid.*

163. *Yomiuri Shinbun* (A Japanese daily newspaper) of 28 August 2000 reported that the number of the cases of exploration by the Chinese vessels in the Japanese side of the hypothetical equidistance line is 4 in 1997, 14 in 1998, 30 in 1999 and 19 in 2000 as of the end of the August 2000.

164. *Sankei Shimbun* (A Japanese daily newspaper) of 29 August 2000.

165. *Ibid.*

166. Article 122 of the LOS Convention defines the enclosed or semi-enclosed seas as follows: "For the purpose of this Convention, "enclosed sea or semi-enclosed sea' means a gulf, basin or sea surrounded by two or more States and connected to another sea or the ocean by a narrow outlet or consisting entirely or primarily of the territorial seas and exclusive economic zones of two or more coastal States". The Mediterranean, Baltic, Caribbean, North Sea, Persian Gulf, and Red Sea are all examples of enclosed or semi-enclosed seas, and have developed regional treaties. The provision of Article 122 of the LOS Convention can be said as an attempt to define one category of region, of which concept determined by reference to the geography of the surrounding landmass. A.E. Boyle, "Globalism and Regionalism in the Protection of the Marine Environment", and Budislav Vukas, "United Nations Convention on the Law of the Sea and the Polar Marine Environment", in Davir Vidas, ed., *Protecting the Polar Marine Environment*, Cambridge University Press (2000) at pp. 26-27 and pp. 35-37.

167. It is said that the possibilities are very high that this waste will be corroded in ten to thirty years and will then produce radio-activity: Jong-bum Choi and Ki-Soon Kim, *Natural Environment and International Law* (written in Korean), Bumyangsa Press (1994), p. 129. As the Soviet Union joined the London Dumping Convention on the Prevention on Marine Pollution by Dumping of Wastes and

Other Matter in December 1975 and the Russian Federation succeeded the membership in December 1991, the Soviet Union's (later the Russian Federation) dumping of nuclear wastes into the Sea of Japan since December 1975 is generally in violation of the London Convention even if the Soviet Union (and the Russian Federation) made a reservation regarding the Annexes III (Regulation for the Prevention of Pollution by Harmful Substances Carried by Sea in Package Forms, or in Freight Containers, Portables Tanks or Road and Rail Tank Wagons), IV (Regulations for the Prevention of Pollution by Garbage from Ships). However, it has been said that as the nuclear waste dumped by the USSR (later Russia) mostly fall into the category of Annexes III, and since there is no provisions in the London Dumping Convention text on nuclear waste it is difficult to rely on the Convention to argue against Russia. However, it seems to this writer, it does not follow that the USSR (Russia) has not been in violation of the customary international law rule of good neighbourhood or the general provisions of the London Dumping Convention. For example, Article 1 of the Convention provides that Contracting Parties shall individually and collectively promote the effective control of all sources of pollution of marine environment, and pledge themselves especially to take all practicable steps to prevent the pollution of the sea by the dumping of waste and other matter that is liable to create hazard to human health, to harm living resources and marine life, to damage amenities or to interfere with other legitimate uses of the sea. Geneva Convention on the High Seas (Article 25) also provides that "Every State shall take measures to prevent pollution of the seas from the dumping of radioactive waste . . ."

168. The Yellow Sea is said to be one of the seven most contaminated seas in the World, as it is contaminated by 7510,000 tons of heavy metals from the Chinese Yellow River every year and by 21,000 tonnes of crude oil from exploitation activities every year. Source: *World Watch 1995*. See also, for marine pollution problem in the Yellow Sea, Y.H. Seung and Y.C. Park, *Physical and Environmental Character of the Yellow Sea,* East and West Studies Series 11, Institute of East and West Studies, Yonsei University (1990).

169. Article 123 of the LOS Convention. Korea, China, Japan and Russia are Parties to the LOS Convention. North Korea is a Signatory to the Convention, but has not yet ratified the Convention.

170. The three environmental Agreements are reproduced in the appendix to this book.

171. Preamble to the Korea-Russia Agreement.

172. Preambles to the Korea-Japan Agreement and to Korea-China Agreement.

173. A.E. Boyle, "Globalism and Regionalism in the Protection of the Marine Environment", in Davir Vidas ed., *Protecting the Polar Marine Environment,* Cambridge University Press (2000), pp. 19-33.

174. L. Alexander, "Regional Arrangements in the Oceans", 71 *A.J.I.L* (1977), p. 84.

175. L. Alexander, *idem.*

176. The environmental community of the North East Asia comprise North and South Korea, Japan, Mongol, China, and the eastern part of Russia. The community is separated from the Pacific by the Japanese Islands to the south and east, to the west from the Europe and West Asia by the Himalayan Mountains and the Pamirs, and to the north from the Western Russia by the Ural and Altai Mountains. The environmental community of the North East Asia comprises North and South Korea, Japan, Mongol, China, and eastern part of Russia. The environmental community in North East Asia in term of air mass is under the influence of the monsoon originating from the Mongol plateau and the Lake Baikal and travelling to the south. For characteristics of the North East Asian environmental community, see Sang-Gyu Lee, "Co-operational Approach for the Environmental Protection of North East Asia" (written in Korean), 13 *Journal of Environment* (1991).

177. Professor Boyle used the term "liberal model of regionalism" for a regional action to set its own standard for the land-based sources under the general requirements of Article 194. He also noted that the function of regional rules or treaties is relatively limited with regard to dumping and pollution from ships because derogation from the global standards is not permitted except for more stringent standards: See A.E. Boyle, "Globalism and Regionalism in the Protection of the Marine Environment", in Davir Vidas ed., *Protecting the Polar Marine Environment*, Cambridge University Press (2000), pp. 19-33.

178. The 5th Paragraph of the Preamble to the Korea-China Agreement. The Korea-Russia Agreement also has similar provision in the 5th paragraph of the Preamble. However, there is no direct mention of the preventive measures in the Korea-Japan Agreements.

179. Principle 15 of the Rio Declaration on Environment and Development. It is also notable here that the UN Framework Convention on Climate Change also adopted precautionary principle by providing in the Article 3(3) that. "The Parties should take precautionary measures to anticipate, prevent or minimise the cause of climate change and mitigate its adverse effects . . . lack of full scientific certainty should not be used as a reason for posting such measures."

180. P. Birnie and A.E. Boyle, *International Law and the Environment*, Oxford University Press (1992), pp. 89-92.

181. E.B. Weiss ed., *Environmental Change and International Law: New Challenges and Dimensions*, UN University Press, p. 18.

182. Article 2 of each of the three Agreements.

183. As Article 194 of the UNCLOS implies a general obligation on the parts of States to act with diligence and due diligence requires the introduction of legislation and administrative controls applicable to public and private conduct which are capable of effectively protecting other States and the global environment, and it

can be expressed as the conduct to be expected of a good government. P.W. Birnie and A.E. Boyle, *International Law and the Environment*, Oxford (1992), pp. 92-93.

184. There is no provision on environmental impact assessment in Korea-Japan and Korea-Russia agreements. But the possibility of implementing the environmental impacts assessment is not excluded because both agreements provide for other forms of co-operation as may be mutually agreed upon.

185. Phoebe N. Okowa, "Procedural Obligations in International Environmental Agreements", 68 *B.Y.I.L.* (1997), p. 275.

186. World Commission on Environment and Development, *Our Common Future*, Oxford University Press (1987), p. 8; P. Birnie and A.E. Boyle, *International Law and the Environment*, Oxford University Press (1992), p. 96.

187. E. B. Weiss, ed., *Environmental Change and International law: New Challenges and Dimensions*, p. 92: *UNEP G.C. Decision* (14/25) June 1987.

188. Article 2(5) of each of the three Agreements. In fact, Korea and Japan agreed at the Joint Environmental Committee which was held in February 1994 to jointly monitor NOx and SO2 in the air even though there is no provision on monitoring in the Agreements; *Dong-ah Ilbo* (a Korean newspaper) 11 Feb. 1995.

189. Ministry of Foreign Affairs and Trade of Korea, "Current Situations of the Environmental Co-operation in North East Asia" (written in Korean); at <www.mofat.go.kr/main/top.htma>.

190. Article 1(2) of Korea-China Agreement.

191. Agenda 21 (chap. 34.1-3) explains the environmentally sound technologies as follows: "Environmentally sound technologies protect the environment, are less polluting, use all resources in a more sustainable manner, recycle more of their ways and products and handle residual waste in a more acceptable manner than the technologies for which they were substitutes. . . . Environmentally sound technologies are not just individual technologies but total systems which include know-how, procedures, goods, and services and equipment as well as organisational and managerial procedure".

192. Article 200 of the LOS Convention.

193. Agenda 21 Chapter 34(4). The Convention on Biological Diversity, Article 16 (2).

194. Agenda 21, Chapter 34(9).

195. As State-of-the-art technologies are usually owned by private enterprises and covered by intellectual property rights and thus commercial channelling is an important vehicle for technology transfer. Commercial channelling occurs in many ways, from selling the intellectual property rights to exporting the technology-based products or services to the host country. Yet the most common way is to give licences

in return for royalties, a practice which is often enhanced through the establishment of joint ventures in the host country with whom licensing agreements are subsequently concluded. Recognising that the royalties for those State-of-the-art technologies are usually high, UNCED recommended to supplier countries the purchase of patents and licenses on commercial terms for their transfer to developing countries on non-commercial terms as part of development co-operation for sustainable development. And there are some multilateral treaties which provide technological co-operation on preferential terms. For example, the Convention on Biological Diversity provides that, Access to and transfer of technology . . . to developing countries shall be provided and/or facilitated under fair and most favourable terms, including on confessional and preferential terms where mutually agreed . . . Agenda 21 Chapter 34 17(e)-iii; The Convention on Biological Diversity, Article 16 (2); Inge Govaere, "The Impact of Intellectual Property Protection on Technology Transfer between EC and Central and Eastern European Countries". 25 *Journal of World Trade*, p. 58.

196. Article 1 of the three Agreements.

197. The Parties to the Convention amended the Convention so as to extend its *ratione materiae* in accordance with the development, *inter alia*, in the United Nations Conference on Environment and Development of 1992, the Declaration of Tunis of 1994 on the Sustainable Development of the Mediterranean and the LOS Convention. In line with the amendment, the Protocols adopted in the 1980s were amended and new protocols were adopted 1990s. See Tullio Scovazzi, "Regional Co-operation in the Field of the Environment", in Tullio Scovazzi ed., *Marine Specially Protected Areas: The General Aspects and the Mediterranean Regional System*, Kluwer Law International (1999), pp. 82-99.

198. Peter H. Sand, *Marine Environmental Law in the United Nations Environment Programme*, 1988, pp. 37-.

199. Ministry of Foreign Affairs and Trade of Korea, "Current Situations of the Environmental Co-operation in North East Asia" (written in Korean), at <www. mofat.go. kr/main/top.htma>.

200. The geographical scope of the Action Plan is agreed as the maritime area between Lat. 33–52 N, and Lon. 121-143 E. in 1996: Ministry of Foreign Affairs and Trade of Korea, *idem.*

201. North Korea has taken a very limited part in the Action Plan; North Korea has participated in the experts meeting before the Action Plan was launched in 1994, but after had been launched, the North Korean government has not participated: Ministry of Foreign Affairs and Trade of Korea, *idem.*

202. UNEP, *Synergies*, Issue 3, September 2000, p. 3.

203. UNEP, *idem.*

204. UNEP, *idem.*

205. Hee Kwon Park, *The Law of the Sea and North East Asia*, Kluwer Law International (2000), p. 48.

206. Jorge E. Illueca, the Assistant Executive Director at the UNEP's Division of Environmental Conventions, wrote that: "If the level of co-operation and commitment continues, NOWPAP will find itself at the forefront of the world's regional seas conventions and action plans and a leader in the sustainable management of marine and coastal areas". UNEP, *Synergies*, Issue 3, September 2000, p. 3.

207. The Meeting of Senior Officials on Environmental Co-operation in North East Asia was initiated by Korea and its annual meetings have been held since 1993: Environment Congress for Asia and the Pacific (the so-called "Eco-Asia") was initiated by Japan and its annual meeting have been held since 1991. Ministry of Foreign Affairs and Trade of Korea, Current Situations of the Environmental Co-operation in North East Asia (written in Korean) at <www.mofat.go.kr/main/top.htma>.

208. Ministry of Foreign Affairs and Trade of Korea, *idem.*

209. For example, in the grey zone agreement of 1976 between Norway and the Soviet Union, they agreed to share almost equally the total allowable catch in the grey area between them. And flag-State jurisdiction has been applied *vis-à-vis* each other's fishing vessels. But fishing quotas can be given to third Parties out of their own respective quotas and where the quota is given to a third Party, then the Party can exercise enforcement jurisdiction over vessels of the third Party to which the Party gave its fishing permit. In the 1977 Protocol, Canada and the United States agreed to apply. However, they agreed that each of them shall exercise enforcement jurisdiction over any fishing vessels of third Parties in the North Pacific and the Bearing Sea.

210. See, for instance, H. Fox, eds., *Joint Development of Offshore Oil and Gas: A Model Agreement for States or Joint Development with Explanatory Commentary*, British Institute of International Law (1989), pp. 89-97.

Conclusion

There have been constant battles between the concepts of *mare liberum* and *mare clausum* throughout the history of the law of the sea.[1] However, since the second half of the 20th century *mare clausum* has been dominant and as a result of that the space of the high seas has significantly dwindled after World War II. Nowadays more than 100 States have claimed extended maritime zones such as 200 N.M. EEZ, fisheries zone and the continental shelf.[2]

Disputes are bound to arise in the overlapping claimed areas of the world, because the task of making maritime boundaries is highly time-consuming or even politically impossible. There are some factors which make the task of achieving delimitation of maritime boundaries elusive.[3] First of all, the task of delimitation involves a number of political decisions, which make policymakers very cautious in agreeing on maritime boundaries. Second, the provisions of the LOS Convention on delimitation of boundaries of the EEZ and the continental shelf, which simply contain references to international law and an equitable solution, are not of much help in solving delimitation disputes. Thirdly, the awareness of policymakers of the stability of maritime boundaries and of the possible future of maritime zones makes them very cautious in negotiations on maritime boundaries. Fourthly, maritime boundaries disputes in many cases are involved in sovereignty disputes over islands in disputed waters. Finally, there are loopholes in the dispute settlement procedures in the LOS Convention, through which Parties to maritime boundaries disputes can avoid the compulsory judicial procedures of the LOS Convention.

In a circumstance where overlapping claims are made but delimitation of the area of overlapping claims is not made, an obvious need arises to search for rules applicable between neighbouring coastal States pending ultimate maritime boundaries which might never take place.

Articles 74(3) and 83(3) of the LOS Convention are the only provisions which manifestly deal with the legal problems pending the delimitation of EEZ/continental shelf. They provide, in identical terms, that "pending agreement as provided for in paragraph 1, the States concerned, in a spirit of understanding and co-operation, shall make every effort to enter into provisional arrangements of a practical nature."

Even if there is a critical view on the use of Articles 74(3) and 83(3) of the LOS Convention,[4] these provisions are the result of long negotiations and contain substantial rules applicable to disputed areas.[5] This provision creates new rules which are in not accordance with provisions in the Geneva Convention. Note that there was great room for a State to justify a unilateral provisional line in the absence of agreement if it is a median line under Article 6 of the 1958 Geneva Convention on the continental shelf.[6] It provides that:

> . . . ***In the absence of agreement***, and unless another boundary line is justified by special circumstances, ***the boundary is the median line*** . . .

However, unlike under Article 6 of the Geneva Convention the justification for a unilateral median line cannot be found in the provisions of Articles 74(3) and 83(3) of the LOS Convention.[7] These provisions are useful in some respects, at least, by prohibiting a unilateral provisional line and obliging States concerned to make every effort to enter into mutual arrangements.[8] These provisions also provide for an obligation of mutual restraint not to jeopardise or hamper the reaching the final agreement on delimitation.[9]

Provisional arrangements share some common aspects with interim measures ordered by domestic or international courts pending their final judgments on the merits. Although the term "arrangement" is used frequently to indicate documents less formal than treaties, an arrangement can take a formal treaty form and can be a treaty under the Vienna Convention on the Law of Treaties.[10] Through such arrangements, coastal States can promote some exploitation activities in disputed areas, while preserving their positions on the final delimitation. With regard to the obligation of mutual restraint, it can be safely argued that there is a customary law rule. However, it does not necessarily follow that all activities are prohibited in disputed areas. It remains to be seen if or when a black list of prohibited activities in disputed areas can be completed through development of case law and State practice.[11]

There are a great number of instances where provisional arrangements are in place pending ultimate delimitation, even though Articles 74(3)/83(3) do not provide illustrative examples of such arrangements. They are joint development of oil and gas zones, joint fishing zones, comprehensive joint exploitation zones, adoption of provisional lines, and recognition of *de facto* maritime boundaries.[12] There are common characteristics in provisional arrangements. First, formal treaties have been usually utilised for the arrangements. Second, conflicting positions of neighbouring coastal States on delimitation of boundaries have been accommodated in shaping joint development zones or joint fishing zones. Third, a "without prejudice clause" has normally been utilised to preserve parties' positions on the ultimate delimitation of maritime boundaries. Finally, usually joint commissions have been set up to implement arrangements.[13]

It is notable that coastal States have taken great caution in shaping joint development zones or joint fishing zones and management regime therein, although these are provided for in provisional arrangements with a without prejudice clause. Coastal States appear to have been concerned about possible impacts of the provisional arrangements on the final delimitation. It is clear that provisional arrangements would not affect parties' positions on delimitation of ultimate boundaries when there is a clearly formulated without prejudice clause in the arrangements. It would be normal for negotiators of State A sitting at a table with their counterparts of State B for provisional arrangements to be very careful in shaping a joint fishing zone or a joint development zone because some of the negotiators from State A and State B would also sit at a table for negotiations on delimitation of boundaries and as a matter of fact, it would be hard, for them in delimitation negotiations to successfully argue for a position which is not in line with those put forward in negotiations on provisional arrangements.

A careful analysis shows that management regime in the joint fishing zone could affect the timing of the ultimate boundaries.[14] A coastal State might choose a white zone approach in the expectation that an early agreement on delimitation would take place, because when there is the effective management of fisheries through a grey zone approach then the urgent need for a final boundary would not be felt. By way of contrast, there would arise an urgent need for a final boundary where there is no effective management of fisheries resources.[15] For example, the Soviet Union concluded two fisheries agreements in the late 1970s: one is a grey zone agreement with Norway and the other is a white zone approach with Sweden. In 1988 the Soviet Union and Sweden reached an agreement on final delimitation, whereas an agreement on delimitation between the Soviet and Norway is still to be concluded.[16] Perhaps delimitation is unlikely to take place, as far as the effective resources management scheme is in place because there is no urgent need to delimit there as far as fisheries resources are concerned.

The disputes in North East Asia arose in a serious manner after the three countries ratified the LOS Convention in 1996, and then subsequently proclaimed their respective exclusive economic zones.[17] The adoption of straight baselines for measuring the breadth of the territorial seas by Japan and China in 1996 further complicated the legal disputes on the law of the sea in the region. Sovereignty issues over islands in the disputed waters are also factors that contribute to the complexity of the legal disputes in the region.

The littoral States in North East Asia took great caution in shaping joint fishing zones and management regimes therein. It took two and half years for Korea and Japan to agree on their provisional fisheries agreement, more than seven years for Korea and China and about four years for China and Japan. It might demonstrate that the coastal States in North East Asia are worried too much about possible impacts of the provisional arrangements on the delimitation of boundaries. However, it is clear that their provisional fisheries agreements negotiated and concluded under the provisions of Article 74(3) of the LOS Convention cannot, as a matter of law, affect their positions on delimitation, although the management regimes in the joint fishing zones might affect the timing of the conclusion of agreements on ultimate boundaries.

The provisional fisheries agreements in North East Asia solve some of the disputes between them on a provisional basis. However, it is also true that these agreements created conflicts between the three littoral States because some of the fishing zones overlap with each other and there have been protests from within the littoral States regarding some aspects of the agreements. Due to these conflicts and some defects in the fisheries agreement, it can be foreseen that the agreements in North East Asia would not be very instrumental for the effective management of the fisheries resources in the region. Does the fact the agreements are not effective for management of the fisheries resources in the region suggest that the fisheries agreements would further the negotiations on, and result in the conclusion of, boundaries agreements? It would be possibly so, but it remains to be seen when and how the littoral States in the region can overcome the issues of territorial sovereignty over small islands in the disputed waters and achieve the difficult task of drawing boundaries.

Lastly, it is advisable that the issues of the protection of the marine environment and marine scientific research in disputed waters should be dealt with in order to avoid jurisdictional conflicts and uncertainty.[18] It is to be recalled that there is a possibility of jurisdictional conflicts at sea between coastal States in disputed areas with regard to dumping, and pollution from ships when these are suspected of substantial discharge or major damage to the environment.[19] These issues can be dealt by provisional arrangements provided for in Articles 74(3)/83(3) of the LOS Convention, although co-operation on the marine environment and scientific research can take place regardless of whether there is a maritime boundary in place or not.[20] The co-operation among the littoral States in the region on the protection of the marine environment is in the initial stage and there is a need to strengthen multi-lateral co-operation in the region by developing regimes for dealing with specific types of pollution.[21]

NOTES

1. See "1. *Mare Clausum* in Dispute" in Chapter One.

2. *Ibid.*

3. See "2. Inherent Difficulties in Delimiting Boundaries of the EEZ/Continental Shelf" in Chapter 1.

4. See "1.3. Rules under the LOS Convention" in Chapter 2.

5. See "2. Drafting History of Article 74(3)/83(3) of the LOS Convention" in Chapter 2.

6. See "1.2. Rules under the Geneva Convention on the Continental Shelf" in Chapter 2.

7. See "1.3. Rules under the LOS Convention" in Chapter 2.

8. See "3.4.2 Obligations to Negotiate" in Chapter 2.

9. See "3.1.2. Provisional Arrangements by Mutual Consent" and "3.4.3. Obligation of Mutual Restraint" in Chapter 2.

10. See "3.2.1. Practice in Use of the Term Arrangements" in Chapter 2.

11. See "3.4.3. Obligation of Mutual Restraint" in Chapter 2.

12. See generally Chapter Three: Practice on Provisional Arrangements in Disputed Areas.

13. See "8. Observations" in Chapter 3.

14. See "3.1.1. The Soviet Union and Norway in the Barents Sea" in Chapter Three. Alex G. Oude Elferink, *The Law of Maritime Boundary Delimitation: A Case Study of the Russian Federation*, Martinus Nijhoff Publishers (1994), p. 201.

15. See "3.3. White Zones" in Chapter 3.

16. See 3.1.1. "The Soviet Union and Norway in the Barents Sea" and "3.3.1 Sweden and the Soviet Union in the Baltic Sea" in Chapter 3.

17. See generally Chapter Four.

18. See "4. Protection of the Marine Environment in Disputed Areas" and "5. Marine Scientific Research in Disputed Waters" in Chapter 2.

19. See "4. Protection of the Marine Environment in Disputed Areas" in Chapter 2.

20. See "6. Protection and Preservation of the Marine Environment in North East Asia" in Chapter 5.

21. See "6.2. Co-operation at the Multilateral Level on the Marine Environmental Issues" in Chapter 5.

APPENDIX

1. Agreement on Fisheries between the Republic of Korea and Japan of 1965

The Republic of Korea and Japan,

Desiring that the maximum sustained productivity of the fishery resources in the waters of their common interests be maintained,

Being convinced that the conservation of the said resources and their rational exploitation and development will serve the interests of both countries,

Confirming that that the principle of the freedom of the high seas shall be respected unless otherwise specifically provided in the present Agreement,

Recognising the desirability of eliminating causes of disputes which may arise from their geographical proximity and the intermingling of their fisheries, and

Desiring to co-operate mutually for the development of their fisheries,

Have agreed as follows:

Article I

1. The Contracting Parties mutually recognise that each Contracting Party has the right to establish within twelve nautical miles measured from its coastal baseline a sea zone in which it will have exclusive jurisdiction with respect to fisheries (hereinafter referred to as "fishery zone"). However, in case where either Contracting Party uses a straight baseline in establishing its fishery zone, such straight baseline will be determined upon consultation with the other Contracting Party.

2. The Contracting Party will not raise against each other any objection to the exclusion by either Contracting Party of the fishing vessels of the other Contracting Party from engaging in fishing operation in the fishery zone of either Contracting Party.

3. The overlapping part of the fishery zones of the Contracting Parties shall be divided into two by the straight lines joining the ends of the part with the midpoint of the straight line drawn across that area at its widest point.

Article II

The Contracting Parties establish a joint regulation zone enclosed by the lines described below (excluding any territorial seas and the Republic of Korea's fishery zone).

(a) Meridian 124° East Longitude north of 37° 30' North Latitude.
(b) Lines connecting the following points in order:
  (i) Intersection of 37° 30' North Latitude and 124° East Longitude
  (ii) Intersection of 36° 45' North Latitude and 124° 30' East Longitude
  (iii) Intersection of 33° 30' North Latitude and 124° 30' East Longitude
  (iv) Intersection of 32° 30' North Latitude and 126° East Longitude
  (v) Intersection of 32° 30' North Latitude and 127° East Longitude
  (vi) Intersection of 34° 34' 30" North Latitude and 129° 2' 50" East Longitude
  (vii) Intersection of 34° 44' 10" North Latitude and 129° 8' East Longitude
  (viii) Intersection of 34° 50' North Latitude and 129° 14' East Longitude
  (ix) Intersection of 35° 30' North Latitude and 130° East Longitude
  (x) Intersection of 37° 30' North Latitude and 131° 10' East Longitude
  (xi)
  (xii) High peak of Uamyong

Article III

The Contracting Parties shall implement in the joint regulation zone, until such time as conservation measures necessary for the maintenance of the maximum sustained productivity of fishery resources are implemented on the basis of sufficient scientific surveys, the provisional regulation measures for fisheries described in the Annex, which constitutes an integral part of the present Agreement with respect to drag-net fishing and seine fishing and to mackerel-angling fishing vessels of not less than 60 tons. (Tonnage is in gross tonnage and is indicated by deducting the tonnage allowed for improving living quarters of the vessels.)

Article IV

1. Enforcement (including halting and boarding of vessels) and jurisdiction in the waters outside the fishery zone shall be carried out and exercised only by the Contracting Party to which the fishing vessel belongs.

2. Either Contracting Party shall give and exercise pertinent guidance and supervision in order to ensure that its nationals and fishing vessels will faithfully observe the provisional regulation measures for fisheries, and shall carry out domestic measures, including appropriate penalties against violations thereof.

Article V

Joint resources survey zone shall be established outside the joint regulation zone. The extent of the said survey zone and the survey to be conducted within these zones shall be determined upon consultation between the two Contracting Parties on the basis of recommendations to be made by the Joint Fisheries Commission provided for in Article VI of the present Agreement.

Article VI

1. In order to realise the objectives of the present Agreement, the Contracting Parties shall establish and maintain the Korea-Japan Joint Fisheries Commission (hereinafter referred to as "the Commission").

2. The Commission shall be composed of two national sections, each consisting of three members appointed by the Governments of the respective Contracting Parties.

3. All regulations, recommendations, and other decisions of the Commission shall be made only by agreement between the national sections.

4. The Commission may decide upon and amend, as occasion may require, rules for the conduct of its meetings.

5. The Commission shall meet at least once each year and such other times as may be requested by either of the national sections. The date and place of the first meeting shall be determined by agreement between the Contracting Parties.

6. At its first meeting, the Commission shall select a chairman and a vice-chairman from different national sections. The chairman and the vice-

chairman shall hold office for a period of one year. Selection of the chairman and the vice-chairman from the national sections shall be made in such a manner as will provide in turn each Contracting Party with representation in these offices.

7. A standing secretariat shall be established under the Commission to carry out the business of the Commission.

8. The official languages of the Commission shall be Korean and Japanese. Proposal and data may be submitted in either official language, or if necessary, in English.

9. In the event that the Commission concluded that joint expenses are necessary, such expenses shall be paid by the Commission through contributions made by the Contracting Parties in the form and proportion recommended by the Commission and approved by the Contracting Parties.

10. The Commission may delegate the disbursement of funds for the joint expenses.

Article VII

1. The Commission shall perform the following functions:

(a) Recommend to the Contracting Parties with respect to scientific survey to be conducted for the purpose of study of the fishery resources in waters of their common interests and to the regulation measures to be taken within the joint regulation zone on the basis of the results of such survey and study;

(b) Recommend to the Contracting Parties with respect to the extent of the joint resource survey zone;

(c) Deliberate, when necessary, on the matters concerning the provisional regulation measures for fisheries and recommend to the Contracting Parties with respect to measures, including the revision of the provisional regulation measures, to be taken on the basis of the results of such deliberation;

(d) Deliberate on necessary matters concerning the safety and order of operation among the fishing vessels of the Contracting Parties and on general principles of measures for handling accidents at sea between the fishing vessels of the Contracting Parties, and recommend to the Contracting Parties with respect to measures to be taken on the basis of the results of such deliberation;

(e) Compile and study data, statistics and records to be provided by the Contracting Parties at the request of the Commission;

(f) Consider and recommend to the Contracting Parties with respect to the enactment of schedule of equivalent penalties for violations of the present Agreement;

(g) Submit annually to the Contracting Parties a report on the operations of the Commission;

(h) In addition to the foregoing, deliberate on various technical questions arising from the implementation of the present Agreement, and recommend, when deemed necessary, to the Contracting Parties with respect to measures to be taken.

2. The Commission, in order to perform its function, may, when necessary, establish subordinate organs composed of experts.

3. The Governments of the Contracting Parties shall respect to the extent possible the recommendations made by the Commission under the provisions of paragraph 1 above.

Article VIII

1. The Contracting Parties shall take measures deemed pertinent toward their respective nationals and fishing vessels in order to have them observe international practices concerning navigation, to ensure safety and maintain proper order in operation among the fishing vessels of the Contracting Parties and to seek smooth and speedy settlements of accidents at sea between the fishing vessels of the Contracting Parties.

2. For the purposes set forth in paragraph 1, the authorities concerned of the Contracting Parties shall, to the extent possible, maintain close contact and co-operate with each other.

Article IX

1. Any dispute between the Contracting Parties concerning the interpretation and implementation of the present Agreement shall be settled primarily through diplomatic channels.

2. Any dispute which fails to be settled by the provisions of paragraph 1 above shall be referred for decision to an arbitration board composed of three arbitrators, one to be appointed by the Government of each Contracting Party within a period of thirty days from the date of receipt by the Government

of either Contracting Party from the Government of the other of a note requesting arbitration of the dispute, and the third arbitrator to be agreed upon by the two arbitrators so chosen within a further period of thirty days or the third arbitrator to be appointed by the Government of a third country agreed upon within such further period by the two arbitrators, provided that such third arbitrator shall not be a national of either Contracting Party.

3. If, within the periods respectively referred to, the Government of either Contracting Party fails to appoint an arbitrator, or the third arbitrator or a third country is not agreed upon, the arbitration board shall be composed of the two arbitrators to be designated by each of the governments of the two countries respectively chosen by the Governments of the Contracting Parties within a period of thirty days and the third arbitrator to be nominated by the government of a third country to be determined upon consultation between the governments so chosen.

4. The Governments of the Contracting Parties shall abide by any award made by the arbitration board under the provisions of the present Article.

Article X

1. The present Agreement shall be ratified. The instrument of ratification shall be exchanged at Seoul as soon as possible. The present Agreement shall enter into force on the date of the exchange of the instruments of ratification.

2. The present Agreement shall continue in force for a period of five years and thereafter until one year from the day on which either Contracting Party shall give notice to other of an intention of terminating the present Agreement.

IN WITNESS WHEREOF the undersigned, being duly authorised thereto by their respective Governments, have signed the present Agreement.

DONE in duplicate at Tokyo, in the Korean and Japanese language, both equally authentic, this twenty-second day of June, 1965.

For the Republic of Korea:	For Japan:
(Signed) Tong Won Lee	(Signed) Dong Jo Kim
(Signed) Etsusaburo Shiina	(Signed) Shinichi Takasugi

Note that Annex, Exchange of Notes Concerning the Straight Baselines for the Fishery Zone of the Republic of Korea, Exchange of Notes Concerning the Fishing Zone of the Republic of Korea, Agreed Minutes Regarding the Agreement on Fisheries between the Republic of Korea and Japan, which were adopted in addition to the Fisheries Agreements of 1965 are not reproduced here.

2. Agreement on Fisheries between the Republic of Korea and Japan of 1999

The Republic of Korea and Japan,

Conscious of the importance of rational conservation, management and optimum utilisation of marine living resources,

Recalling the tradition of co-operative relations in the field of fisheries between the two countries maintained on the basis of the "Agreement on Fisheries between the Republic of Korea and Japan" signed at Tokyo on 22 June 1976,

Bearing in mind that the two countries are States Parties to the United Nations Convention on the Law of the Sea (hereinafter referred to as "the LOS Convention"), and

Desiring to establish a new fisheries order and further the co-operative relations in the field of fisheries between the two countries on the basis of the LOS Convention,

Have agreed as follows:

Article 1

This Agreement applies to the exclusive economic zone of the Republic of Korea and the exclusive economic zone of Japan (hereinafter referred to as "the Agreement Zone").

Article 2

Each Contracting Party, on the basis of the principle of reciprocity, shall permit nationals and fishing vessels of the other Contracting Party to harvest in its exclusive economic zone, pursuant to this Agreement and its relevant laws and regulations.

Article 3

1. Each Contracting party shall annually determine species allowed for harvesting, quotas of catch, areas of fishing and other specific conditions in its exclusive economic zone for nationals and fishing vessels of the other Contracting Party and notify in writing the other Contracting Party of such determination.

2. Each Contracting Party, in making such determination as provided for in paragraph 1, shall respect the outcome of the consultation of the Korea-Japan Joint Fisheries Committee to be established under the provisions of Article 12, and shall take into account the state of marine living resources in its exclusive economic zone, its harvesting capacity, circumstances of mutual access and other relevant factors.

Article 4

1. After receiving a written notification from the other Contracting Party on the determination as provided for Article 3, the competent authorities of each Contracting party shall apply to the competent authorities of the other Contracting Party for the issuance of permit for its nationals and fishing vessels wishing to harvest in the economic zone of the other Contracting Party. The competent authorities of the other Contracting Party shall issue the permits in accordance with this Agreement and its relevant laws and regulations on fisheries.

2. The fishing vessels which obtain such permits shall display the permits in an easily visible place in the wheelhouse and clearly show identification markings while fishing.

3. The competent authorities of each Contracting Party shall notify in writing the competent authorities of the other Contracting Party of procedural rules including those on application and issuance of permit, reporting of actual harvest, identification of markings of fishing vessels and recording of fishing logs.

4. The competent authorities of each Contracting Party may collect fishing fees and reasonable permit issuance fee.

Article 5

1. Nationals and fishing vessels of each Contracting Party shall comply with this Agreement and the relevant laws and regulations concerning

the fisheries of the other Contracting Party, when harvesting in the exclusive economic zone of the other Contracting Party.

2. Each Contracting Party shall take necessary measures to ensure that its nationals and fishing vessels harvesting in the economic zone of the other Contracting Party, comply with specific conditions on fishing in the exclusive economic zone of the other Contracting Party, as determined by the other Contracting Party, in accordance with the provisions of Article 3 and the provisions of this Agreement. Such measures shall not include boarding, stopping and other enforcement measures against its nationals and fishing vessels in the exclusive economic zone of the other Contracting party.

Article 6

1. Each Contracting Party may take necessary measures in accordance with international law in its exclusive economic zone to ensure that nationals and fishing vessels of the other Contracting Party harvesting in its exclusive economic zone comply with the provisions of this Agreement and specific conditions on fishing in its exclusive economic zone as determined by it in accordance with the provisions of Article 3.

2. In cases of the arrest or detention of fishing vessels and their crews of the other Contracting Party as a measure under paragraph 1, the competent authorities of each Contracting Party shall promptly inform, through diplomatic channels, the other Contracting Party of the measures taken and the penalties imposed thereafter.

3. Arrested or detained fishing vessels and their crews shall be promptly released upon the posting of adequate bond or the submission of documents guaranteeing such posting.

4. Each Contracting Party shall notify without delay the other Contracting Party of the conservation measures for marine living resources and other conditions, stipulated in its laws and regulations on fishing.

Article 7

1. Each Contracting Party shall exercise sovereign rights on fisheries in its coastal part of the Agreement Zone up to the straight lines connectingthe following points in the sequence given below, and shall deem such part to be its exclusive economic zone in applying the provisions of Articles 2 to 6.

a. point at Lat. 32° 57.0' N. Long. 127° 41.1' E
b. point at Lat. 32° 57.5' N. Long. 127° 41.9' E
c. point at Lat. 33° 01.3' N. Long. 127° 44.0' E
d. point at Lat. 33° 08.7' N. Long. 127° 48.3' E
e. point at Lat. 33° 13.7' N. Long. 127° 51.6' E
f. point at Lat. 33° 16.2' N. Long. 127° 52.3' E
g. point at Lat. 33° 45.1' N. Long. 128° 21.7' E
h. point at Lat. 33° 47.4' N. Long. 128° 25.5' E
i. point at Lat. 33° 50.4' N. Long. 128° 26.1' E
j. point at Lat. 34° 08.2' N. Long. 128° 41.3' E
k. point at Lat. 34° 13.0' N. Long. 128° 47.6' E
l. point at Lat. 34° 18.0' N. Long. 128° 52.8' E
m. point at Lat. 34° 18.5' N. Long. 128° 53.3' E
n. point at Lat. 34° 24.5' N. Long. 128° 57.3' E
o. point at Lat. 34° 27.6' N. Long. 128° 59.4' E
p. point at Lat. 34° 29.2' N. Long. 129° 00.2' E
q. point at Lat. 34° 32.1' N. Long. 129° 00.8' E
r. point at Lat. 34° 32.6' N. Long. 129° 00.8' E
s. point at Lat. 34° 40.3' N. Long. 129° 03.1' E
t. point at Lat. 34° 49.7' N. Long. 129° 12.1' E
u. point at Lat. 34° 50.6' N. Long. 129° 13.0' E
v. point at Lat. 34° 52.4' N. Long. 129° 15.8' E
w. point at Lat. 34° 54.3' N. Long. 129° 18.4' E
x. point at Lat. 34° 57.0' N. Long. 129° 21.7' E
y. point at Lat. 34° 57.6' N. Long. 129° 22.6' E
z. point at Lat. 34° 58.6' N. Long. 129° 25.3' E
aa. point at Lat. 35° 01.2' N. Long. 129° 32.9' E
bb. point at Lat. 35° 04.1' N. Long. 129° 40.7' E
cc. point at Lat. 35° 06.8' N. Long. 130° 07.5' E
dd. point at Lat. 35° 07.' N. Long. 130° 16.4' E
ee. point at Lat. 35° 18.2' N. Long. 130° 23.3' E
ff. point at Lat. 35° 33.7' N. Long. 130° 34.1' E
gg. point at Lat. 35° 42.3' N. Long. 130° 42.7' E
hh. point at Lat. 36° 03.8' N. Long. 131° 08.3' E
ii. point at Lat. 36° 03.8' N. Long. 131° 15.9' E

2. Each Contracting Party shall not exercise sovereign rights on fisheries in the part at the other Contracting Party's side of the Agreement Zone joining the lines provided for in paragraph 1, and shall deem such part to be the exclusive economic zone of the other Contracting Party in applying the provisions of Articles 2 to 6.

Article 8

The provisions of Articles 2 to 6 shall not apply to the following zones:

a. the zone provided for in Article 9, Paragraph 1.

b. the zone provided for in Article 9, paragraph 2.

Article 9

1. The provisions of Annex I. paragraph 2 shall apply to the area surrounded by the line formed by straight lines joining the points of the following sub-paragraphs in sequence:

a. point at Lat. 36° 10.0' N. Long. 131° 15.9' E
b. point at Lat. 35° 33.75' N. Long. 131° 46.5' E
c. point at Lat. 35° 59.5' N. Long. 132° 13.7' E
d. point at Lat. 36° 18.5' N. Long. 132° 13.7' E
e. point at Lat. 36° 56.2' N. Long. 132° 55.8' E
f. point at Lat. 36° 56.2' N. Long. 135° 30.0' E
g. point at Lat. 38° 37.0' N. Long. 135° 30.0' E
h. point at Lat. 39° 51.75' N. Long. 134° 11.5' E
i. point at Lat. 38° 37.0' N. Long. 132° 59.8' E
j. point at Lat. 38° 37.0' N. Long. 131° 40.0' E
k. point at Lat. 37° 25.5' N. Long. 131° 40.0' E
l. point at Lat. 37° 08.0' N. Long. 131° 34.0' E
m. point at Lat. 36° 52.0' N. Long. 131° 10.0' E
n. point at Lat. 36° 52.0' N. Long. 130° 22.5' E
o. point at Lat. 36° 10.0' N. Long. 130° 22.5' E
p. point at Lat. 36° 10.0' N. Long. 127° 15.9' E

2. The provisions of Annex I, paragraph 3 shall apply to the area north of the southernmost parallel of the exclusive economic zone of the Republic of Korea, in the zone surrounded by lines of the following sub-paragraphs:

a. the straight line joining the point at Lat. 32° 57.0' N, Log. 127° 41.1' E and the point at Lat. 32° 34.0' N. Log. 127° 9.0 'E

b. the straight line joining the point at Lat. 32° 34.0' N, Log. 127° 9.0' E and the point at Lat. 31° 0.0 'N. Log. 125° 51.5' E

c. the straight line running from the point at Lat. 31° 0.0' N, Log. 125° 51.5' E and passing through the point at Lat. 30° 56.0'N., Log. 125° 52.0'E

d. the straight line joining the point at Lat. 32° 57.0' N Log. 127° 41.1' E and the point at Lat. 31° 20.0' N., Log. 127° 13.0' E

e. the straight line running from the point at Lat. 31° 20.0' N, Log. 127° 13.0' E and passing through the point at Lat. 31° 0.0' N., Log. 127° 5.0' E

Article 10

The Contracting Party shall co-operate with each other for rational conservation and management, and optimum utilisation of marine living resources in the Agreement Zone. Such co-operation shall include exchange of relevant statistical information on marine living resources and fishing industry data.

Article 11

1. The Contracting Parties shall take appropriate measures to ensure that their respective nationals and fishing vessels comply with international rules and regulations on navigation, maintain the safety and order of fishing operations among their fishing vessels and smoothly and promptly settle accidents at sea between fishing vessels of the Contracting Parties.

2. For the purpose enumerated in paragraph 1, the competent authorities of the Contracting Parties shall communicate with each other as closely as possible.

Article 12

1. The Contracting Parties shall establish the Korea-Japan Fisheries Committee (hereinafter referred to as "the Committee") to effectively achieve the objectives of this Agreement.

2. The Committee shall be composed of one Representative and one Member appointed respectively by the Governments of the Contracting Parties and, when necessary, may establish subsidiary bodies composed of specialists.

3. The Committee shall meet once each year in the two countries alternately and may hold *ad hoc* meetings upon agreement between the Contracting Parties. If a subsidiary body under paragraph 2 is established, such subsidiary body may hold meetings at any time upon agreement between the Government Representatives of the Contracting Parties of the Committee.

4. The Committee shall consult on the following matters and recommend the outcome of the consultations to the Contracting Parties. The Contracting Parties shall respect the recommendations of the Committee:

a. matters relating to specific conditions on fishing as stipulated under Article 3;

b. matters relating to maintenance of fishing order;

c. matters relating to the state of marine living resources;

d. matters relating to co-operation between the two countries in the field of fisheries;

e. matters relating to the conservation and management of marine living resources in the zone provided for in Article 9, paragraph 1; and

f. other matters relating to the implementation of this Agreement.

5. The Committee shall consult and decide on matters related to conservation and management of marine living resources in the zone provided in Article 9, paragraph 2.

6. All recommendations and decisions of the Committee shall be made only through agreement between the Representatives of the Contracting Parties.

Article 13

1. Any dispute between the Contracting Parties concerning the interpretation or application of this Agreement shall be settled through negotiations.

2. If a dispute referred to in paragraph 1 is not settled through negotiations, such dispute shall, upon agreement between the Contracting Parties, be settled through the following procedures:

a. When the Government of either Contracting Party, upon receipt of an official document specifying the cause of a dispute and requesting arbitration thereof from the Government of the other Contracting Party, notifies the Government of the other Contracting Party of the acceptance of such request, the dispute shall be submitted for a decision to an committee composed of three arbitrators: one arbitrator appointed by the Government of each Contracting Party within a 30-day period from the receipt of such notification, and the third arbitrator either to be agreed upon within 30 days following the 30-day period by the two arbitrators appointed, or to be nominated by the Government of a third country agreed upon by the two arbitrators with 30 days following the 30-day period. The third arbitrator shall not be a national of either Contractor.

b. If the Government of either Contracting Party fails to appoint an arbitrator within the period specified in sub-paragraph a. or the third arbitrator or third country is not agreed upon within the period specified in sub-paragraph a., the Committee shall be composed of two arbitrators each nominated by the Government of a country selected by the Governments of each Contracting Party within 30 days following the periods specified in sub-paragraph a., and the third arbitrator to be nominated by third country decided upon through consultations between the two Governments so selected.

c. Each Contracting Party shall bear the expenses in relation to the arbitrator appointed by it or the arbitrator nominated by the Government of the country selected by it, and the expenses necessary for the participation of its Government in the arbitration. The expenses for the performance of duties of the third arbitrator shall be equally borne by the Contracting Parties.

d. The Governments of the Contracting Parties shall abide by the majority decision of the arbitral committee in accordance with the provisions of this Article.

Article 14

Annex I and Annex II to this Agreement shall form an integral part of this Agreement.

Article 15

Nothing in this Agreement shall be deemed to prejudice the position of each Contracting Party relating to issues in international law other than matters on fisheries.

Article 16

1. This Agreement shall be ratified. The instruments of ratification shall be exchanged at Seoul as soon as possible. This Agreement shall enter into force from the date of the exchange of the instruments of ratification.

2. This Agreement shall remain in force for a period of three years from the date of its entry into force. Thereafter, either Contracting Party may notify, in writing, the other Contracting Party of its intention to terminate this Agreement, which shall terminate six months after the date of such notification. This Agreement shall continue to remain in force, unless so terminated.

Article 17

The Agreement on Fisheries between the Republic of Korea and Japan signed at Tokyo, 22 June 1965 shall lose its effect on the date of entry into force of this Agreement.

IN WITNESS WHEREOF, the undersigned, being duly authorised thereto by their respective Governments, have signed this Agreement.

DONE in duplicate at Kagoshima, this 28th day of November 1998, in the Korean and Japanese languages, both texts being equally authentic.

FOR THE REPUBLIC OF KOREA FOR JAPAN

ANNEX I

1. The Contracting Parties shall continue to negotiate in good faith for an early delimitation of the exclusive economic zone.

2. The Contracting Parties shall co-operate in accordance with the provisions of the following sub-paragraphs to ensure that the maintenance of the marine living resources in the zone provided for in Article 9, paragraph 1 of this Agreement is not endangered from over-exploitation.

a. Each Contracting Party shall not apply its relevant laws and regulations on fisheries to nationals and fishing vessels of the other Contracting Party in the zone.

b. Each Contracting Party shall respect the recommendations made following the outcome of consultations of the Korea-Japan Fisheries Committee (hereinafter referred to as the "Committee") established under the provisions of Article 12 of this Agreement, and take such necessary measures for its nationals and fishing vessels as the conservation of marine living resources and appropriate management thereof, including a ceiling on the number of fishing vessels per mode of fishing in the zone.

c. Each Contracting Party shall notify the other Contracting Party of the measures which it implements on its nationals and fishing vessels in the zone. The Contracting Parties shall take such notification into full account in having their Representatives of the Committee participate in consultations for recommendations under sub-paragraph b.

d. Each Contracting Party shall provide the other Contracting Party with information on catches per species and modes of fishing by its nationals and fishing vessels harvesting in the zone, and other relevant information.

e. If a Contracting Party discovers that nationals and fishing vessels of the other Contracting Party are violating the measures implemented in the zone by the other Contracting Party in accordance with the provisions of sub-paragraph b, it may inform the other Contracting Party of the fact and relevant circumstances. The other Contracting Party, in regulating its nationals and fishing vessels, shall, after having verified the fact relating to such notification and taken necessary measures, notify the results thereof to the other Contracting Party.

3. The Contracting Parties shall co-operate in accordance with the provisions of the following sub-paragraphs to ensure that the maintenance of marine living resources in the zone provided for in Article 9, paragraph 2 of this Agreement is not endangered by over-exploitation.

a. Each Contracting Party shall not apply its relevant laws and regulations on fisheries to nationals and fishing vessels of the other Contracting Party in the zone.

b. Each Contracting Party shall, in accordance with the decisions of the Committee, take such necessary measures for its nationals and fishing vessels as the conservation of marine living resources and appropriate management thereof, including a ceiling on number of fishing vessels per mode of fishing in the zone.

c. Each Contracting Party shall notify the other Contracting Party of the measures which it implements on its nationals and fishing vessels in the zone. The Contracting Parties shall take such notification into full account in having their Representatives of the Committee participate in consultations for recommendations under sub-paragraph b.

d. Each Contracting Party shall provide the other Contracting Party with information on catches per species and modes of fishing by its na-tionals and fishing vessels harvesting in the zone, and other relevant information.

e. If a Contracting Party discovers that nationals and fishing vessels of the other Contracting Party are violating the measures implemented in the zone by the other Contracting Party in accordance with the provisions of sub-paragraph b, it may inform the other Contracting Party of the fact and relevant circumstances. The other Contracting Party, in regulating its nationals and fishing vessels, shall, after having verified the fact relating to such notification and taken necessary measures, notify the results thereof to the other Contracting Party.

Annex II

1. Each Contracting Party shall exercise sovereign rights on fisheries in the part of the Agreement Zone situated on its side from the line provided for in Article 9, paragraphs 1 and 2, and shall deem such part to be its exclusive economic zone in applying the provisions of Articles 2 to 6 of this Agreement.

2. Each Contracting Party shall not exercise sovereign rights on fisheries in the part of the Agreement Zone of the other Contracting Party from the zones provided for in Article 9, paragraphs 1 and 2 and shall deem such

part to be the exclusive economic zone of the other Contracting Party in applying the provisions of Article 2 to 6 of this Agreement.

3. The provisions of paragraphs 1 and 2 shall not be applied to a certain part of the Agreement Zone situated to the north west of the line formed by the straight lines joining the following points in sequence. Each Contracting Party shall not apply its relevant laws and regulations to nationals and fishing vessels of the other Contracting Party in such part.

a. point at Lat. 38° 37.0' N, Long. 131° 40.0' E.
b. point at Lat. 38° 37.0' N, Long. 132° 59.8' E.
c. point at Lat. 39° 51.75' N, Long. 134° 11.5' E.

AGREED MINUTE

The representative of the Government of the Republic of Korea and Japan have agreed to record the following matters relating to the relevant provisions of the Agreement on Fisheries between the Republic of Korea and Japan (hereinafter referred to as the "Agreement") signed today:

1. The Governments of both countries shall closely co-operate to maintain good fisheries order in the East China Sea.

2. The Government of the Republic of Korea in relation to the establishment of the zone provided for in Article 9, Paragraph 2 of the Agreement, has an intention to co-operate with the Government of Japan in order not to damage fisheries relations established by Japan and a third country in a certain area of the East China Sea. However, this shall not be deemed to prejudice the position of the Republic of Korea on fisheries agreements concluded by Japan with such third country.

3. The Government of Japan in relation to the establishment of the zone provided for in Article 9, paragraph 2 of the Agreement, has an intention to co-operate with the Government of such a third country in order to make possible certain fishing activities by nationals and fishing vessels of the Republic of Korea in other certain area of the East China Sea under the fisheries relations established by Japan and the third country.

4. Governments of both countries have the intention to negotiate specific ways to maintain a good fisheries order in the East China Sea on the basis of fisheries agreements concluded or to be concluded by either Party with

a third country and through the Korea-Japan Joint Fisheries Committee which is to be established under Article 12 of the Agreement and committees of such a kind which are to be established under fisheries agreements with such third country.

Kagoshima, 28 November 1998

FOR THE REPUBLIC OF KOREA FOR JAPAN

3. Agreement on Fisheries between the People's Republic of China and Japan

The Government of the People's Republic of China and the Government of Japan,

Recalling the Joint Communique of the Government of the People's Republic of China and the Government of Japan issued on 29 September 1972,

Considering the traditional co-operative relationship in the field of fisheries including the relations based on the Agreement on Fisheries between the People's Republic of China and Japan signed on 15 August 1975,

In order to establish a new fisheries order between the two countries in accordance with the spirit of the United Nations Convention on the Law of the Sea made on 10 December 1982, conserve and rationally utilise marine living resources of common concerns, and maintain a proper order of maritime fishing operations,

After friendly consultations,

Have agreed as follows:

Article 1

The area to which this Agreement is applied (hereinafter referred to as "the Agreement Zone") shall consist of the exclusive economic zones of both the People's Republic of China and Japan.

Article 2

1. Each Contracting Party shall, based on the principle of mutual interest, in accordance with this Agreement and the respective relevant laws and

regulations, permit the nationals and fishing vessels of the other Contracting Party to conduct fishing operations within its own exclusive economic zone.

2. The competent authorities of each Contracting Party, under Annex I of this Agreement, shall issue permits to the nationals and fishing vessels of the other Contracting Party to authorise them to conduct fishing operations. The said competent authorities may collect reasonable fees for issuing permits.

3. The nationals and fishing vessels of each Contracting Party shall conduct fishing operations within the exclusive economic zone of the other Contracting Party in accordance with this Agreement and the relevant laws and regulations of the other Contracting Party.

Article 3

Each Contracting Party shall determine each year, taking into account the situation of resources in its exclusive economic zone, its capacity to harvest, the traditional fishing activities, the implementation of mutual permission of fishing and other related factors, such operational conditions for the nationals and fishing vessels of the other Contracting Party fishing within its exclusive economic zone as the species which may be caught, quotas of catch and the area in which the fishing operation is permitted. This determination shall be made with respect for the result of the consultations in China-Japan Joint Fisheries Committee, which is established under Article 11 of this Agreement.

Article 4

1. Each Contracting Party shall take necessary measures to ensure that its nationals and fishing vessels conducting fishing operations within the exclusive economic zone of the other Contracting Party, comply with the provisions of this Agreement, the conservation measures of marine living resources and other terms and conditions established in the relevant laws and regulations of the other Contracting Party.

2. Each Contracting Party shall notify without delay the other Contracting Party of the conservation measures of marine living resources and other terms and conditions established in its relevant laws and regulations.

Article 5

1. Each Contracting Party may, subject to international law, take necessary measures within its exclusive economic zone to ensure that nationals

and fishing vessels of the other Contracting Party comply with the conservation measures of marine living resources and other terms and conditions established in its relevant laws and regulations.

2. Arrested or detained vessels and their crews shall be promptly released upon the posting of appropriate bond or other security.

3. In case of arrest or detention of the vessels of the other Contracting Party of the conservation measures of marine living resources and other terms and conditions established in its relevant laws and regulations.

Article 6

The provisions from Article 2 to Article 5 shall be applied to the Agreement Zone except the areas referred to in the following sub-paragraphs of (a) and (b):

(a) the area prescribed in paragraph 1 of Article 7; and

(b) the part of the Agreement Zone in the East China Sea below Lat. 27 N. and the part of the Agreement Zone to the west of Long. 125 E. below the East China Sea (excluding the exclusive economic zone of the People's Republic of China in the South China Sea)

Article 7

1. Paragraphs 2 and 3 of this Article shall be applied to the Area (hereinafter referred to as "the Provisional Measures Zone") surrounded by the line joining the following co-ordinates in sequence.

(a) point at Lat. 30.40° N. Long. 124° 10.1' E

(b) point at Lat. 30° N. Long. 123° 56.4' E

(c) point at Lat. 29° N. Long. 123° 25.5' E

(d) point at Lat. 28° N. Long. 122° 47.9' E

(e) point at Lat. 27° N. Long. 121° 57.4' E

(f) point at Lat. 27° N. Long. 125° 58.3' E

(g) point at Lat. 28° N. Long. 127° 15.1' E

(h) point at Lat. 29° N. Long. 128° 0.9' E

(i) point at Lat. 30° 37.0' N. Long. 128° 32.2' E

(j) point at Lat. 30° 40' N. Long. 128° 26.1' E

(k) point at Lat. 30° 40' N. Long. 124°10.1' E

2. Both Contracting Parties, subject to the determination of the China-Japan Joint Fisheries Committee established under Article 11 of this Agreement, taking into account the effect on the traditional fishing of each

Contracting Party, shall take proper conservation measures and quantitative management measures in the Provisional Measures Zone in order to ensure that the maintenance of marine living resources in not endangered by over-exploitation.

3. Each Contracting Party shall apply the enforcement measures and other necessary measures to its nationals and vessels fishing in the Provisional Measures Zone. Each Contracting Party may not apply the enforcement measures and other necessary measures to the nationals and fishing vessels of the other Contracting Party fishing in the Provisional Measures Zone. Nevertheless, either Contracting Party may, in case of finding an offence of the nationals and fishing vessels of the other Contracting Party against the operational regulations determined by the China-Japan Joint Fisheries Committee established under Article 11 of this Agreement, call attention of the said nationals and fishing vessels to that fact, and notify the other Contracting Party of the fact with its related situation. The other Contracting Party shall take necessary measures with respect for that notification, and subsequently notify the former of the result.

Article 8

Each Contracting Party shall apply necessary measures such as instructions and others to its nationals and fishing vessels in order to secure safe navigation and fishing operations, maintain a proper order of fishing operations, and deal with maritime accidents smoothly and promptly.

Article 9

1. In cases where any national or fishing vessel of either Contracting Party runs across maritime casualties or other emergency situations on the coast of the other Contracting Party, the other Contracting Party shall provide assistance and protection to the extent possible, and notify the authorities concerned of the former Contracting Party of the related situations.

2. Any national or fishing vessel of either Contracting Party, in cases where they need to take refuge due to rough weather or other emergency situations, may take refuge in a harbour or other places of of the other Contracting Party after they give notification to the authorities concerned of the other Contracting Party in accordance with the provisions of Annex II to this Agreement.

Article 10

Both Contracting Parties shall co-operate for scientific research on fisheries and conservation of marine living resources.

Article 11

1. Both Contracting Parties shall, for the purpose of achieving the objectives of this Agreement, establish the China-Japan Fisheries Committee (hereinafter referred to as "the Fisheries Committee"). The Fisheries Committee shall consist of four members, two of whom will be appointed respectively by the Government of each Contracting Party.

2. The Fisheries Committee shall discharge the following duties:

(1) The Fisheries Committee will consult on the matters related to the provisions of Article 3 and the area referred to in sub-paragraph (b) of Article 6, and make recommendations to the Government of each Contracting Party. Subjects of consultation may include, *inter alia*:

(a) the species which may be caught, quotas of catch and other conditions of the fishing operation of the nationals and fishing vessels of the other Contracting Party prescribed in the provisions of Article 3;

(b) the maintenance of an order of fishing operations;

(c) the circumstances and conservation of marine living resources; and

(d) co-operation on fishing between the Contracting Parties.

(2) The Fisheries Committee will consult on and determine matters related to the provisions of Article 7.

(3) The Fisheries Committee will, as necessary, make recommendations to the Governments of the Contracting Parties on the revision of the Annex to this Agreement.

(4) The Fisheries Committee will review the implementations of this Agreement and other matters related to this Agreement.

3. All the recommendations and determinations of the Fisheries Committee shall be reached only through the agreement of the members of both Contracting Parties.

4. The Governments of Contracting Parties shall, with respect for the recommendations referred to in sub-paragraph (1) of paragraph 2 of this Article and subject to determinations referred to in sub-paragraph (2) of paragraph 2 of this Article, take necessary measures.

5. The Fisheries Committee shall meet once a year alternately in the People's Republic of China and Japan. The Fisheries Committee may, as necessary, meet on *ad hoc* basis upon agreement of the Contracting Parties.

Article 12

Nothing in this Agreement shall be deemed to prejudice the position of either Contracting Party in regard to any question on the law of the Sea.

Article 13

1. The Annex, including the revised one in accordance with paragraph 2 of this Article, forms an integral part of this Agreement.

2. The Governments of both Contracting Parties may revise the Annex to this Agreement by written agreement.

Article 14

1. This Agreement shall enter into force on a date to be agreed upon by an exchange of notes between the Governments of the Contracting Parties, following the completion of their procedures required by respective internal laws and regulations for the entry intro force of this Agreement. This Agreement shall remain in force for a period of five years, and shall continue in force thereafter subject to termination in accordance with paragraph 2 of this Article.

2. Either Contracting Party may terminate this Agreement at any time on or after the date of expiration of an initial period of five years by giving six months' written notice to the other Contracting Party.

3. The Agreement on Fisheries between the People's Republic of China and Japan signed on 15 August 1975 shall expire on the date of entry into force of this Agreement.

DONE in duplicate, in the Japanese and Chinese languages, both texts being equally authentic, at Tokyo, this eleventh day of November 1997.

For the Government of the People's Republic of China

For the Government of Japan

Annex I

Each Contracting Party shall, under paragraph 2 of Article 2 of this Agreement, take the following measures concerning permits:

1. The competent authorities of each Contracting Party shall, after receiving from the competent authorities of the other Contracting Party the written notice on the determination prescribed in the provisions of Article 3 of this Agreement, submit applications for permits to the competent authorities of the other Contracting Party for its nationals and fishing vessels that wish to conduct fishing operations in the latter's exclusive economic zone. The competent authorities of the other Contracting Party shall issues permits pursuant to this Agreement and its relevant laws and regulations.

2. The competent authorities of each Contracting Party shall give written notice to the competent authorities of the other Contracting Party on the procedural rules on fishing, including application for and issuance of permits, submission of information on fishing, signs of fishing vessels and other procedural rules on fishing operations diaries.

3. The fishing vessels permitted shall post the permit in a place easily accessible in the wheelhouse, and prominently display the sign assigned by the other Contracting Party.

Annex II

The implementation of paragraph 2 of Article 9 of this Agreement shall be carried out in the manner stipulated as follows:

1. The contact point designated by the Government of the People's Republic of China shall be the Harbour Management Authorities, which exercises jurisdiction over the harbours concerned. The contact point designated by the Government of Japan shall be each Regional Maritime Safety Headquarters, which exercise jurisdiction over harbours for refuge and other places.

2. The detailed means of communication will be mutually informed at the China-Japan Joint Fisheries Committee.

3. The contents which either Contracting Party shall inform the contact point designated by the other Contracting Party shall be set out below:

the name of the fishing vessel, the identification signal, the current position (latitude, longitude), the harbour where the fishing vessel is registered, the tonnage and the length, the name of the captain, the number of crews, the destination of refuge, the reason for refuge, the scheduled arrival time and the means of communication.

AGREED MINUTE

The representatives of the Government of the Peoples' Republic of China and of the Government of Japan, in relation to the relevant provisions of the Agreement on Fisheries between the Peoples' Republic of China and Japan (hereinafter referred to as "the Agreement") agreed to record the following:

1. Both Governments expressed their intentions to continue the consultations on the delimitation of the exclusive economic zones and the continental shelves of both countries, and make efforts so that a mutually acceptable agreement be reached. Both Governments also expressed their intentions that the establishment of the Provisional Measures Zone prescribed in the provisions of paragraph 1 of Article 7 should not be deemed to prejudice the positions of both Parties in regard to the delimitation of the exclusive economic zones and the continental shelves.

2. Both Governments expressed their intentions that, given the traditional and co-operative fishing relations between the two countries, in implementing this Agreement and establishing fishing relations with a third country, both Parties should respect the current fishing operations in a certain area of the East China Sea in the north of the area prescribed in the provisions of paragraph 1 of Article 7, pay due consideration to the traditional fishing of the other country and the resource situation of the said area, and not unduly damage the interest of fishing of the other country in the said area.

11 November 1997 in Tokyo

For the Government of the People's Republic of China	For the Government of Japan

EXCHANGE OF LETTERS concerning the area referred to in sub-paragraph (b) of Article 6 of the Agreement on Fisheries between the People's Republic of China and Japan

(Chinese Letter)

I have the honour to refer to the Agreement on Fisheries between the People's Republic of China and Japan, which was signed today, and to state as follows:

> The Government of the People's Republic of China has the intention not to apply its relevant laws and regulations on fisheries to Japanese nationals on the condition that both the People's Republic of China and Japan are in co-operative relations in order to ensure that the maintenance of marine living resources is not endangered by over-exploitation.

I avail myself of this opportunity to extend to Your Excellency the assurances of my highest consideration.

11 November 1977 in Tokyo

Ambassador Extraordinary and Plenipotentiary of the People's Republic of China

(Japanese Letter)

I have the honour to refer to the Agreement on Fisheries between Japan and the People's Republic of China, which was signed today, and to state as follows:

> The Government of Japan has the intention not to apply its relevant laws and regulations on fisheries to Chinese nationals on the condition that both Japan and the People's Republic of China are in co-operative relations in order to ensure that the maintenance of marine living resources is not endangered by over-exploitation.

I avail myself of this opportunity to extend to Your Excellency the assurances of my highest consideration.

11 November 1977 in Tokyo
Minister for Foreign Affairs of Japan

JAPANESE LETTER concerning Chinese Fishing for Squid

I have the honour to refer to the Agreement on Fisheries between Japan and the People's Republic of China, which was signed today, and to state as follows:

1. The Government of the People's Republic of China requested, under Article 3 of the said Agreement, that Chinese fishing vessels for squid be permitted to carry out fishing operations, without being charged any fees, for five years after the entry into force of the Agreement in the exclusive economic zone of Japan in the Japan Sea (East Sea) and the Pacific Ocean, and that the number of fishing vessels and the amount of catch will not exceed the actual number of the year 1996.

2. The Government of Japan has the intention to bear in mind such request by the Chinese side, accept it in principle and, taking into account the situation of fishery resources in the said area, consult on and determine the concrete measures of implementation in the China-Japan Fisheries Committee established under Article 11 of the Agreement.

I avail myself of this opportunity to extend to Your Excellency the assurances of my highest consideration.

11 November 1977 in Tokyo

Minister for Foreign Affairs of Japan

Ambassador Extraordinary and Plenipotentiary
of the People's Republic of China

4. Agreement on Fisheries between the Republic of Korea and the People's Republic of China

The Government of Korea and the Government of the People's Republic of China,

In accordance with the provisions of the United Nations Convention on the Law of the Sea made on 10 December 1982,

In order to promote conservation and rational utilisation of marine living resources of common concerns, maintain a proper order of fishing operations at sea, strengthen and expedite co-operations in the field of fisheries,

Through friendly consultations,

Have agreed as follows:

Article 1

The area to which this Agreement is applied (hereinafter referred to as "the Agreement Zone") shall consist of the exclusive economic zones of Korea and the People's Republic of China.

Article 2

1. Each Contracting Party shall, in accordance with this Agreement and provisions of its relevant laws and regulations, permit the nationals and fishing vessels of the other Contracting Party to conduct fishing operations within its own exclusive economic zone.

2. The competent authorities of each Contracting Party, in accordance with the provisions of Annex I to this Agreement and its relevant laws and regulations, shall issue fishing permits to the nationals and fishing vessels of the other Contracting Party.

Article 3

1. Each Contracting party shall annually determine species allowed for harvesting, quotas of catch, areas of fishing and other specific conditions in its exclusive economic zone for nationals and fishing vessels of the other Contracting Party and notify in writing the other Contracting Party of such determination.

2. Each Contracting Party, in making such determination as provided for in paragraph 1, shall take into account the state of marine living resources in its exclusive economic zone, its harvesting capacity, circumstances of mutual access and other relevant factors, and shall respect the outcome of the consultation of the Korea-China Joint Fisheries Committee which is to be established under the provisions of Article 13.

Article 4

1. Nationals and fishing vessels of each Contracting Party shall comply with this Agreement and the relevant laws and regulations concerning the fisheries of the other Contracting Party, when harvesting in the exclusive economic zone of the other Contracting Party.

2. Each Contracting Party shall take necessary measures to ensure that its nationals and fishing vessels harvesting in the economic zone of the other Contracting Party, comply with the provisions of this Agreement, the conservation measures of marine living resources and other terms and conditions established in the relevant laws and regulations of the other Contracting Party.

3. Each Contracting Party shall notify without delay the other Contracting Party of the conservation measures of marine living resources and other conditions provided for in its laws and regulations.

Article 5

1. Each Contracting Party may, in accordance with international law, take necessary measures within its exclusive economic zone to ensure that nationals and fishing vessels of the other Contracting Party comply with the conservation measures of marine living resources and other conditions provided for in its laws and regulations.

2. Arrested or detained vessels and their crews shall be promptly released upon the posting of appropriate bond or other security.

3. The Contracting Party, in cases of arrest or detention of fishing vessels and their crews of the other Contracting Party, shall promptly inform through diplomatic channels the other Contracting Party of the measures taken and the penalties imposed thereafter.

Article 6

The provisions in Article 2 to Article 5 shall be applied to the Agreement Zone except the areas designated in Articles 7, 8 and 9.

Article 7

1. Paragraphs 2 and 3 of this Article shall be applied to the Area (hereinafter referred to as "the Provisional Measures Zone") surrounded by the line joining the following co-ordinates in sequence.

a. point at Lat. 37° 00' N. Long. 123° 40' E
b. point at Lat. 36° 22' 23" N. Long. 123° 10" 52" E
c. point at Lat. 35° 30' N. Long. 122° 11° 54" E
d. point at Lat. 35° 30' N. Long. 122° 01°54" E
e. point at Lat. 34° 00' N. Long. 122° 01° 54" E
f. point at Lat. 34° 00' N. Long. 122° 11° 54" E
g. point at Lat. 33° 20' N. Long. 122° 41' E

h. point at Lat. 32° 20' N. Long. 123° 45' E
i. point at Lat. 32° 11' N. Long. 123° 49' 30" E
j. point at Lat. 32° 11' N. Long. 125° 25' E
k. point at Lat. 33° 20' N. Long. 124° 08' E
l. point at Lat. 34° 00' N. Long. 124° 00' 30" E
m. point at Lat. 35° 00' N. Long. 124° 07' 30" E
n. point at Lat. 35° 30' N. Long. 124° 30' E
o. point at Lat. 36° 45' N. Long. 124° 30' E
p. point at Lat. 37° 00' N. Long. 124° 20' E
q. point at Lat. 37° 00' N. Long. 123° 40' E

2. Both Contracting Parties, following the determination of the Korea-China Joint Fisheries Committee established under Article 13 of this Agreement, taking into account the effect on the traditional fishing of each Contracting Party, shall take joint conservation measures and quantitative management measures in the Provisional Measures Zone.

3. Each Contracting Party shall apply the management measures and other necessary measures to its nationals and vessels fishing in the Provisional Measure Zone. Each Contracting Party may not apply the management measures and other measures to the nationals and fishing vessels of the other Contracting Party fishing in the Provisional Measure Zone. Either Contracting Party may, in case of finding an offence of the nationals and fishing vessels of the other Contracting Party against the operational regulations determined by the Korea-China Joint Fisheries Committee, call attention of the said nationals and fishing vessels to that fact, and notify the other Contracting Party of the fact with its related situation. The other Contracting Party shall take necessary measures with respect for that notification, and subsequently notify the former of the result.

Article 8

1. For four years from the date of entry into force of this Agreement, paragraphs 2 to 4 of this Article shall be applied to the Area (hereinafter referred to as "the Transitional Zones") surrounded by the line joining the following co-ordinates in sequence.

(1) Co-ordinates of the Transitional Zone in the Korean side
a. point at Lat. 35° 30' N. Long. 124° 30' E (K1)
b. point at Lat. 35° 00' N. Long. 124° 07' 30" E (K2)
c. point at Lat. 34° 00' N. Long. 124° 00' 30" E (K3)
d. point at Lat. 33° 20' N. Long. 124° 08' E (K4)

e. point at Lat. 32° 11' N. Long. 125°25' E (K5)
f. point at Lat. 32° 11' N. Long. 126° 45' E (K6)
g. point at Lat. 32° 40' N. Long. 127° 00' E (K7)
h. point at Lat. 32° 24' 30" N. Long. 126' 17' E (K8)
i. point at Lat. 32° 29' N. Long. 125° 57' 30" E (K9)
j. point at Lat. 33° 20' N. Long. 125° 28' E (K10)
k. point at Lat. 34° 00' N. Long. 124° 35' E (K11)
l. point at Lat. 34° 25' N. Long. 124° 33' E (K12)
m. point at Lat. 35° 30' N. Long. 124° 48' E (K13)
n. point at Lat. 35° 30' N. Long. 124° 30' E (K14)

(2) Co-ordinates of the Transitional Zone in the Chinese side
a. point at Lat. 35° 30' N. Long. 121° 55' E (C1)
b. point at Lat. 35° 00' N. Long. 121° 30' E (C2)
c. point at Lat. 34° 00' N. Long. 121° 30' E (C3)
d. point at Lat. 33° 20' N. Long. 122° 00' E (C4)
e. point at Lat. 31° 50' N. Long. 123° 00' E (C5)
f. point at Lat. 31° 50' N. Long. 124° 00' E (C6)
g. point at Lat. 32° 20' N. Long. 123° 45' E (C7)
h. point at Lat. 33° 20' N. Long. 122° 41' E (C8)
i. point at Lat. 34° 00' N. Long. 122° 11' 54" E (C9)
j. point at Lat. 34° 00' N. Long. 122° 01' 54" E (C10)
k. point at Lat. 35° 30' N. Long. 122° 01' 54" E (C11)
l. point at Lat. 35° 30' N. Long. 121° 55' E (C12)

2. Each Contracting Party shall take appropriate measures in order to gradually implement exclusive economic zone regimes in the Transitional Zones and make efforts to reach a balance by adjusting and reducing the level of fishing activities of its nationals and fishing vessels in the Transitional Zone in the other Contracting Party's side to reach a balance.

3. Both Contracting Parties shall take the same conservation and management measures as provided in paragraphs 2 and 3 of Article 7 in the Transitional Zones, and may take joint surveillance measures including joint boarding, stopping and inspection.

4. Each Contracting Party shall issue fishing permits to its fishing vessels engaging in fishing activities in the Transitional Zone at the other Contracting Party's side and exchange the list of such fishing vessels with the other Contracting Party.

5. The provisions of Articles 2 to 5 shall be applied in Transitional Zones after 4 years from the date of entry into force of this Agreement.

Article 9

Both Contracting Parties, in a certain area to the north of the Provisional Measures Zone provided for in Article 7, paragraph 1 and the other certain area to the south of the Provisional Measure Zone provided for in Article 7, paragraph 1 and Transitional Zones provided for in Article 8, paragraph 1 shall not apply their laws and regulations on fisheries against nationals and fishing vessels of the other Party and shall maintain the current fishing order, unless agreed otherwise between the two Parties.

Article 10

Each Contracting Party shall apply necessary measures such as instructions and others to its nationals and fishing vessels in order to secure safe navigation and fishing operations, maintain a proper order of fishing operations, and deal with maritime accidents smoothly and promptly.

Article 11

1. In cases where any national or fishing vessel of either Contracting Party runs across maritime casualties or other emergency situations on the coast of the other Contracting Party, the other Contracting Party shall provide assistance and protection to the extent possible, and notify the authorities concerned of the former Contracting Party of the related situations.

2. Any national or fishing vessel of either Contracting Party, in cases where they need to take refuge due to rough weather or other emergency situations, may take refuge in a harbour or other places of the other Contracting Party after they give notification to the authorities concerned of the other Contracting Party in accordance with the provisions of Annex II to this Agreement.

Article 12

Both Contracting Parties shall strengthen the co-operation for scientific research (including exchange of necessary data) on conservation and reasonable exploitation of marine living resources.

Article 13

1. Both Contracting Parties shall, for the purpose of facilitating the achievement of the objectives of this Agreement, establish the Korea-China

Fisheries Committee (hereinafter referred to as "the Fisheries Committee"). The Fisheries Committee shall be composed of one Representative and a few members appointed respectively by the Governments of the Contracting Parties, and, when necessary, may establish expert sub-committees.

2. The Fisheries Committee shall discharge the following duties:

(1) The Fisheries Committee will consult on the following matters and make recommendations to the Government of each Contracting Party

(a) the species which may be caught, quotas of catch and other conditions for the fishing operation of the nationals and fishing vessels of the other Contracting Party prescribed in the provisions of Article 3;

(b) the maintenance of the order of fishing operations;

(c) the circumstances and conservation of marine living resources; and

(d) co-operation on fishing between the Contracting Parties.

(2) The Fisheries Committee may, as necessary, make recommendations to the Governments of the Contracting Parties on the revision of the Annex to this Agreement.

(3) The Fisheries Committee will consult on and determine matters related to the provisions of Articles 7 and 8.

(4) The Fisheries Committee will review the implementations of this Agreement and other matters related to this Agreement.

3. All the recommendations and determinations of the Fisheries Committee shall be reached only through the agreement of the members of the Contracting Parties.

4. The Governments of Contracting Parties shall respect the recommendations referred to in sub-paragraph (1) of paragraph 2 of this Article and implement necessary measures with the determinations referred to in sub-paragraph (3) of paragraph 2 of this Article.

5. The Fisheries Committee shall meet once a year alternately in Korea and the People's Republic of China. The Fisheries Committee may, as necessary, meet on *ad hoc* basis upon agreement of the Contracting Parties.

Article 14

No provision in this Agreement shall be interpreted in such a way as to prejudice the position of either Contracting Party on issues in the law of the Sea.

Article 15

The Annex to this Article forms an integral part of this Agreement.

Article 16

1. This Agreement shall enter into force from the date of an exchange of notes between the Governments of the Contracting Parties, following the completion of their procedures required by respective internal laws.

2. This Agreement shall remain in force for a period of five years from the date of its entry into force. Thereafter, this Agreement shall continue to remain in force, unless so terminated in accordance with paragraph 3 of this Article.

3. Either Contracting Party may terminate this Agreement at any time on or after the date of expiration of an initial period of five years by giving a one year written notice to the other Contracting Party.

IN WITNESS WHEREOF, the undersigned, being duly authorised thereto by their respective Governments, have signed this Agreement.

DONE in duplicate at Beijing this day of 3rd of August of 2000, in the Korean and Chinese languages, both texts being equally authentic.

FOR THE REPUBLIC OF KOREA	FOR THE PEOPLE'S REPUBLIC OF CHINA

Annex I

Each Contracting Party shall, under paragraph 2 of Article 2 of this Agreement, take the following measures concerning permits:

1. The competent authorities of each Contracting Party shall, after receiving from the competent authorities of the other Contracting Party the written notice on the determination prescribed in the provisions of Article 3 of this Agreement, submit applications for permits to the competent authorities of the other Contracting Party for its nationals and fishing vessels that wish to conduct fishing operations in the latter's exclusive economic zone. The competent authorities of the other Contracting Party may impose appropriate fees in issuing permits.

2. The competent authorities of each Contracting Party shall give written notice to the competent authorities of the other Contracting Party on the procedural rules on fishing, including application for and issuance of permits, submission of information on fishing, signs of fishing vessels and other procedural rules on fishing operations diaries.

3. The fishing vessels permitted shall post the permit in a place easily accessible in the wheelhouse, and prominently display the sign assigned by the other Contracting Party.

ANNEX II

The implementation of paragraph 2 of Article 11 of this Agreement shall be carried out in the manner stipulated as follows:

1. The contact point designated by the Government of the Republic of Korea shall be each Korean Maritime Police Station and the contact point designated by the Government of the People's Republic of China shall be the Harbour Management Authorities, which exercises jurisdiction over the harbours concerned.

2. The detailed means of communication will be mutually informed at the China-Japan Joint Fisheries Committee.

3. The contents which either Contracting Party shall inform the contact point designated by the other Contracting Party shall be set out below:

> the name of the fishing vessel, the identification signal, the current position (latitude, longitude), the harbour where the fishing vessel is registered, the tonnage and the length, the name of the captain, the number of crews, the destination of refuge, the reason of refuge, the scheduled arrival time and the means of communication.

MEMORANDUM OF UNDERSTANDING

The representatives of the Government of the Republic of Korea and the Government of the Peoples' Republic of China, in relation to the relevant provisions of the Agreement on Fisheries between the Republic of Korea and the Peoples' Republic of China (hereinafter referred to as "the Agreement") signed on 3 August 2000, reached the following understanding:

1. Both Parties shall respect the laws and regulations which the coastal States are now enforcing in a certain area in the Korean side to the north of the northern limits of Provisional Measures Zone provided for in Article 7, paragraph 1 of the Agreement and in other certain area in the Chinese side to the southern limits of the Provisional Measure Zone provided for in Article 7, paragraph 1 and the Transitional Measure Zones provided for in Article 8, paragraph 1, and should take necessary measures to ensure that their respective nationals and fishing vessels observe such laws and regulations.

2. Both Parties shall notify each other the laws and regulations mentioned in Paragraph 1, and consult, through the Korea-China Joint Fisheries Committee established under Article 13 of the Agreement, the ways for efficient implementations of such laws and regulations.

This Understanding is signed at Beijing on 3 August 2000 and will be valid from the date of the entry into force of the Agreement.

5. Agreement between the Government of the Republic of Korea and the Government of Japan on Co-operation in the Field of Environmental Protection

The Government of the Republic of Korea and the Government of Japan,

Noting that global environmental deterioration would pose serious threats to the survival of mankind,

Recognising the urgent need of regional and global efforts to prevent such deterioration and to ensure environmentally sound and sustainable economic and social development,

Believing that co-operation between the two Governments is of mutual advantage in coping with similar problems of environmental protection in each country and will further promote such regional and global efforts,

Noting that the framework of the co-operation between the two Governments in the field of science and technology has been provided for by the Agreement between the Government of the Republic of Korea and the Government of Japan on Co-operation in the Field of Science and Technology signed on December 20, 1985, and Desiring to strengthen the co-operation between the two Governments in the field of environmental protection,

Have agreed as follows:

Article 1

The two Governments shall, on the basis of equality and mutual benefit, maintain and promote co-operation in the field of environmental protection.

Article 2

Co-operative activities under this Agreement may take the following forms:

(a) Exchange of information and data on research and development activities, policies, practices, legislation and regulations, and on technology related to environmental protection;

(b) Exchange of scientists, technical personnel and other experts;

(c) Joint seminars and meetings by scientists, technical personnel and other experts;

(d) Implementation of agreed co-operative projects, including joint research; and

(e) Other forms of co-operation as may be mutually agreed upon.

Article 3

1. In order to co-ordinate and facilitate co-operative activities under this Agreement, a Joint Committee on Environmental Co-operation (hereinafter referred to as "the Committee") consisting of the representatives of the two Governments shall be established.

2. The Committee shall meet, in principle once a year, alternately in the Republic of Korea and in Japan.

3. The Committee shall have the following functions:

(a) to discuss any matter relating to the implementation of this Agreement;

(b) to review the progress of the implementation of this Agreement; and

(c) to propose to the two Governments specific measures to ensure the enhancement of co-operation under this Agreement.

4. Contacts between the two Governments concerning the implementation of this Agreement when the Committee is not in session shall be made through diplomatic channels.

Article 4

Co-operative activities may be undertaken in mutually agreed areas pertaining to environmental protection and improvement, such as:

(a) Pollution abatement and control, which comprise: air pollution control, including control of emissions from mobile and stationary sources; water pollution control, including municipal and industrial waste-water treatment; marine pollution control; soil pollution control, including agricultural runoff and pesticide control; waste management and resource recovery; control and disposal of toxic substances; noise abatement

(b) Conservation of ecosystems and biological diversity;

(c) Prevention of dangerous anthropogenic interference with the climate system; and

(d) Other areas of environmental protection and improvement as may be mutually agreed upon.

Article 5

Implementing arrangements setting forth the details and procedures of the specific co-operative activities under this Agreement may be made between the two Governments or their agencies, whichever is appropriate.

Article 6

Each Government shall accord to the nationals of the other country facilities necessary for the implementation of the co-operative activities under this Agreement.

Article 7

This Agreement shall be implemented within the scope of laws and regulations in force in each country.

Article 8

1. Scientific and technological information of a non-proprietary nature arising from the co-operative activities under this Agreement may be made available to the public by either Government through customary channels and in accordance with the normal procedures of the participating agencies.

2. The disposition of patents, designs, and other industrial property arising from the co-operative activities under this Agreement shall be provided for in the implementing arrangements referred to in Article 5.

Article 9

No provision in this Agreement shall be interpreted as to affect other arrangements for co-operation between the two Governments, existing at the date of signature of this Agreement or concluded thereafter.

Article 10

1. This Agreement shall enter into force on the date of signature and remain in force for a period of two years.

2. This Agreement shall be automatically extended for successive periods of two years, unless either Government notifies the other Government in writing of its intention to terminate this Agreement at least six months prior to the expiry of any one period.

3. The termination of this Agreement shall not affect the completion of any project and program undertaken in accordance with the implementing arrangements referred to in Article 5 and not fully executed at the time of the termination of this Agreement.

DONE in duplicate at Seoul on June 29, 1993 in the English language.

FOR THE GOVERNMENT OF THE REPUBIC OF KOREA	FOR THE GOVERNMENT OF JAPAN

6. Agreement on Environmental Co-operation between the Government of the Republic of Korea and the Government of the People's Republic of China

The Government of the Republic of Korea and the Government of the People's Republic of China (hereinafter referred to as "the Parties"),

Noting that the global environmental degradation poses serious threats to the survival of mankind,

Recognising the urgent need of the global efforts to prevent environmental deterioration and to achieve the environmentally sound and sustainable development,

Believing that the co-operation between the Parties in the field of environment is of mutual benefit for addressing the environmental challenges and is essential for the protection and improvement of the regional and global environment, and

Considering that preventive measures should serve as important elements in the co-operative activities of the Parties to minimise the possible adverse effects of environmental damages,

Have agreed as follows:

Article 1

1. On the basis of equality and mutual benefit, the Parties shall encourage and promote co-operation in the field of environmental protection.

2. The principal objective of this co-operation is to provide better opportunities to exchange information, technology and experience related to the environmental protection and to collaborate on the issues of mutual interest.

Article 2

Co-operative activities under this Agreement shall include the following forms:

1. Exchange of data, information, technology and materials relating to the environmental protection;

2. Exchange of environmental experts and officials;

3. Organisation of joint seminars, symposia and meetings on general or specific environmental issues;

4. Implementation of joint research on subjects of mutual interest, including joint assessment of environmental impact;

5. Other forms of co-operation as may be mutually agreed upon.

Article 3

Co-operation, if mutually agreed upon, may be undertaken in the following areas pertaining to environmental protection and improvement:

1. Pollution abatement and control, which comprise: air pollution control, including control of emissions from mobile and stationary sources;

water pollution control, including municipal and industrial waste water treatment and total load control of water pollutants; coastal and marine pollution control; agricultural runoff and pesticide control; solid waste management and resource recovery; control of transboundary movement and disposal of hazardous solid wastes; management of toxic chemicals; noise abatement; conservation of biodiversity; management of environment and natural resources;

2. Contribution to the protection and improvement of sub-regional, regional, and global environment;

3. Other areas of environmental protection and improvement.

Article 4

1. In order to co-ordinate and facilitate co-operative activities under this Agreement, the Parties shall establish a Joint Committee on Environmental Co-operation (hereinafter referred to as "the Committee") consisting of the representatives designated by each Party.

2. The Committee shall meet, in principle, annually where appropriate in Korea and in China alternately on the date to be agreed upon through diplomatic channels. The frequency of the Committee's meeting may be reduced or increased according to the actual situation.

3. The Committee shall perform the following functions:

a. to discuss matters relating to the implementation of this Agreement;

b. to monitor and review the progress of the implementation of this Agreement; and

c. to recommend to the Parties specific measure for enhancement of the co-operation under this Agreement.

4. Appropriate consultations, when the Committee is not in session, shall be conducted through diplomatic channels.

Article 5

1. With a view to facilitating the bilateral co-operation, the Parties shall encourage, where appropriate, the conclusion of supplementary arrangements between the government agencies, research institutes, universities and enterprise specifying the terms and conditions of particular co-operative

programmes and projects, the procedures to be followed, and other appropriate matters.

2. The treatment of intellectual property rights arising from the co-operative activities under this Agreement shall be provided for in such arrangements.

Article 6

1. The Parties shall bear the expenses to be incurred in conjunction with the implementation of the co-operative programmes and projects between their government agencies or institutes under this Agreement on the basis of equality and in accordance with the availability of assets.

2. Each Party shall provide the nationals of the other Party with the appropriate assistance essential for the implementation of the co-operative activities under this Agreement.

Article 7

1. Nothing in this Agreement shall affect the obligations of the Parties deriving from any treaty, convention, regional or international agreement relating to the protection of environment.

2. Co-operative activities under this Agreement including the conclusion of such supplementary arrangements as referred to in Article 5 shall be undertaken in accordance with applicable laws and regulations of each country.

Article 8

1. This Agreement shall enter into force thirty days after the date of signature and remain in force for a period of five years.

2. This Agreement shall be automatically renewed for successive periods of five years, unless either Party notifies the other Party six months in advance in writing of its intention to terminate this Agreement.

3. The termination of this Agreement shall not affect the completion of any project or program being undertaken under this Agreement and not fully executed at the time of the termination of this Agreement.

IN WITNESS WHEREOF, the undersigned, being duly authorised by their respective Governments, have signed this Agreement.

DONE in duplicate at Beijing this 28th day of October 1993 in the Korean, Chinese and English languages, all three versions being equally authentic. In case of any divergence of interpretation, the English version shall prevail.

FOR THE GOVERNMENT OF THE REPUBLIC OF KOREA	FOR THE GOVERNMENT OF THE PEOPLE'S REPUBLIC OF CHINA

7. Agreement between the Government of the Republic of Korea and the Government of the Russian Federation on Co-operation in the Field of the Environment

The Government of the Republic of Korea and the Government of the Russian Federation (hereinafter referred to as "the Parties"),

Noting that global environmental degradation poses serious threats to the survival of mankind,

Recognising the urgent need of the global efforts to prevent environmental deterioration and to achieve the environmentally sound and sustainable development,

Believing that the co-operation between the Parties in the field of environment is of mutual advantage for addressing the environmental challenges and is essential for the protection and improvement of the regional and global environment, and

Considering that preventive or precautionary measures should serve as important elements in the co-operative activities of the Parties to minimise the possible adverse effects of environmental damages,

Have agreed as follows:

Article 1

1 On the basis of equality and mutual benefit, the Parties shall encourage and promote co-operation in the field of environmental protection and sustainable development.

2. The principal objective of this co-operation is to provide better opportunities to exchange information, technology and experience related to the sustainable development and to collaborate on the issues of mutual interest.

Article 2

Co-operative activities under this Agreement shall include the following forms:

1. Exchange of data, information, technology and materials relating to the environmental protection and sustainable development;

2. Exchange of environmental experts and officials;

3. Organisation of joint seminars, symposia and meetings on general or specific environmental issues;

4. Implementation of joint research projects on subjects of mutual interest, including joint assessment of environmental impact;

5. Other forms of co-operation as may be mutually agreed upon.

Article 3

Co-operation, if mutually agreed upon, may be undertaken in the following areas pertaining to environmental protection and improvement:

1. Pollution abatement and control, which comprise: air pollution control, including control of emissions from mobile and stationary sources; water pollution control, including municipal and industrial waste-water treatment; coastal and marine pollution control; agricultural runoff and pesticide control; solid waste management and resource recovery; control and disposal of toxic substances; noise abatement; management of environment and natural resources;

2. Contribution to the protection and improvement of sub-regional, regional, and global environment;

3. Other areas of environmental protection and improvement.

Article 4

1. In order to co-ordinate and facilitate co-operative activities under this Agreement, the Parties shall establish a Joint Committee on the Environmental Co-operation (hereinafter referred to as "the Committee") consisting of the representatives designated by each Party.

2. The Committee shall meet, in principle, in Korea and in Russia alternately once in two years.

3. The Committee shall perform the following functions:

a. to discuss matters relating to the implementation of this Agreement;

b. to monitor and review the progress of the implementation of this Agreement; and

c. to recommend to the Parties specific measures, including the establishment of working groups, for the enhancement of the co-operation under this Agreement.

4. Appropriate consultations, when the Committee is not in session, shall be conducted through diplomatic channels.

Article 5

With a view to facilitating the bilateral co-operation, the Parties shall encourage, where appropriate, the conclusion of supplementary arrangements between the government agencies, research institutes, universities and enterprises specifying the terms and conditions of particular co-operative programmes and projects, the procedures to be followed, and other appropriate matters.

The treatment of intellectual property arising from the co-operative activities under this Agreement shall be provided for in such arrangements.

Article 6

1. The Parties shall bear the expenses to be incurred in conjunction with the implementation of the co-operative programmes and projects between their government agencies or institutes under this Agreement on the basis of equality and in accordance with the availability of assets.

2. Each Party shall provide the nationals of the other Party with the appropriate assistance essential for the implementation of the co-operative activities under this Agreement.

Article 7

1. Nothing in this Agreement shall affect the obligations of the Parties deriving from any treaty, convention, regional or international agreement relating to the protection of environment.

2. Co-operative activities under this Agreement including the conclusion of such supplementary arrangements as referred to in Article 5 shall be undertaken in accordance with applicable laws and regulations of each country.

Article 8

1. This Agreement shall enter into force thirty days after the date of signature and remain in force for a period of five years.

2. This Agreement shall be automatically renewed for successive periods of five years, unless either Party notifies the other Party six months in advance in writing of its intention to terminate this Agreement.

3. The termination of this Agreement shall not affect the completion of any project or program undertaken in accordance with the supplementary arrangements referred to in Article 5 and not fully executed at the time of the termination of this Agreement.

IN WITNESS WHEREOF, the undersigned, being duly authorised by their respective Governments, have signed this Agreement.

DONE in duplicate at Moscow on June 2, 1994 in the Korean, Russian and English languages, all texts being equally authentic. In case of any divergence of interpretation, the English text shall prevail.

FOR THE GOVERNMENT OF THE REPUBLIC OF KOREA	FOR THE GOVERNMENT OF THE RUSSIAN FEDERATION

Selected Bibliography

1. BOOKS

Adede, A.O., *The System for Settlement of Disputes under the United Nations Convention on the Law of the Sea*, Martinus Nijhoff Publishers, 1987.

AL-Bahara, H.M., *The Legal Status of the Arabian Gulf States: A Study of their Treaty Relations and their International Problems*, Manchester University Press, 1968.

American Law Institute, *Restatement of the Law: The Foreign Relations Law of the United States*, third Edition, 1986, vol. 1.

Armstrong, P. and V. Forbes, *The Falkland Islands and their Adjacent Maritime Area*, Maritime Briefing Vol. 2 No. 3, IBRU, University of Durham, 1997.

Attard, D.H., *The Exclusive Economic Zone*, Oxford Clarendon Press, 1987.

Aust, A., *Modern Treaty Law and Practice*, Cambridge University Press, 2000.

Beazley, P.B., *Maritime Limits and Baselines: A Guide to their Delineation*, The Hydrographic Society Special Publication No. 2, 1978.

Bentwich, N. and A. Martin, *A commentary on the Charter of the United Nations*, Routledge & Kegan Paul L.T.D., 1950.

Bernhardt, R., ed., *Encyclopaedia of International Law*, vol 7.

Birnie, P.W. and A.E. Boyle, *International Law and the Environment*, Oxford Clarendon Press, 1992.

Blake, G., ed., *IBRU Workshop on International Maritime Boundaries*, University of Durham, 4-9 April 1999.

——, ed., *World Boundaries Vol. 5: Maritime Boundaries*, Routledge, 1994.

Bowett, D.W., *The Law of the Sea*, Manchester University Press, 1967.

——, *The Law of International Institutions*, Stevens & Sons, 1982.

Brittin, B.H., *International Law for Seagoing Officers*, US Naval Institute, 1956.

Brown, E.D., *The UN Convention on the Law of the Sea, 1982: A Guide for National Policy Making Legislation and Administration*, Book 1, UK Government Commonwealth Secretariat, 1987.

——, *The Sea-Bed Energy and Mineral: The International Legal Regime, Vol. 1: The Continental Shelf*, Martinus Nijhoff Publishers, 1992.

——, *The International Law of the Sea*, 2 Vols, Dartmouth, 1994.

Brownlie, I., *Principles of International Law*, Oxford University Press, 1998.

Burke, W.T., *The New International Law of Fisheries*, Oxford University Press.

Bynkershoek, *De Dominio Maris Dissertatio* (1737), translated into English by Ralph Van Deman Magoffin, Oxford University Press, 1923.

Charney, J.I., and L. Alexander, ed., *International Maritime Boundaries*, Vol. I and Vol. II, Martinus Nijhoff Publishers, 1993.

——, ed., *International Maritime Boundaries*, Vol. III, Martinus Nijhoff Publishers, 1998.

Cheng, Bin, *General Principles of Law as Applied by International Courts and Tribunals*, Stevens & Sons, 1953.

Chinkin, C., *Third Parties in International Law*, Oxford University Press, 1993.

Chiyuki, M., *Japan and the Law of the Sea* (written in Japanese), Yooshindou Press, 1995.

Churchill, R.R and A.V. Lowe, *The Law of the Sea*, Manchester University Press, 1999.

Churchill, R., and G. Ulfstein, *Marine Management in the Disputed Areas: The Case of the Barents Sea*, Routledge, 1992.

Davir Vidas, ed., *Protecting the Polar Marine Environment*, Cambridge University, 2000.

Dipla, H., *Le regime juridique des iles dans dans le droit international de la mer*, 1984.

El-Hakim, A.A., *The Middle Eastern States and the Law of the Sea*, Manchester University Press, 1979.

Elferink, A.G.O., *The Law of Maritime Boundary Delimitation: A Case Study of the Russian Federation*, Martinus Nijhoff Publishers, 1994.

Evans, M.D., *Relevant Circumstances and Maritime Delimitation*, Oxford Clerendon Press, 1989.

Falklands Government Office, *Falklands Islands' Interim Conservation & Management Zone: Fisheries Report 87/88.*

FAO, *World Fisheries 2000.*

Fox, H., *et al.* eds., *Joint Development of Offshore Oil and Gas: A Model Agreement for States for Joint Development with Explanatory Commentary*, London: British Institute of International and Comparative Law, 1989.

Foreign & Commonwealth Office, *Press Release* No. 241, 28 November 1990.

Franckx, E., *Maritime Boundaries in the Baltic Sea: Past, Present and Future*, Maritime Briefing Vol. 2, No. 2, University of Durham, 1996.

Fulton, T.W., *The Sovereignty of the Sea*, William Blackwood and Sons, 1911.

Gidel, B., *Le droit international public de la mer*, 1934.

Grotius, H., *Mare Liberum*; translated by Ralph Van Deman Magoffin, *The Freedom of the Sea or The Right which Belongs to the Dutch to Take Part in the East Indian Trade*, New York Oxford University Press, 1916.

Gustafsen, L.S., *The Sovereignty Dispute over the Falklands (Malvinas) Islands*, New York Oxford University Press, 1988.

Hakapää, K., *Marine Pollution in International Law*, Suomalinen Tiedeakatemia, 1981.

Harris, D.J., *Cases and Materials on International Law*, Sweet & Maxwell, 1998.

Hey, E., *The Regime for the Exploitation of Transboundary Marine Fisheries Resources*, Martinus Nijhoff, 1989.

Higgins, R., *Problems and Process: International Law and How We Use It*, Oxford University Press, 1995.

Hodgson, R. and L., Alexander, "Towards an Objective Analysis of Special Circumstance of Oceanic Archipelagos and Atolls", *Law of the Sea Institute Occasional Paper* No. 13, 1971.

Hollick, Ann L., *US Foreign Policy and the Law of the Sea*, Princeton University Press, 1981.

Ji, G., *Maritime Jurisdiction in the Three China Seas*, University of California Institute on Global Conflict and Co-operation, Policy Paper No. 19, 1995.

Jeanette Greenfield, *China's Practice in the Law of the Sea*, Oxford Clarendon Press.

Jennings, R. and Sir A. Watts, *Oppenheim's International Law*, Longman, 1992, vol. 1.

Jessup, P.C., *The Law of Territorial Waters and Maritime Jurisdiction*, G.A. Jennings Co, 1927.

Johnston, D.M., *The Theory and History of Ocean Boundary-Making*, McGill-Queen's University Press, 1988.

Kim, H., *The Impact of the U.N. Convention on the Law of the Sea on the Territorial Sea Regime of Selected East Asian Countries*, Ph.D. thesis, University of Wales College of Cardiff, 1993.

Lapradelle, P., "La Frontiere: Etude de droit international", Les Editions Internationales, 1928.

Lauterpacht, L., ed., *Oppenheim's International Law*, Longman, 1955.

Lay, H.S., *et al.* eds., *New Directions in the Law of the Sea*, Oceana Publication, 1973, Vol. 1.

Ma, Y., *Legal Problems of Seabed Boundary Delimitation in the East China Sea*, Occasional Papers/Reprint Series in Contemporary Asia Studies Inc., 1984.

Makarczysk, J., ed., *Essays in International Law in Honour of Judge Lachs*, Martinus Nijhoff Publishers, 1984.

MacDonald, C.G., *Iran, Saudi Arabia, and the Law of the Sea: Political Integration and Legal Development in the Persian Gulf*, Greenwood Press, Connecticut USA, 1980.

McNair, *The Law of the Treaties*, Oxford University Press, 1961.

——, ed., Oppenheim's *International Law*, 4th edn., 1928, Vol. I.

Merrills, J.G., *International Dispute Settlement*, Cambridge University Press, 1998.

Myoshi, M., *The Joint Development of Offshore Oil and Gas in Relation to Maritime Boundary Delimitation*, Maritime Briefing IBRU, University of Durham, 1999.

Neyadi, M.H.H., *The Maritime Zones of the United Arab Emirates with Particular Reference to Delimitation*, Ph.D. Thesis, Edinburgh University, 1997.

Nordquist, M., S.H. Lay and K.R. Simmonds, eds., *New Directions in the Law of the Sea*, Oceana Publications, 1980.

Nordquist, M.H., *et al.* eds., *United Nations Convention on the Law of the Sea 1982: A Commentary*, Martinus Nijhoff Publishers, 1993.

O'Connell, D.P., *International Law*, Stevens & Sons, 1970.

——, *The Influence of Law on the Sea Power*, Manchester University Press, 1975.

Park, C., *The Law and Practice Relating to the International Regulation of Fisheries in Asia with Particular Reference to the Korea-Japanese Dispute*, University of Edinburgh, Ph.D. Thesis, 1971.

——, *Continental shelf issues in the Yellow Sea and the East China Sea*, Law of the Sea Institute Occasional Paper No. 15, University of Rhode Island, 1972.

Park, H.K., *The Law of the Sea and North East Asia: A Challenge for Co-operation*, Kluwer Law International, 2000.

Platzöder, *Third Untied Nations Conference on the Law of the Sea: Document.*

Prescott, J.R.V., *Maritime Jurisdiction in East Asian Seas*, East-West Environment and Policy Institute Occasional Paper No. 4, 1987.

Reisman, W.M. and G.S. Westerman, *Straight Baselines in International Maritime Boundary Delimitation*, Macmillan, 1992.

Roach, A. and R.W. Smith, *United States Responses to Excessive Maritime Claims*, Martinus Nijhoff Publishers, 1996.

Rosenne, S., *Development in the Law of Treaties 1945-1986*, Cambridge University Press, 1989.

——, *The Law and Practice of the International Court*, A.W. Sijthoff-Leyden (1965).

Recueil des Cours, 121 Academie de Droit International, 1967.

Shaw, M.N., *International Law*, Cambridge University Press, 1994.

Schwarzenberger, G., *A Manual of International Law*, Stevens & Sons, 1967.

Simma, B., ed., *The Charter of the United Nations: A Commentary*, Oxford University Press, 1995.

Smith, R.W. and B.L. Thomas, *Island Disputes and the Law of the Sea: An Examination of Sovereignty and Delimitation Disputes*, Maritime Briefing Vol. 2, No. 4, University of Durham, 1998.

Smith, R.W., *Exclusive Economic Zone Claims: An Analysis and Primary Document*, Nijhoff Publisher, 1986.

Soons, A., *Artificial Islands and Installations in International Law*, 17 Law of the Sea Occasional Paper No. 22, 1977.

——, *Marine Scientific Research and the Law of the Sea*, Kluwer Law and Taxation Publishers, 1982.

U.N., *The Law of the Sea: Status of the United Nations Convention on the Law of the Sea*, 1985.

——, *The Law of the Sea Bulletin* No. 5, 1985.

——, The Law of the Sea, Regime of Islands – Legislative History of Part VII (Article 121) of the United Nations Convention on the Law of the Sea, 1988.

——, *National Maritime Claims*, 1989.

——, *The Law of the Sea: National Claims to Maritime Jurisdiction*, 1992.

——, "National Claims over Maritime Zones", *Law of the Sea Bulletin* No. 23, June 1993.

——, *The Law of the Sea, Regime of Islands-Legislative History of Part VII (Article 121) of the United Nations Convention on the Law of the Sea*, 1988.

——, *Law of the Sea Bulletin* No. 35, 1997.

——, *Law of the Sea Bulletin* No. 38, United Nations, 1999.

——, *The Law of the Sea: National Legislation on the Territorial Sea, the Right of Innocent Passage and the Contiguous Zone*, United Nations, 1995.

UNEP, *Synergies*, Issue 3, September 2000.

US State Department, *Limits in the Seas, No. 117 – Straight Baseline Claim: China*, 1996.

——, *Limits in the Seas No. 121 – Straight Baseline and Territorial Sea Claims: South Korea*, 1998.

——, "Straight Baselines Claim: China", *Limits in the Seas* No. 117, 9 July 1996.

Vicuna, F.O., *The Exclusive Economic Zone*, Cambridge University Press, 1989.

Valencia, M.J., *A Maritime Regime for North East Asia*, Oxford University Press, 1996.

Weil, P., *The Law of Maritime Delimitation-Reflection*, translated in to English by M. Mac Glashman Grotius Publication Limited, 1989.

Whiteman, M.M., *Digest of International Law*, US Department of State (1965), Vol. 4, p. 1018.

Welwood, W., *An Abridgement of all Sea-Lawes, gathered forth of all Writing and Monument, which are to be found among any people or Nation upon the coasts of the great Ocean and Mediterranean Sea: And specially ordered and disposed for the use and benefit of all benevolent Sea-farers, within his Maisties Dominions of Great Britanne, Ireland, and the adjacent Isles thereof*, London, 1613.

World Commission on Environment and Development, *Our Common Future*, Oxford University Press.

Weiss, E.B., ed., *Environmental Change and International law: New Challenges and dimensions.*

2. ARTICLES

Akaha, T., "Fishery Relations in North East Asia", a Paper presented at the International Conference on the UN Convention on the Law of the Sea and East Asia, held in Seoul on 29-30 November, 1995, p. 9.

Adede, A.D., "Toward the Formation of the Rules of Delimitation of Sea Boundaries between States with Adjacent or Opposite Coasts", 19 *V.J.I.L.* (1979), p. 213.

——, "Report No. 4-3: Guinea-Guinea Bissau", *International Maritime Boundaries*, pp. 857-865.

——, "Report Number 4-4: Guinea-Bissau-Senegal", *International Maritime Boundaries*, pp. 872-874.

Alexander, L., "Regional Arrangements in the Oceans", 71 *A.J.I.L.* (1977), p. 84.

——, "Report Number 1-1: Canada-Denmark (Greenland)", *International Maritime Boundaries*, p. 372.

Anderson, D.H., "Report Number 9-2: France-Spain", *International Maritime Boundaries*, pp. 1728-1734.

——, "Report No. 9-4: Iceland-Norway (Jan Mayen)", *International Maritime Boundaries*, pp. 1756-1760.

——, "Report Number 9-5: Ireland-United Kingdom", *International Maritime Boundaries*, pp. 1767-1780.

——, "Report No. 9-10: Denmark-UK", *International Maritime Boundaries.*

——, "Report No. 9-11: Federal Republic of Germany-The Netherlands", *International Maritime Boundaries.*

——, "Report No. 9-12: Federal Republic of Germany-UK", *International Maritime Boundaries.*

——, "Report No. 9-13: The Netherlands-UK", *International Maritime Boundaries*, pp. 1825-1869.

——, "Report Number 9-15: Norway-United Kingdom", *International Maritime Boundaries*, pp. 1881-1882.

——, "The Straddling Stocks Agreement of 1995 – An Initial Assessment", 45 *I.C.L.Q.* (1996), pp. 463-475.

——, "British Accession to the UN Convention on the Law of the Sea", 46 *I.C.L.Q.* (1997), pp. 778-779.

Aust, A., "The Theory and Practice of Informal International Instruments", 35 *I.C.L.Q.* (1986), pp. 797-799.

Barston, R.P. and P.W. Birnie, "The Falkland Islands/Islas Malvinas conflict: A question of zones", 7 *M.P.* (1983), p. 20.

Bergin, A., "The Australian-Indonesian Timor Gap Boundary Agreement", 5 *I.J.E.C.L.* (1990), pp. 383-385.

Birnie, P., "The Legal Framework Relating to Ocean Resources and Their Development, Management and Protection: Implications for Marine Scientific Research (provisional version)", a paper delivered at the International Conference on Oceanography, held in Lisbon, 14-19 November 1994, p. 12.

Bowett, D.B., "The *Arbitration* between the United Kingdom and France concerning the Continental Shelf Boundary in the English Channel and South-Western Approach", 49 *B.Y.I.L.* (1978), p. 14.

——, "Jurisdiction: Changing Patterns of Authority over Activities and Resources", 52 *B.Y.I.L.* (1983), p. 14.

——, "Islands, Rocks, Reefs and Low-tide Elevations", *International Maritime Boundaries*, pp. 131-151.

Boyle, A.E., "Problems of Compulsory Jurisdiction and the Settlement of Dispute Relating to Straddling Fish Stocks", *The International Journal of Maritime and Coastal Law* Vol. 14, no. 1 (1999), pp. 13-15.

——, "Dispute Settlement and the Law of the Sea Convention; Problems of Fragmentation and Jurisdiction", 46 *I.C.L.Q.* (1997), p. 36.

——, "The Law of Treaties and the Anglo-French Continental Shelf", 1980 *I.C.L.Q.*, vol. 29, pp. 498-508.

——, "Globalism and Regionalism in the Protection of the Marine Environment", in Davir Vidas ed., *Protecting the Polar Marine Environment*, Cambridge University Press (2000), pp. 19-33.

——, "UNCLOS, ITLOS and the Settlement of Maritime Boundary Disputes between Taiwan and Japan", in Taiwan Law Society and Taiwan Institute of International Law ed., *ibid.* pp. 143-144.

Charney, J.I., "Central East Asian Maritime Boundaries and the Law of the Sea", 89 *A.J.I.L.* (1995), p. 741.

——, "The Exclusive Economic Zone and Public International Law", 15 *O.D.I.L.* (1985), p. 239.

——, "Rocks that cannot sustain human habitation", 93 *A.J.I.L*, 1999, p. 862 and p. 872.

——, "Progress in International Maritime Boundary Delimitation Law", 88 *A.J.I.L.* (1994), pp. 227-242.

——, "Report Number 1-2: Canada-France (St. Pierre and Miqueon)", *International Maritime Boundaries*, pp. 396-399.

——, "International Maritime Boundary Delimitation", 88 *A.J.I.L.* (1994), p. 247.

——, "The Diaoyu/Senkaku Islands Maritime and Territorial Dispute", Taiwan Law Society and Taiwan Institute of International Law eds., *International Law Conference on the Dispute over Dioya/Senkaku Islands* (2-3 April 1997, Taiwan), pp. 128-129.

——, "The Dioyu/Senkaku Islands Maritime and Territorial Dispute", *ibid.*, p. 126.

Cheng, T., "Communist China and the Law of the Sea", 63 *A.J.I.L.* (1969), pp. 54-56.

Churchill, R.R., "Current Developments: The Law of the Sea", 38 *I.C.L.Q.* (1989), pp. 413-417.

——, "Joint Development Zones: International Legal Issues", H. Fox *et al.* eds., *Joint Development of Offshore Oil and Gas*, Vol. II, British Institute of International Law, 1990, p. 64.

——, "Maritime Delimitation in the Jan Mayen Area", *M.P.* (January 1985), pp. 26-27.

——, "Falkland Island – Maritime Jurisdiction and Co-operation Arrangement with Argentina" 46 *I.C.L.Q.* (1997), p. 464.

——, "The Greenland/Jan Mayen Case and Its Significance for the International Law of Maritime Boundary Delimitation", 9 *IJMCL* (1994), pp. 1-29.

——, "Joint Development Zones: International Legal Issues", *Joint Development of Offshore Oil and Gas*, H. Fox, eds., British Institute of International Law (1990), pp. 55-67.

——, "Fisheries issues in maritime boundary delimitation", *M.P.* (1993), pp. 48-49.

Colson, D., "Report Number 1-3: Canada-United States", *International Maritime Boundaries*, pp. 401-408.

——, "The Legal Regime of Maritime Boundary Agreements", *International Maritime Boundary*, pp. 61-62.

Davis, P.G. and C. Redgwell, "The International Legal Regulation of Straddling Fish Stocks", *B.Y.I.L.* (1997), pp. 265-272.

El Ghoneimy, M.T., "The Legal Status of the Saudi-Kuwaiti Neutral Zone", 15 *I.C.L.Q.* (1966), pp. 690-717.

Ely N. and R.W. Petrowski Jr., "Boundaries of Seabed Off the Pacific Coast of Asia", 8 *N.R.L.* (1975), pp. 620-623.

Evans, M.D., "Delimitation: The Common Maritime Boundaries", 64 *B.Y.I.L.* (1994), p. 294.

——, "The Restoration of Diplomatic Relations between Argentina and the United Kingdom", *40 I.C.L.Q.* (1991), pp. 476-481.

Fitzmaurice-Lachs, M. "The Legal Regime of the Baltic Sea Fisheries", 24 *Netherlands International Law Review* (1982), p. 205.

Franckx, E., "Report Number 10-9: Sweden-Soviet", *International Maritime Boundaries*, p. 2061.

——, "Report Number 10-3: Finland-Sweden", *International Maritime Boundaries*, pp. 1945-1957.

——, "Report Number 10-10: Poland-Sweden", *International Maritime Boundaries*, p. 2080.

Gao, Z., "The Legal Concept and Aspects of Joint Development in International Law", *Ocean Yearbook* 13, The University of Chicago Press, p. 120.

Gilmore, W.C., "The Newfoundland Continental Shelf dispute in the Supreme Court of Canada", 8 *M.P.* (1984), pp. 323-329.

——, *Combating International Drugs Trafficking: The 1988 United Nations Convention Against Illicit Traffic in Narcotic Drugs and Psychotropic Substances*, Commonwealth Secretariat (1991), p. 35.

Govaere, I., "The Impact of Intellectual Property Protection on Technology Transfer between EC and Central and Eastern European Countries", 25 *Journal of World Trade*, p. 58.

Hayashi, M., "The Management of Transboundary Fish Stocks under the LOS Convention", 8 *I.J.M.C.L.* (1993), p. 249.

Herrisman, M. and M. Tsamenyi, "The 1997 Austria-Indonesia Maritime Boundary Treaty: A Secure Legal Regime for Offshore Resource Development?", 1998 *O.D.I.L.* vol. 27, pp. 361-396.

Hodgson, R.D. and R. Smith, "The Informal Single Negotiating Next (Committee II): A Geographical Perspective", 3 *O.D.I.L.* (1976), p. 225.

Hollic, A.L., "The Origin of 200-mile offshore zones 2", 71 *A.J.I.L.* (1977), pp. 492-498.

Hsu, R.T.S., "A Rational Approach to Maritime Delimitation", 13 *OD.I.L.* (1983), pp. 108-112.

Hutchinson, D.N., "The Limits to Continental Shelf Jurisdiction in Customary International Law", *56 B.Y.I.L.* (1985), p. 181.

Jhe, S.H., "Some Delimitation Issues in the Maritime Areas Surrounding the Korean

Peninsula", 11 *Hae-yang-jung-chak-yon-ku* (Studies on Marie Policy) No. 1, pp. 109-110.

Jimenez de Arechaaga, E., "Report Number 3-1: Argentina-Chile", *International Maritime Boundaries*, pp. 719-721.

——, "Report Number 3-2: Argentina-Uruguay", *International Maritime Boundaries*, pp. 757-762.

——, "Report Number 3-5: Chile-Peru", *International Maritime Boundaries*, p. 795.

Knauss, J.A., "Development of the Freedom of Scientific Research Issue of the Third Law of the Sea Conference", 1 *O.D.I.L.* (1973), p. 94.

Kwiatowska, B., "The Eritrea-Yemen *Arbitration*: Landmark Progress in THE Acquisition of Territorial Sovereignty and Equitable Maritime Boundary Delimitation", 32 *O.D.I.L.* (2001), pp. 1-25.

Kwiatkowska, B. and A.H. Soons, "Entitlement to maritime areas of rocks which cannot sustain human habitation or economic life of their own", 21 *N.Y.I.L* (1990), p. 165.

Lagoni, R., "Interim Measures Pending Maritime Delimitation", 78 *A.J.I.L.* (1984), pp. 350-360.

——, "Oil and Gas Deposits Across National Frontiers", 73 *A.J.I.L.* (1979), pp. 233-239.

Lauterpacht, H., "Sovereignty over Submarine Areas", 27 *B.Y.I.L.* (1950), p. 420.

Leiner, F.C., "Maritime Security Zones: prohibited yet perpetuated", 24 *VJIL* (1984), p. 980.

Lowe, A.V., "The *M/V Saiga Case*: The First Case in the International Tribunal for the Law of the Sea", 48 *I.C.L.Q.* (1999), pp. 186-199.

Legault. L., and B. Hankey, "Method, Opposite and Adjacency, and Proportionality in Maritime Boundary Delimitation", *International Maritime Boundaries*, p. 204.

McDorman, T.L., "Number 5-19: Malaysia-Vietnam", *International Maritime Boundaries*, Vol. III on CD Rom.

Manner, E.J., "Settlement of Sea-Boundary Delimitation Dispute According to the Provisions of the 1982 Law of the Sea Convention", J. Makarczysk, ed., *Essays in International Law in Honour of Judge Lachs*, Martinus Nijhoff Publishers (1984), p. 639.

Marston, G., "The Stability of Land and Sea Boundary Delimitation International Law", in G.H. Blake ed., *World Boundaries Vol. 5: Maritime Boundaries*, Routledge (1994), p. 144.

Miyoshi, M., "New Japan-China Fisheries Agreement", 41 *The Japanese Annual of International Law* (1998), pp. 32-34.

——, "Japan's Straight Baseline System", 111 *The Journal of International Affairs*, Aichi University, Japan (1999), pp. 1-17.

——, "New Japan-China Fisheries Agreement," 41 *Japanese Annual of International Law* (1998), pp. 31-33.

——, "The Food and Agriculture Organisation of the United Nations Compliance", 10 *I.J.M.C.L.* (1995), pp. 412-416.

Moowoong, P., "China's Oil Development in the East China Sea" (written in Japanese), *Dongah Monthly* (June 2000).

Nakauci, K., "Problems of Delimitation in the East China Sea and Sea of Japan", 6 *O.D.I.L.* (1979), pp. 312-314.

Nandan, S.N. and Anderson, D.H., "Straits Used for International Navigation", 61 *B.Y.I.L.* (1990), pp. 176-178.

Nelson, L.D.M., "The Delimitation of Maritime Boundaries in the Caribbean", Johnston and Saunders, eds., *Ocean Boundaries Making: Regional Issues and Development* (1988), p. 167.

Nordholt, H.S., "Delimitation of the Continental Shelf in the East China Sea", 32 *Netherlands International Law Review* (1985), p. 136.

Nweihed, K.G., "Report Number 2-14: United States (Puerto Rico and the Virgin Islands)-Venezuela", *International Maritime Boundaries*, p. 695.

——, "Report Number 2-2: Colombia-Dominican Republic", *International Maritime Boundaries*, pp. 479-480.

——, "Report Number 2-12: The Netherlands(Antilles)-Venezuela", *International Maritime Boundaries*, p. 623.

——, "Report Number 2-13(2): Trinidad and Tobago-Venezuela", *International Maritime Boundaries*, pp. 656-658.

——, "EZ (uneasy) Delimitation in the Semi-enclosed Caribbean Sea: Recent Agreement between Venezuela and Her Neighbours", 8 *O.D.I.L.* (1980), p. 17.

Oda, S., "The Normalization of Relations between Japan and the Republic of Korea", 61 *A.J.I.L.* (1967), pp. 51-52.

Oda, S. and H. Owada, "Annual Review of Japanese Practice in International Law XV" (1976-77), 28 *J.A.I.L.* (1985), pp. 132-134.

Ohira, Z. and T. Kuwahara, "Fisheries Problems between Japan and People's Republic of China", 3 *Japanese Annual of International Law* (1959), p. 109.

Okowa, P. N., "Procedural Obligations in International Environmental Agreements", 68 *B.Y.I.L.* (1997).

Onorato, W.T., "Apportionment of an International Common Petroleum Deposit", 17 *I.C.L.Q.* (1968), pp. 85-101.

Oxman, B.H., "Political, Strategic, and Historical Considerations", *International Maritime Boundaries*, p. 12.

——, "The Third United Nations Conference on the Law of the Sea: The Ninth Session (1980)", 75 *A.J.I.L.* (1981), pp. 211-256.

——, "Political, Strategic, and Historical Considerations", *International Maritime Boundaries*, pp. 10-11.

——, "The Third United Nations Conference on the Law of the Sea: The 1976 New York Sessions", 71 *A.J.I.L.* (1977), pp. 261-267.

——, "The Third United Nations Conference on the Law of the Sea: The Seventh Session (1978)", 73 *A.J.I.L.* (1978), p. 23.

Park, C., "Oil under Troubled Waters," 14 *Harvard International Law Journal* (1973), No. 2, pp. 212-260.

——, "Report Number 5-3: Australia-Papua New Guinea", *International Maritime Boundaries*, pp. 930-934.

——, "Report Number 5-12: Agreement between Japan and Republic of Korea Concerning the Establishment of Boundary in the Northern Part of the Continental Shelf Adjacent to two Countries", *International Maritime Boundaries*, pp. 1058-1089.

——, "Report Number 5-13(2): Malaysia-Thailand (Gulf of Thailand Continental Shelf)", *International Maritime Boundaries*, pp. 1111-1123.

——, "Fishing under Troubled Waters: The North East Asia Controversy", 2 *O.D.I.L.* (1974), pp. 116-118.

——, "Fishing under Troubled Water: The North East Asia Fishery Controversy", 2 *O.D.I.L* (1974), pp. 93-135.

——, "The Sino-Japanese-Korean Sea Resources Controversy and the Hypothesis of a 200-Mile Economic Zone", 16 *H.I.L.J.* (1975), pp. 41-42.

——, "The South China Sea Disputes: Who owns the islands and the natural resources", 5 *O.D.I.L.* (1978), pp. 27-59.

——, "China and Maritime Jurisdiction", 22 *German Yearbook of International Law* (1979), pp. 119-141.

Park, H.K., "*Case Reports – Japan v. Kim Sun-Ki Case* No. 35. 1977, Matsue

District Court, Shimane Prefecture, Japan, August 15, 1977", 92 *A.J.I.L.* No. 2 (1998), pp. 301-305.

Piertrowski Jr., R.F., "Report Number 7-7: Iran-Saudi Arabia", *International Maritime Boundaries*, pp. 1519-1532.

——, "Report Number 7-10: Sharjah-Umm al Qaywayn", *International Maritime Boundaries*, p. 155.

Prescott, J.R.V., "Report No. 4-4 (4) & (5): Guinea-Bissau-Senegal", *International Maritime Boundaries.*

——, "Report Number 6-2 (2): Australia-Indonesia (Timor and Arafura Seas)", *International Maritime Boundaries*, p. 1213.

——, "Report Number 6-2 (4): Australia-Indonesia (Fisheries)", *International Maritime Boundaries*, p. 1232.

——, "Report Number 6-2 (5); Australia-Indonesia", *International Maritime Boundaries*, p. 1246.

——, "Report Number 6-5: France (Reunion)-Mauritius", *International Maritime Boundaries*, pp. 1354-1358.

——, "Report No. 6-7: India-Indonesia-Thailand", *International Maritime Boundaries*, pp. 1374-1388.

——, "Report Number 6-9: India-Maldives-Sri Lanka", *International Maritime Boundaries*, pp. 1401-1408.

——, "Report Number 6-12: Indonesia-Malaysia-Thailand", *International Maritime Boundaries*, pp. 1443-1454.

Roach, J.A. and R.W. Smith, "Straight Baselines: The Need for a Universally Applied Norm", 31 *O.D.I.L.* (2000), pp. 31-50.

Scott, S.V., "The Inclusion of Sedentary Fisheries within the Continental Shelf Doctrine", 41 *ICLQ* (1992), pp. 788-807.

Scovazzi, T., "Report Number 8-7 (2): Italy-Yugoslavia", *International Maritime Boundaries*, pp. 1640-1641.

Scovazzi, T. and G. Francalanci, "France-Italy (Report Number 8-2)", *International Maritime Boundaries*, pp. 1571-1576.

Scrivener, D., "Oil, Fish and Strategy: The USSR, Svalbard and the Barents Sea", *Aberdeen Studies in Defence Economics* No. 27 (1987).

Shearer, I., "Problems of Jurisdiction and law enforcement against delinquent vessels", 35 *I.C.L.Q.* (1986), p. 334.

Siddle, J., "Anglo-American Co-operation in the Suppression of Drug Smuggling", 1982 *I.C.L.Q.* Vol. 31, p. 738.

Smith, R.W., "Report Number 1-4: Cuba-United States", *International Maritime Boundary*, pp. 416-425.

Soeparjadi, P.A., *et al.*, "World Oil Development: Far East", 70 *AAPG Bulletin* (1985), pp. 1479-1485.

Soons, A.H.A., "The International Legal Regime of Marine Scientific Research", 24 *Netherlands International Law review* (1977), p. 401.

——, "The International Legal Regime of Marine Scientific Research", 24 *Netherlands International Law Review* (1977), pp. 412-429; United Nations, *Y.I.L.C.* 1956, Vol. II, p. 10.

Symmons, C., "The Rockall Dispute Deepens: An Analysis of Recent Danish and Icelandic Actions", 35 *I.C.L.Q.* (1986), pp. 164-165, and p. 261.

——, "The Maritime Zones around the Falkland Islands", 37 *I.C.L.Q.* (1988), pp. 289-291.

Taiwan Law Society and Taiwan Institute of International Law eds., *International Law Conference on the Dispute over Diaoyu/Senkaku Islands* (2-3 April 1997, Taiwan).

Takabayashi, H., "Normalisation of Relations between Japan and the Republic of Korea", 10 *Japanese Annual of International Law* (1966), pp. 16-23.

Takeyama, M. "Japan's Foreign Negotiations over Offshore Petroleum Development", in R.L. Friedheim *et al.* eds., *Japan and new Ocean Regime*, 1984, p. 287.

Toshiro M., "Legal Bases and Analysis of Japan's Claims to Diaoyu Islands", in *International Law Conference on the Dispute over Diaoyu/Senkaku Islands*, pp. 21-45.

UK Foreign & Common Wealth Office, "Background Brief: Britain's Accession to the United Nations Convention on the Law of the Sea", London, January 1998, p. 7. available at <www.Faco.gov.uk/news/archive.asp>.

Valencia, M., and J. Van Dyke, "Southeast Asian Seas: Joint Development of Hydrocarbons in the Overlapping Claims Areas", 16 *O.D.I.L.* (986), pp. 218-219.

Valencia, M.J., "North East Asia: Petroleum Potential, Jurisdictional Claims and International Relations", 20 *O.D.I.L.* (1989), p. 47.

Vallat, "The Continental Shelf", 23 *B.Y.I.L.* (1946), pp. 333-338.

Van Dyke, J.M., "The Aegean Sea dispute: option and avenues", 20 *M.P.* (1996), pp. 397-403.

Van Dyke J.M., *et al.*, "The Exclusive Economic Zone of the Northwestern Hawaiian Islands: When Do Uninhabited Islands Generate an EEZ", 25 *San Diego Law Review* (1988), pp. 449-454.

Widdows, K., "The Unilateral Denunciation of Treaties Containing No Denunciation", 53 *B.Y.I.L.* (1982), p. 113.

Wooster, W.S., "Research in Troubled Waters: US Research Vessel Clearance Experience, 1972-1978", 9 *O.D.I.L.* (1981), pp. 219-231.

Young, "Recent Development with Regard to the Continental Shelf", 42 *A.J.I.L.* (1948), pp. 851-857.

Zou Keyuan, "China's exclusive economic zone and continental shelf: developments, problems, and prospects", 25 *M.P.* (2001), pp. 77-78.

3. U.N. OFFICIAL DOCUMENTS

Official Records of the First United Nations Conference on the Law of the Sea, Vol. III, 1958.

ILC Yearbook Vol. II, 1950.

ILC Yearbook Vol. I, 1953.

ILC Yearbook Vol. II, 1956.

ILC Yearbook Vol. II, 1966.

United Nations Conference on the Law of Treaties, Official Records of the First Session, UN Doc. A/Con, 1968.

Third United Nations Conference on the Law of the Sea.

UN Doc. NG 7/2 of 20 April 1978.

UN Doc. NG 7/10 of 1 May 1978.

A/CONF.62/C.2/L.14 (1974).

A/CONF.62/C.2/L.31/REV.1 (1974).

A/CONF.62/C.2/L.43 (1974).

A/CONF.62.WP.8/PART II (ISNT, 1975).

A/CONF.62/WP.8/Rec.1 (1976).

A/CONF.62/WP.8/Rev.1/Part II (ISNT, 1976).

A/CONF.62/WP.109ICNT, (1977).

A/CONF.62/91 1979.

A/CONF.62/WP.10/Rev.29ICNY/Rev.2 (1980).

U.N. Doc. ST/LEG/SER.B/6 (1957).

U.N. Doc. ST/LEG/SER.B/8 (1959).

United Nations Legislative Series, ST/LEG/SER.B/19.

U.N. Sea-Bed Committee Doc. A/AC.138/SR.57.

Publications on Ocean Development

1. R.P. Anand: *Legal Regime of the Sea-Bed and the Developing Countries.* 1976 ISBN 90-286-0616-5
2. N. Papadakis: *The International Legal Regime of Artificial Islands.* 1977 ISBN 90-286-0127-9
3. S. Oda: *The Law of the Sea in Our Time.* Volume I: *New Developments, 1966-1975.* 1977 ISBN 90-286-0277-1
4. S. Oda: *The Law of the Sea in Our Time.* Volume II: *The UN Seabed Committee, 1968-1973.* 1977 ISBN 90-286-0287-9
5. C.O. Okidi: *Regional Control of Ocean Pollution.* Legal and Institutional Problems and Prospects. 1978 ISBN 90-286-0367-0
6. N.S. Rembe: *Africa and the International Law of the Sea.* A Study of the Contribution of the African States to the 3rd UN Conference on the Law of the Sea. 1980 ISBN 90-286-0639-4
7. R.P. Anand: *Origin and Development of the Law of the Sea.* History of International Law Revisited. 1983 ISBN 90-247-2617-4
8. A.M. Post: *Deepsea Mining and the Law of the Sea.* 1983 ISBN 90-247-3049-X
9. S.P. Jagota: *Maritime Boundary.* 1985 ISBN 90-247-3133-X
10. A.O. Adede: *The System for Settlement of Disputes under the UN Convention on the Law of the Sea.* A Drafting History and a Commentary. 1987 ISBN 90-247-3324-3
11. M. Dahmani: *The Fisheries Regime of the Exclusive Economic Zone.* 1987 ISBN 90-247-3374-X
12. S. Oda: *International Control of Sea Resources.* Reprint with a New Introduction. 1989 ISBN 90-247-3800-8
13. D.G. Dallmeyer and L. DeVorsey, Jr. (eds.): *Rights to Oceanic Resources.* Deciding and Drawing Maritime Boundaries. 1989 ISBN 0-7923-0019-X
14. B. Kwiatkowska: *The 200 Mile Exclusive Economic Zone in the New Law of the Sea.* 1989 ISBN 0-7923-0074-2
15. H.W. Jayewardene: *The Regime of Islands in International Law.* 1990 ISBN 0-7923-0130-7
16. D.M. Johnston and M.J. Valencia: *Pacific Ocean Boundary Problems.* Status and Solutions. 1990 ISBN 0-7923-0862-X
17. J.A. de Yturriaga: *Straits Used for International Navigation.* A Spanish Perspective. 1991 ISBN 0-7923-1141-8

18. C.C. Joyner: *Antarctica and the Law of the Sea.* 1992 ISBN 0-7923-1823-4
19. D. Pharand and U. Leanza (eds.): *The Continental Shelf and the Exclusive Economic Zone: Delimitation and Legal Regime/Le Plateau continental et la Zone économique exclusive: Délimitation et régime juridique.* 1993 ISBN 0-7923-2056-5
20. F. Laursen: *Small Powers at Sea.* Scandinavia and the New International Marine Order. 1993 ISBN 0-7923-2341-6
21. J. Crawford and D.R. Rothwell (eds.): *The Law of the Sea in the Asian Pacific Region.* 1995 ISBN 0-7923-2742-X
22. M. Munavvar: *Ocean States.* Archipelagic Regimes in the Law of the Sea. 1995 ISBN 0-7923-2882-5
23. A. Strati: *The Protection of the Underwater Cultural Heritage: An Emerging Objective of the Contemporary Law of the Sea.* 1995 ISBN 0-7923-3052-8
24. A.G. Oude Elferink: *The Law of Maritime Boundary Delimitation.* A Case Study of the Russian Federation. 1994 ISBN 0-7923-3082-X
25. Y. Li: *Transfer of Technology for Deep Sea-Bed Mining.* The 1982 Law of the Sea Convention and Beyond. 1994 ISBN 0-7923-3212-1
26. T.O. Akintoba: *African States and Contemporary International Law.* A Case Study of the 1982 Law of the Sea Convention and the Exclusive Economic Zone. 1996. ISBN 90-411-0144-6
27. J.A. Roach and R.W. Smith: *United States Responses to Excessive Maritime Claims.* Second Edition. 1996 ISBN 90-411-0225-6
28. T. Treves (ed.): *The Law of the Sea.* The European Union and its Member States. 1997 ISBN 90-411-0326-0
29. A. Razavi: *Continental Shelf Delimitation and Related Maritime Issues in the Persian Gulf.* 1997 ISBN 90-411-0333-3
30. J.A. de Yturriaga: *The International Regime of Fisheries.* From UNCLOS 1982 to the Presential Sea. 1997 ISBN 90-411-0365-1
31. M.J. Valencia, J.M. Van Dyke and N.A. Ludwig: *Sharing the Resources of the South China Sea.* 1997 ISBN 90-411-0411-9
32. E.C. Farrell: *The Socialist Republic of Vietnam and the Law of the Sea.* An Analysis of Vietnamese Behavior within the Emerging International Oceans Regime. 1997 ISBN 90-411-0473-9
33. P.B. Payoyo: *Cries of the Sea.* World Inequality, Sustainable Development and the Common Heritage of Humanity. 1997 ISBN 90-411-0504-2
34. H.N. Scheiber (ed.): *Law of the Sea.* The Common Heritage and Emerging Challenges. 2000 ISBN 90-411-1401-7

35. D.R. Rothwell and S. Bateman (eds.): *Navigational Rights and Freedoms and the New Law of the Sea.* 2000 ISBN 90-411-1499-8
36. M.J. Valencia (ed.): *Maritime Regime Building.* Lessons Learned and their Relevance for Northeast Asia. 2001 ISBN 90-411-1580-3
37. A.G. Oude Elferink and D.R. Rothwell (eds.): *The Law of the Sea and Polar Maritime Delimination and Jurisdiction.* 2001 ISBN 90-411-1648-6
38. Robert Kolb, *Case Law on Equitable Maritime Delimitation/Jurisprudence sur les délimitations maritimes selon l'équité*: Digest and Commentaries/ Répertoire et commentaires. 2002. ISBN 90-411-1976-0
39. Simon Marr, *The Precautionary Principle in the Law of the Sea:* Modern Decision Making in International Law. 2002. ISBN 90-411-2015-7
40. Sun Pyo Kim: *Maritime Delimitation and Interim Arrangements in North East Asia.* 2003. ISBN 90-04-13669-X

www.ingramcontent.com/pod-product-compliance
Lightning Source LLC
LaVergne TN
LVHW010542300926
849162LV00002B/206

9789004136694